INTRODUCTION TO NUCLEAR AND PARTICLE PHYSICS

INTRODUCTION TO NUCLEAR AND PARTICLE PHYSICS

ASHOK DAS
THOMAS FERBEL
University of Rochester

JOHN WILEY & SONS, INC.
NEW YORK CHICHESTER BRISBANE TORONTO SINGAPORE

Acquisitions Editor	Cliff Mills
Marketing Manager	Catherine Faduska
Production Supervisor	Bonnie Cabot
Designer	David Levy/Kevin Murphy
Manufacturing Manager	Andrea Price
Illustration	Jaime Perea
Copyeditor	Trumbull Rogers

This book was set in 10/12 Palatino by Publication Services and
printed and bound by Courier Stoughton. The cover was printed by Phoenix Color Corp.

Recognizing the importance of preserving what has been written, it is a
policy of John Wiley & Sons, Inc. to have books of enduring value published
in the United States printed on acid-free paper, and we exert our best
efforts to that end.

Library of Congress Cataloging in Publication Data:
Das, Ashok, 1953–
Introduction to nuclear and particle physics / Ashok Das, Thomas Ferbel.
p. cm.
Includes index.
ISBN 0-471-57132-6
1. Nuclear physics. 2. Particles (Nuclear physics). I. Ferbel, Thomas. II. Title.
QC776.D37 1994 93-19307
539.7—dc20 CIP

Printed in the United States of America

10 9 8 7 6 5 4 3 2

To
Our Teachers
and
Our Students

PREFACE AND INTRODUCTION

This book is based on a one-semester course on Nuclear and Particle Physics that we have taught to undergraduate juniors and seniors at the University of Rochester. Naturally, the previous experience and background of our students determined to a large extent the level at which we presented the material. This ranged from a very qualitative and hand-waving exposition to one class that consisted of a mix of about six engineering and math majors, to relatively formal and quantitative developments for classes that were composed of about ten to fifteen well-prepared physics majors. It will not come as a great surprise to find out that, independent of the degree of sophistication of our students, they were invariably fascinated by the subject matter, which provided great wonderment and stimulation to them. In class we always strove to stress the general underlying ideas of nuclear and particle physics, and we hope that in transforming our lecture notes into this more formal text, we have not committed the common sin of sacrificing physical content and beauty for difficulty and rigor.

APOLOGIES

This book is intended primarily for use in a senior undergraduate course, and particularly for students who have had previous contact with Quantum Mechanics. In fact, more than just slight contact is required in order to appreciate many of the subtleties we have infused into the manuscript. A one-semester course in Quantum Mechanics should be of great help to the student in navigating through the fantastic world of nuclear and particle phenomena. Although, in principle, our book is self-contained, there are parts of several chapters that will be daunting. For example, the sections on Relativistic Variables and Quantum Treatment of Rutherford Scattering in Chapter I, some of the more formal material in Chapters X, XI, and XIII, and the section on Time Development and Analysis of the K^0–$\overline{K}^0$ System in Chapter XII, are all especially demanding. Although the treatment of the mass matrix for the kaon system may be considered as too advanced, and not essential for the overall development of the material in the book, we believe that the other difficult sections are quite important. (Also, we felt that mathematically advanced students would appreciate some of these more challenging excursions.) Nevertheless, if deemed necessary, the formal concepts in these harder sections can be de-emphasized in favor of the phenomenological content.

Having chosen a somewhat historical development for particle physics, we had difficulty in infusing the quark structure of hadrons early into our logical development. We felt that this early introduction was important for familiarizing students with the systematics of hadrons and their constituents. To achieve this goal, we introduced the properties of quarks in the Problems section of Chapter IX, well before the discussion of their relevance in the Standard Model in Chapter XIII. Although this might not be considered the best approach, it should nevertheless provide students, through problems, with the valuable experience of interpreting hadrons in terms of their quark content, and in reducing the possible confusion and frustration caused by keeping track of so many different hadrons.

UNITS AND TABLES OF NUCLEAR AND PARTICLE PROPERTIES

We use the cgs system of units throughout the text, except that energy, mass, and momentum are specified in terms of eV. This often requires the use of $\hbar c$ to convert from cgs to the mixed system. Whenever possible, we have shown explicitly in the text how such changes in units are made. Periodically, when we depart from our normal convention, as we do for the case of magnetic moments, we warn the reader of this change, and again offer examples or problems to ease the transition between different sets of units.

We have found that the best source of information on properties of nuclei and particles, as well as on fundamental constants, is the all-inclusive *CRC Handbook of Chemistry and Physics* (CRC Press, Inc.). Because every library has copies of this work, we have not provided such detailed information in our manuscript, and urge students to consult the CRC tables when need arises. We have included, however, some useful physical constants in an appendix to this book.

OTHER REFERENCES

The subjects of nuclear and particle physics share a common heritage. The theoretical origins of the two fields and their reliance on quantum mechanics, as well as the evolution of their experimental techniques, provide much overlap in content. It is therefore sensible to present these two areas of physics, especially at the undergraduate level, in a unified manner. And, in fact, there are several excellent texts that have recently been published, or extensively revised, that provide the kind of combined exposition that we have presented. The books *Subatomic Physics* by Hans Frauenfelder and Ernest Henley (Prentice-Hall, Inc.) and *Nuclear and Particle Physics* by W. S. C. Williams (Oxford University Press) are particularly worthy of noting, because they also offer a panoramic view of nuclear and particle physics of the kind that we have attempted to give in our book. We believe that the emphasis in all three of these works is sufficiently different and original to make them all complementary and of value to students learning these two exciting fields of physics.

ACKNOWLEDGMENTS

It gives us great pleasure to acknowledge the superb typing (and seemingly endless retyping) of this manuscript by Ms. Judy Mack. Her great care and grace under pressure were vital to the ultimate success of our project. We also thank David Rocco for the artwork, the reviewers from Wiley for their valuable suggestions concerning both style and content of the book, and Richard Hagen for pointing out several typos and possible sources of confusion in the first printing of this book. Finally, we thank our students for their stimulating and often refreshing comments, and our colleagues for their patience in responding to some of our trivial questions.

A. DAS AND T. FERBEL
University of Rochester

CONTENTS

Chapter I

RUTHERFORD SCATTERING

INTRODUCTORY REMARKS

Matter has distinct levels of structure. For example, atoms, once considered the ultimate building blocks, are themselves composed of nuclei and electrons. The nucleus, in turn, consists of protons and neutrons, which we now believe are made of quarks and gluons. Gaining an understanding of the fundamental structure of matter has not been an easy achievement, primarily because the dimensions of the constituents are so small. For example, the typical size of an atom is about 10^{-8} cm, the average nucleus is about 10^{-12} cm in diameter, neutrons and protons have radii of about 10^{-13} cm, while electrons and quarks are believed to be without structure down to distances of at least 10^{-16} cm (namely, they behave as particles of $\lesssim 10^{-16}$ cm in size).

The study of the structure of matter presents formidable challenges both experimentally and theoretically, simply because we are dealing with the submicroscopic domain, where much of our classical intuition regarding the behavior of objects fails us. Experimental investigations of atomic spectra provided our first insights into atomic structure. These studies ultimately led to the birth of quantum mechanics, which beautifully explained, both qualitatively and quantitatively, not only the observed spectra and the structure of the atom, but also clarified the nature of chemical bonding, and a host of phenomena in condensed matter. The remarkable success of quantum theory in explaining atomic phenomena was mainly due to two reasons. First, the interaction responsible for holding the atom together is the long-ranged electromagnetic force, whose properties were well understood in the classical domain, and whose principles carried over quite readily to the quantum regime. Second, the strength of the electromagnetic coupling is weak enough (recall that the dimensionless coupling constant is represented by the fine structure constant, $\alpha = (e^2/\hbar c) \simeq \frac{1}{137}$), so that the properties of even complex atomic systems can be estimated reliably using approximations based on perturbative quantum mechanical calculations. Peering beyond the atom into the nuclear domain, however, the situation changes drastically. The force that holds the nucleus together—the nuclear force as we will call it—is obviously very strong since it holds the positively charged protons together inside a small nucleus, despite the presence of the Coulomb force that acts to repel them. Furthermore, the nuclear force is short-ranged, and therefore, unlike the electromagnetic force, more difficult to probe. (We know that the nuclear force is short-ranged because its effect can hardly be noticed outside of the nucleus.) There is no classical equivalent for such a force and, therefore, without any intuition to guide us, we are at a clear disadvantage in trying to unravel the structure of the nucleus.

It is because of the lack of classical analogies that experiments play such important roles in deciphering the fundamental structure of subatomic matter. Experiments provide information on properties of nuclei and on their constituents, at the very smallest length scales; these data are then used to construct theoretical models of nuclei and of the nuclear force. Of course, the kinds of experiments that can be performed in this domain present interesting challenges in their own right, and we will discuss some of the techniques used in the field in Chapter VII. In general, much of the experimental information, both in nuclear and particle physics, is derived from scattering measurements—similar, in principle, to those that Ernest Rutherford and his collaborators performed in discovering the nucleus. In such experiments, beams of energetic particles are directed into a fixed target, or, alternatively, two beams of energetic particles are made to collide. In either case, the results of collisions in such scattering experiments provide invaluable, and often the only attainable, information about subatomic systems. Since the basic principles in most of these experiments are quite similar, we will next sketch the ideas behind the pioneering work of Rutherford and his colleagues that was carried out at the University of Manchester, England, around 1910, and which provided the foundation for nuclear and particle physics.

RUTHERFORD SCATTERING

The series of measurements performed by Hans Geiger and Ernest Marsden under Rutherford's direction at Manchester provide a classic example of a "fixed target" experiment. The target was a thin foil of relatively large atomic number, while the projectiles consisted of a collimated beam of low energy α-particles, which, as we see in the next chapter, are nothing more than the nuclei of helium atoms. The basic outcome of these experiments was that most of the α-particles went straight through the gold foil with very little angular deviation. Occasionally, however, the deflections were quite large. A detailed analysis of these observations revealed the structure of the target, which ultimately led to the nuclear model of the atom.

To fully appreciate the beauty of these experiments, it is essential to analyze the results in their proper historical context. Prior to this work, the only popular model of the atom was due to Joseph Thomson, who visualized the electrically neutral atom as a "plum pudding" where negatively charged electrons were embedded, like raisins, within a uniform distribution of positive charge. If this model were correct, one would expect only small deviations in the α-particles' trajectory (primarily due to the scattering from the electrons), unlike what was found by Geiger and Marsden. To see this, let us do a few simple kinematic calculations. Because the velocities of the α-particles in these experiments were well below 0.1 c (where c refers to the speed of light), we will ignore relativistic effects.

Let us assume that an α-particle with mass m_α and initial velocity $\vec{v}_0$ collides head-on with a target particle of mass m_t, which is initially at rest (see Fig. 1.1). After the collision, both particles move with respective velocities $\vec{v}_\alpha$ and $\vec{v}_t$. Assuming that the collision is elastic (namely, that no kinetic energy is converted or lost in the process), momentum and energy conservation yield the following relations.

Momentum conservation:

$$m_\alpha \vec{v}_0 = m_\alpha \vec{v}_\alpha + m_t \vec{v}_t$$

or

$$\vec{v}_0 = \vec{v}_\alpha + \frac{m_t}{m_\alpha} \vec{v}_t \tag{1.1}$$

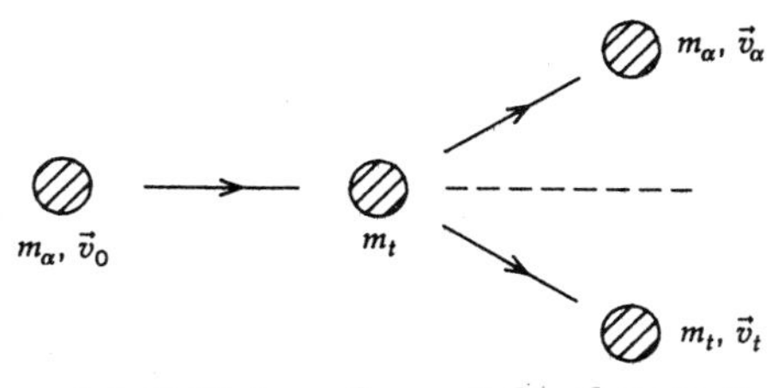

Figure 1.1 Collision of a particle of mass m_α and velocity $\vec{v}_0$ with a target particle of mass m_t.

Energy conservation:

$$\frac{1}{2}m_\alpha v_0^2 = \frac{1}{2}m_\alpha v_\alpha^2 + \frac{1}{2}m_t v_t^2$$

or

$$v_0^2 = v_\alpha^2 + \frac{m_t}{m_\alpha}v_t^2 \tag{1.2}$$

where we have labeled $(\vec{v}_i)^2 = \vec{v}_i \cdot \vec{v}_i$ as v_i^2, for $i = 0$, α, and t. Squaring the relation in Eq. 1.1 and comparing with Eq. 1.2, we obtain

$$v_0^2 = \left(v_\alpha^2 + \left(\frac{m_t}{m_\alpha}\right)^2 v_t^2 + 2\frac{m_t}{m_\alpha}\vec{v}_\alpha \cdot \vec{v}_t\right) = v_\alpha^2 + \frac{m_t}{m_\alpha}v_t^2$$

or

$$v_t^2\left(1 - \frac{m_t}{m_\alpha}\right) = 2\vec{v}_\alpha \cdot \vec{v}_t \tag{1.3}$$

It is clear from this analysis that, if $m_t \ll m_\alpha$, then the left-hand side of Eq. 1.3 is positive and, consequently, from the right-hand side we conclude that the motion of the α-particle and the target must be essentially along the incident direction. In other words, in such a case, one would expect only small deviations in the trajectory of the α-particle. On the other hand, if $m_t \gg m_\alpha$, then the left-hand side of Eq. 1.3 will be negative, which will then imply large angles between the trajectories of the α-particle and the recoiling nucleus, or large-angle scattering. To get a feeling for the magnitude of the numbers involved, let us recall that the masses of the electron and the α-particle have the following approximate values:

$$\begin{aligned} m_e &\simeq 0.5\ \mathrm{MeV}/c^2 \\ m_\alpha &\simeq 4 \times 10^3\ \mathrm{MeV}/c^2 \end{aligned} \tag{1.4}$$

Therefore, if we identify

$$m_t = m_e$$

then

$$\frac{m_t}{m_\alpha} \simeq 10^{-4} \tag{1.5}$$

Now, from Eq. 1.3 it follows that $v_e = v_t \lesssim 2v_\alpha$, and then Eq. 1.2 yields $v_\alpha \simeq v_0$. Therefore, $m_e v_e = m_\alpha(m_e/m_\alpha)v_e \lesssim 2\times10^{-4}m_\alpha v_\alpha \simeq 2\times10^{-4}m_\alpha v_0$, and the magnitude of the momentum transfer to the electron target is therefore $\lesssim 10^{-4}$ of the incident momentum. Consequently, the change in the momentum of the α-particle is quite small and, in the framework of the "plum pudding" model of the atom, we would expect only slight deviations in the α-trajectory after scattering from atomic electrons; thus, the outcome of the experiments, namely the occasional scatters through large angles, would pose a serious puzzle. On the other hand, if we accept the nuclear model, wherein the atom has a positively charged core (the nucleus) containing most of the mass of the atom, and electrons moving around it, then the experimental observations would follow quite naturally. For example, setting the mass of the target to that of the gold nucleus

$$m_t = m_{\text{Au}} \simeq 2 \times 10^5 \text{ MeV}/c^2 \tag{1.6}$$

yields

$$\frac{m_t}{m_\alpha} \simeq 50 \tag{1.7}$$

A simple analysis of Eq. 1.3 gives $v_t \leq (2m_\alpha v_\alpha)/m_t$, and from Eq. 1.2 we again obtain that $v_\alpha \simeq v_0$. Therefore, $m_t v_t \leq 2m_\alpha v_\alpha \simeq 2m_\alpha v_0$. This means that the nucleus can carry away up to twice the incident momentum, which implies that the α-particle can recoil backwards with a momentum essentially equal and opposite to its initial value. Such large momentum transfers to the nucleus can therefore provide large scattering angles. Consequently, in the Rutherford picture, we would expect those α-particles that scatter off the atomic electrons in gold to have only small-angle deflections in their trajectories, while the α-particles that occasionally scatter off the massive nuclear centers to suffer large angular deviations.

The analysis of the scattering process, however, is not this straightforward, and this is simply because we have completely ignored the forces involved in the problem.* We know that a particle with charge Ze produces a Coulomb potential of the form

$$U(\vec{r}) = \frac{Ze}{r} \tag{1.8}$$

We also know that two electrically charged particles separated by a distance $r = |\vec{r}|$ experience a Coulomb force giving rise to a potential energy

$$V(r) = \frac{ZZ'e^2}{r} \tag{1.9}$$

*We have also tacitly assumed, in the context of the Thomson model, that contributions to large-angle scattering from the diffuse positively charged nuclear matter can be ignored. This is, in fact, the case, as discussed by Thomson in his historic paper.

Here Ze and $Z'e$ are the charges of the two particles. An important point to note about the Coulomb force is that it is conservative and central. A force is said to be conservative if it can be related to the potential energy through a gradient, namely

$$\vec{F}(\vec{r}) = -\vec{\nabla}V(\vec{r}) \tag{1.10}$$

and it is defined to be central if

$$V(\vec{r}) = V(|\vec{r}|) = V(r) \tag{1.11}$$

In other words, the potential energy associated with a central force depends only on the distance between the particles and not on their angular coordinates. Because the description of scattering in a central potential is no more complicated than that in a Coulomb potential, we will first discuss the general case.

Let us consider the classical scattering of a particle from a fixed center. We assume that the particle is incident along the z-axis with an initial velocity $\vec{v}_0$. (It is worth noting that, outside the foil, the incident and the outgoing trajectories are essentially straight lines, and that all the deflection occurs at close distances of the order of atomic dimensions, where the interaction is most intense.) If we assume that the potential (force) falls off at infinity, then conservation of energy would imply that the total energy equals the initial energy:

$$E = \frac{1}{2}mv_0^2 = \text{constant} > 0 \tag{1.12}$$

Equivalently, we can relate the incident velocity to the total energy:

$$v_0 = \sqrt{\frac{2E}{m}} \tag{1.13}$$

Let us describe the motion of the particle using polar coordinates with the fixed center as the origin (see Fig. 1.2). If r denotes the radial coordinate of the incident particle, and χ the angle with respect to the z-axis, then the potential (being central) would be independent of χ. Consequently,

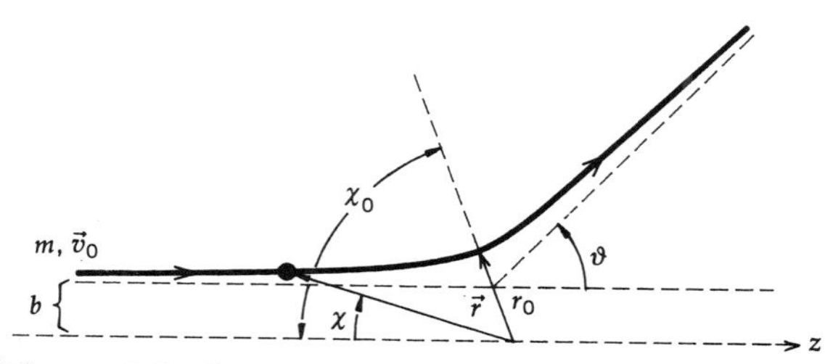

Figure 1.2 The scattering of a particle of mass m, with initial (asymptotic) velocity $\vec{v}_0$, from a center of force at the origin.

the angular momentum will be a constant during the entire motion. (That is, since $\vec{r}$ and $\vec{F}$ are colinear, the torque $\vec{r} \times \vec{F}$ vanishes, and the angular momentum $\vec{r} \times m\vec{v}$ cannot change.) For the incident particle, the angular momentum is clearly perpendicular to the plane of motion and has a magnitude $\ell = mv_0 b$, where b is known as the *impact parameter*. The impact parameter represents the transverse distance that the incident particle would fly by the source if there was no force acting. Using Eq. 1.13, we can obtain the following relation:

$$\ell = m\sqrt{\frac{2E}{m}}\,b = b\sqrt{2mE}$$

or

$$\frac{1}{b^2} = \frac{2mE}{\ell^2} \tag{1.14}$$

From its definition, the angular momentum can also be related to the angular frequency, $\dot{\chi}$, as follows:

$$\ell = |\vec{r} \times m\vec{v}| = \left|m\vec{r} \times \left(\frac{dr}{dt}\hat{r} + r\frac{d\chi}{dt}\hat{\chi}\right)\right| = mr^2\frac{d\chi}{dt} \equiv mr^2\dot{\chi} \tag{1.15}$$

where, as usual, we have defined a unit vector $\hat{\chi}$ perpendicular to $\vec{r} = r\hat{r}$, with $\vec{v}(\vec{r}) = \dot{r}\hat{r} + r\dot{\chi}\hat{\chi}$ expressed in terms of a radial and an angular component of the velocity, and the dot above a variable stands for differentiation with respect to time. Equation 1.15 can be rewritten as

$$\frac{d\chi}{dt} = \frac{\ell}{mr^2} \tag{1.16}$$

At any point along the trajectory, the energy is fixed, and can be written as

$$\begin{aligned} E &= \frac{1}{2}m\left(\frac{dr}{dt}\right)^2 + \frac{1}{2}mr^2\left(\frac{d\chi}{dt}\right)^2 + V(r) \\ &= \frac{1}{2}m\left(\frac{dr}{dt}\right)^2 + \frac{1}{2}mr^2\left(\frac{\ell}{mr^2}\right)^2 + V(r) \end{aligned}$$

or

$$\frac{1}{2}m\left(\frac{dr}{dt}\right)^2 = E - \frac{\ell^2}{2mr^2} - V(r)$$

or

$$\frac{dr}{dt} = -\left[\frac{2}{m}\left(E - V(r) - \frac{\ell^2}{2mr^2}\right)\right]^{1/2} \tag{1.17}$$

The term $\ell^2/2mr^2$ is referred to as the centrifugal barrier, which for $\ell \neq 0$ can be considered as a repulsive contribution to an overall effective potential $V_{\text{eff}}(r) = V(r) + \ell^2/2mr^2$. Both positive and negative roots are allowed in Eq. 1.17, but we have chosen the negative root because the radial coordinate decreases with time until the point of closest approach, and that is the time domain we will be examining.* Rearranging the factors in Eq. 1.17 and using Eq. 1.15, we obtain

$$\frac{dr}{dt} = -\left[\frac{2}{m} \cdot \frac{\ell^2}{2mr^2}\left\{\frac{2mEr^2}{\ell^2}\left(1 - \frac{V(r)}{E}\right) - 1\right\}\right]^{1/2}$$

$$= -\frac{\ell}{mr}\left[\frac{r^2}{b^2}\left(1 - \frac{V(r)}{E}\right) - 1\right]^{1/2}$$

$$= -\frac{\ell}{mrb}\left[r^2\left(1 - \frac{V(r)}{E}\right) - b^2\right]^{1/2} \tag{1.18}$$

From Eqs. 1.16 and 1.18, we now obtain

$$d\chi = \frac{\ell}{mr^2}dt = \frac{\ell}{mr^2}\frac{dt}{dr}dr$$

$$= -\frac{\ell}{mr^2} \cdot \frac{dr}{(\ell/mrb)\left[r^2(1 - (V(r)/E)) - b^2\right]^{1/2}}$$

or

$$d\chi = -\frac{b\,dr}{r\left[r^2(1 - (V(r)/E)) - b^2\right]^{1/2}} \tag{1.19}$$

Integrating this between the initial point and the point of closest approach, we obtain

---

*The motion is completely symmetric about the point of closest approach ($r = r_0$), and consequently the positive and negative roots provide identical information. In fact, if the α-particle approached the target with the velocity v_0 along the exiting trajectory in Fig. 1.2, it would then emerge on the entering trajectory, with the same asymptotic velocity. A simple way to see that this is true is to imagine the collision as observed from both above and below the plane of scattering shown in Fig. 1.2. Viewed from these two perspectives, the motion in Fig. 1.2 appears as the mirror image of the reversed trajectory. This symmetry is a consequence of time-reversal invariance of the equations of motion; a concept to be discussed in Chapter XI.

$$\int_0^{\chi_0} d\chi = -\int_\infty^{r_0} \frac{b\,dr}{r\left[r^2\left(1-(V(r)/E)\right)-b^2\right]^{1/2}}$$

or

$$\chi_0 = b\int_{r_0}^{\infty} \frac{dr}{r\left[r^2\left(1-(V(r)/E)\right)-b^2\right]^{1/2}} \tag{1.20}$$

The point of closest approach is determined by noting that, as the particle approaches from infinity, its velocity decreases continuously (assuming the repulsive potential for the case of an α-particle approaching a nucleus), until the point of closest approach, where the radial velocity (dr/dt) vanishes and subsequently changes sign. That is, beyond this point, the velocity of the particle increases again. Therefore, at the distance of closest approach, when $r = r_0$, both the radial and the absolute velocity attain a minimum, and we have that

$$\left.\frac{dr}{dt}\right|_{r=r_0} = 0$$

which, from Eqs. 1.17 and 1.18, means that

$$E - V(r_0) - \frac{\ell^2}{2mr_0^2} = 0$$

or

$$r_0^2\left(1 - \frac{V(r_0)}{E}\right) - b^2 = 0 \tag{1.21}$$

Thus, given a specific form of the potential, we can determine r_0, and therefore χ_0, as a function of the impact parameter b.* Defining the scattering angle θ as the change in the asymptotic angles of the trajectory, we have that

$$\theta = \pi - 2\chi_0 = \pi - 2b\int_{r_0}^{\infty} \frac{dr}{r\left[r^2\left(1-(V(r)/E)\right)-b^2\right]^{1/2}} \tag{1.22}$$

Consequently, given an impact parameter b, and a fixed energy E, the scattering angle of a particle in a potential can, at least in principle, be completely determined.

---

*We note that, in general, with $\ell \neq 0$ and $E > 0$, that is, for $b \neq 0$, $d\chi/dt$ is maximum at $r = r_0$ (see Eq. 1.16). We also wish to point out that for $\ell \neq 0$, even for an attractive Coulomb potential, there will be a finite result for r_0 as determined from Eq. 1.21. This is because the centrifugal barrier for $\ell \neq 0$ acts as a repulsive potential that dominates over Coulomb attraction at small distances.

As an application of the general result, let us now return to the scattering of a charged particle from a repulsive Coulomb potential, for which the potential energy is given by Eq. 1.9:

$$V(r) = \frac{ZZ'e^2}{r} \tag{1.23}$$

where $Z'e$ represents the charge of the incident particle and Ze the charge of the scattering center. (The scattering of an α-particle from a nucleus would then correspond to $Z' = 2$, with Ze representing the nuclear charge.) The distance of closest approach can be obtained from Eq. 1.21:

$$r_0^2 - \frac{ZZ'e^2}{E} r_0 - b^2 = 0$$

or

$$r_0 = \frac{(ZZ'e^2/E) \pm \sqrt{(ZZ'e^2/E)^2 + 4b^2}}{2} \tag{1.24}$$

Since the radial coordinate can by definition only be positive, we conclude that

$$r_0 = \frac{ZZ'e^2}{2E}\left(1 + \sqrt{1 + \frac{4b^2E^2}{(ZZ'e^2)^2}}\right) \tag{1.25}$$

Consequently, from Eq. 1.22, we obtain

$$\theta = \pi - 2b\int_{r_0}^{\infty} \frac{dr}{r\,[r^2(1 - (ZZ'e^2/r)) - b^2]^{1/2}} \tag{1.26}$$

Let us define a new variable

$$x = \frac{1}{r} \tag{1.27}$$

which gives

$$x_0 = \frac{1}{r_0} = \frac{2E}{ZZ'e^2}\left(1 + \sqrt{1 + \frac{4b^2E^2}{(ZZ'e^2)^2}}\right)^{-1} \tag{1.28}$$

From Eq. 1.27, we obtain

$$dx = -\frac{dr}{r^2} \qquad \text{or} \qquad dr = -\frac{dx}{x^2}$$

and, in terms of this new variable, we can write

$$\theta = \pi - 2b \int_{x_0}^{0} \left(-\frac{dx}{x^2}\right) \frac{x}{[(1/x^2) - (ZZ'e^2/x) - b^2]^{1/2}}$$

$$= \pi + 2b \int_{x_0}^{0} \frac{dx}{(1 - (ZZ'e^2/E)x - b^2x^2)^{1/2}} \tag{1.29}$$

Now, using the following result from the integral tables:

$$\int \frac{dx}{\sqrt{\alpha + \beta x + \gamma x^2}} = \frac{1}{\sqrt{-\gamma}} \cos^{-1}\left(-\frac{\beta + 2\gamma x}{\sqrt{\beta^2 - 4\alpha\gamma}}\right) \tag{1.30}$$

we obtain

$$\theta = \pi + 2b \cdot \frac{1}{b} \cos^{-1}\left(\frac{(ZZ'e^2/E) + 2b^2x}{\sqrt{(ZZ'e^2/E)^2 + 4b^2}}\right)\Bigg|_{x_0}^{0}$$

$$= \pi + 2\cos^{-1}\left(\frac{1 + (2b^2E)/(ZZ'e^2)x}{\sqrt{1 + [(4b^2E^2)/(ZZ'e^2)^2]}}\right)\Bigg|_{x_0}^{0}$$

$$= \pi + 2\cos^{-1}\left(\frac{1}{\sqrt{1 + [(4b^2E^2)/(ZZ'e^2)^2]}}\right) - 2\cos^{-1}(1)$$

$$= \pi + 2\cos^{-1}\left(\frac{1}{\sqrt{1 + [(4b^2E^2)/(ZZ'e^2)^2]}}\right) \tag{1.31}$$

Equivalently, we can write

$$\frac{1}{\sqrt{1 + [(4b^2E^2)/(ZZ'e^2)^2]}} = \cos\left(\frac{\theta}{2} - \frac{\pi}{2}\right)$$

or

$$\frac{1}{1 + [(4b^2E^2)/(ZZ'e^2)^2]} = \cos^2\left(\frac{\theta}{2} - \frac{\pi}{2}\right) = \sin^2\frac{\theta}{2} = \frac{1}{\operatorname{cosec}^2(\theta/2)}$$

or

$$\frac{2bE}{ZZ'e^2} = \cot\frac{\theta}{2}$$

or

$$b = \frac{ZZ'e^2}{2E} \cot\frac{\theta}{2} \tag{1.32}$$

This relates the scattering angle, which is a measurable quantity, to the impact parameter, which cannot be observed directly. Note that, for fixed b, E, and Z', the scattering angle is larger for a larger value of Z. This is consistent with our intuition in that the Coulomb potential is stronger for larger Z, and leads to a larger deflection. Similarly, for a fixed b, Z, and Z', the scattering angle is larger when E is smaller. Qualitatively, we can understand this as follows. When the particle has low energy, its velocity is smaller and, therefore, it spends more time in the potential and suffers a greater amount of scattering. Finally, let us note that, for fixed Z, Z', and E, the scattering angle is larger for smaller b. Namely, when the impact parameter is smaller, the particle feels the force more strongly, and hence the deflection is larger. Equation 1.32 therefore incorporates all the qualitative features that we expect of scattering in the Coulomb field.

SCATTERING CROSS SECTION

As we have seen, the scattering of a particle in a potential is completely determined once we know the impact parameter and the energy of the particle; and therefore, for a fixed incident energy, the deflection is defined by just the impact parameter. To perform an experiment, we prepare an incident flux of beam particles of known energy, and measure the number of particles scattered out of the beam at different θ. Because this number is determined entirely by the impact parameters involved in the collisions, such measurements reflect these impact parameters and thereby the range of the interaction and the effective size of the target atoms.

Let N_0 denote the number of particles incident on the target foil per unit area per unit time. (Because the target density in the foil is low, the flux will always be uniform over the dimensions of any single target center.) Of these, the ones with impact parameters between b and $b + db$ will undergo angular deflections between θ and $\theta - d\theta$, and will scatter into a solid angle $d\Omega$. (Remember, that, as illustrated in Fig. 1.3, the larger the impact parameter, the smaller will be the scattering angle, and consequently a positive db corresponds to a negative $d\theta$.) The number of such

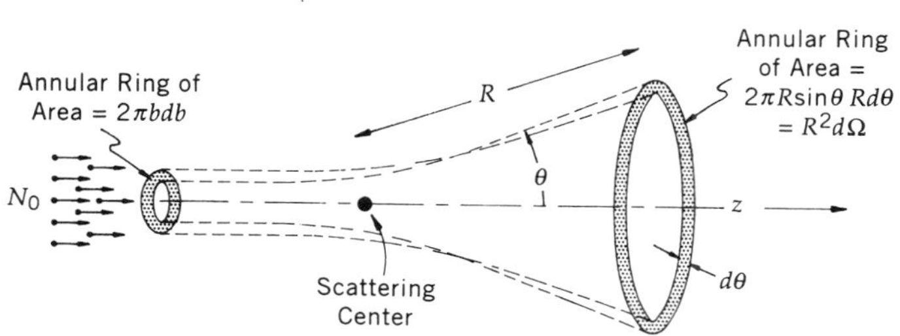

Figure 1.3 Incident particles within the area $2\pi bdb$ of any scattering center are emitted into the annular ring of area $R^2 d\Omega$ at angle θ.

particles scattered per unit time will be $2\pi N_0 b db$, since $2\pi b db$ is the area of the circular ring around the incident direction centered on each target, through which these particles must pass in order to scatter into the solid angle between θ and $\theta - d\theta$. It may be puzzling that we do not have to be concerned with the fact that we have many target particles in our foil, and that any single beam particle, in principle, comes within some impact parameter of all of them! This would clearly provide a great complication to our analysis. What we are assuming, however, is that our foil is exceedingly thin, so that multiple collisions of one beam particle are negligible; and we also have the Rutherford atomic model in mind, which means that the separation between nuclei is vast relative to their size. Normally, very large impact parameters provide very little scattering, so it is the trajectory that comes nearest to the nuclear center that matters most. (And, of course, the effect of electrons, because of their small mass, is also quite negligible.) When the thickness or density of the medium cannot be ignored, other interesting phenomena involving coherence and interference between scattering centers come into play. Cherenkov radiation and the density effect in ionization have their origins in such ramifications (see Chapters VI and VII).

For a central potential, at any impact parameter b, we can think of the scattering source as presenting an effective transverse cross-sectional area $\Delta\sigma = 2\pi b db$ for scattering of an incident particle by an angle θ into $d\Omega$. Because the specific relationship between b and θ (e.g., Eq. 1.32) depends rather explicitly on the nature of the force; for example, whether it has r^{-2} behavior, whether it is central, and so forth. In general, $\Delta\sigma$ can depend on both θ and ϕ, so we can write

$$\Delta\sigma(\theta, \phi) = b db d\phi = -\frac{d\sigma}{d\Omega}(\theta, \phi) \cdot d\Omega = -\frac{d\sigma}{d\Omega}(\theta, \phi) \sin\theta d\theta d\phi \qquad (1.33)$$

which defines the differential cross section $d\sigma/d\Omega$, and where the negative sign reflects the fact that θ decreases as b increases. When there is no azimuthal dependence in the scattering—for example, when the interaction has spherical symmetry—we can integrate over ϕ (as we have already done implicitly in our discussion of the annular rings of area $2\pi b db$) and write

$$\Delta\sigma(\theta) = -\frac{d\sigma}{d\Omega}(\theta) \cdot 2\pi \sin\theta d\theta = 2\pi b db$$

or

$$\frac{d\sigma}{d\Omega}(\theta) = -\frac{b}{\sin\theta}\frac{db}{d\theta} \qquad (1.34)$$

We wish to note that, since the Coulomb potential is central (depends only on distance and not on angle), we have assumed azimuthal symmetry in the scattering. This means that all positions along the annular ring of radius b are equivalent, and that the differential cross section is only a function of the angle θ and not ϕ. It follows, therefore, that measuring the yield as a function of θ, or the differential cross section, is equivalent to measuring the entire effect of the scattering.

In subatomic experiments, the unit, normally used to measure cross-sectional area, is the barn, which is defined as 10^{-24} cm^2. This is a very small quantity, but then we have to remember that the typical size of a nucleus is about 10^{-12} cm and, therefore, the cross-sectional area for a medium-sized nucleus (if we assume it to be a sphere) would be of the order of a barn. This is consequently a relatively natural unit for such measurements. The units of solid angle are steradians, and 4π sr corresponds to a sum over all solid angles, that is, all θ and ϕ. One can also define a total scattering cross section by integrating the differential cross section over all angles:

$$\sigma_{\mathrm{TOT}} = \int d\Omega \frac{d\sigma}{d\Omega}(\theta, \phi) = 2\pi \int_0^{\pi} d\theta \sin\theta \frac{d\sigma}{d\Omega}(\theta) \tag{1.35}$$

where, in the last step, we have again assumed azimuthal symmetry. (If there is any ϕ-dependence observed in the scattering, the last step in Eq. 1.35 would not hold.) The total cross section represents, in some sense, the effective size that the source presents for scattering at all possible impact parameters.

Let us now calculate the cross section for Rutherford scattering. We know from Eq. 1.32 that

$$b = \frac{ZZ'e^2}{2E} \cot\frac{\theta}{2}$$

It follows, therefore, that

$$\frac{db}{d\theta} = -\frac{1}{2} \cdot \frac{ZZ'e^2}{2E} \operatorname{cosec}^2\frac{\theta}{2} \tag{1.36}$$

The negative sign in Eq. 1.36 again reflects the fact that, as b increases, θ decreases, and that there is less deflection for larger impact parameter. Substituting this back into the definition of the scattering cross section, we obtain

$$\frac{d\sigma}{d\Omega}(\theta) = -\frac{b}{\sin\theta}\frac{db}{d\theta} = \left(\frac{ZZ'e^2}{4E}\right)^2 \operatorname{cosec}^4\frac{\theta}{2} = \left(\frac{ZZ'e^2}{4E}\right)^2 \frac{1}{\sin^4(\theta/2)} \tag{1.37}$$

If we now integrate this relation over θ (note from Eq. 1.34 that, because there is no azimuthal dependence, $d\Omega = 2\pi\sin\theta d\theta$), we obtain the total cross section:

$$\sigma_{\mathrm{TOT}} = \int \frac{d\sigma}{d\Omega}(\theta) d\Omega = 2\pi \int_0^\pi d\theta \sin\theta \frac{d\sigma}{d\Omega}(\theta)$$

$$= 8\pi \left(\frac{ZZ'e^2}{4E}\right)^2 \int_0^1 d\left(\sin\frac{\theta}{2}\right) \frac{1}{\sin^3(\theta/2)} \longrightarrow \infty \quad (1.38)$$

This divergence may seem troublesome, but it is consistent with our earlier discussion. Namely, the total cross section reflects the largest values of impact parameter a particle can have and still undergo scattering. In the case of the Coulomb potential, the long-ranged force extends to infinity and, consequently, a particle very far away from the center will still experience the Coulomb force, albeit only very slightly, and this is the origin of the divergence. Because the Coulomb force drops off rapidly with distance, however, and does not lead to any appreciable scattering beyond some finite value of the impact parameter, it is appropriate to cut off the angular integration at some finite $\theta = \theta_0 > 0^0$, corresponding to some realistic cutoff for the impact parameter. This cutoff provides a finite σ_{TOT} for observable scattering angles (that is, for $\theta > \theta_0$), which can be compared with experimental measurements. Finally, we should point out that our results cannot be valid for impact parameters much beyond the innermost electron levels, because such electrons will shield and thereby reduce the effective nuclear charge.

MEASURING CROSS SECTIONS

Let us now see how we would go about performing a measurement in order to extract a cross section. Macroscopically, we have a beam of α-particles (Geiger and Marsden used a collimated source of α-particles from a sample of radioactive radon), a thin foil, and some scintillating material for detecting the scattered particles. This was, initially, a thin coat of ZnS phosphor deposited on a glass screen, and viewed by eye through a telescope. The telescope was able to rotate in one plane, and thereby trace out the counting rate as a function of θ (but not ϕ). Schematically, the apparatus can be represented as in Fig. 1.4.

Now, if we have our flux of N_0 α-particles per unit area per second impinging on the thin foil, then some of these will pass through essentially undeflected, while others will be scattered through an angle between $\theta - d\theta$ and θ, corresponding to impact parameters between $b + db$ and b. Here $d\theta$ can be regarded as the angle subtended by the aperture of the telescope. In fact, the telescope views a small area of the screen given approximately by $Rd\theta \cdot R\sin\theta d\phi = R^2 d\Omega$, where R is the distance from the foil to the point of observation on the screen. The scattered particles that appear in this part of the screen are those that pass through and emerge from the part of the annular ring of impact radius b, width db, and arc length $bd\phi$. Had Geiger

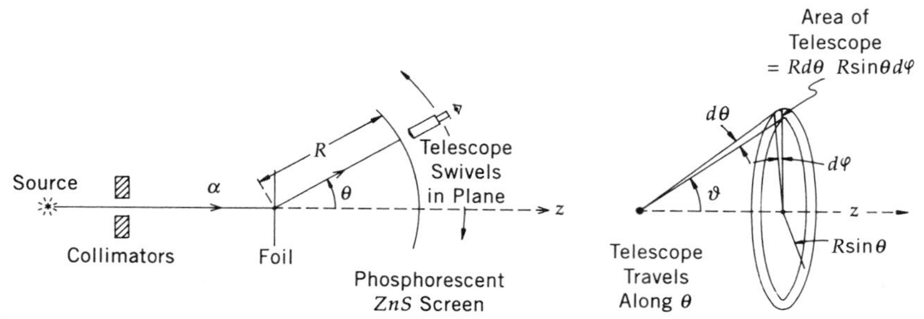

Figure 1.4 Sketch of the macroscopic geometry of Rutherford scattering.

and Marsden constructed a circle of telescopes about the beam center to view the α-particles emitted for all values of ϕ for any particular fixed angle θ, they would have certainly increased their event rate by $2\pi/d\phi$, but at the substantial cost of greatly complicating their experiment.

The next step is to determine the fraction of the incident particles that approach our nuclear targets within the small areas $\Delta\sigma = bd\phi db$ at impact parameters b. This is, in fact, the same fraction $-dn/N_0$ that will be removed from the beam and scattered into the part of the screen of area $R^2 d\Omega$ at angles (θ, ϕ) that is being viewed. This fraction must be identical to the ratio of the sum of the small $bd\phi db$ areas for all N nuclear centers within the foil, divided by the area (S) of the foil.

$$-\frac{dn}{N_0} = \frac{Nbd\phi db}{S} = \frac{N}{S}\Delta\sigma(\theta, \phi) \tag{1.39}$$

For a foil of thickness t, density ρ, atomic weight A, $N = (\rho t S/A)A_0$, where A_0 is Avogadro's number of atoms per mole. Thus, for the number of α-particles scattered per unit time into the detector at angles (θ, ϕ), we can write

$$dn = \frac{N_0 \rho t}{A} A_0 \frac{d\sigma}{d\Omega}(\theta, \phi) d\Omega$$

or

$$\frac{dn}{d\Omega} = \frac{N_0 \rho t A_0}{A} \frac{d\sigma}{d\Omega}(\theta, \phi) \tag{1.40}$$

For any given detector situated at angles (θ, ϕ) relative to the beam axis, and subtending a solid angle $d\Omega$ (which is determined by the transverse dimensions of the detector – its area/R^2), we will observe dn counts per second

$$dn = N_0 \frac{N}{S} \frac{d\sigma}{d\Omega}(\theta, \phi) d\Omega \tag{1.41}$$

This is a general expression, valid for any scattering process, independent of the existence of a theory that might be able to provide an expression for $d\sigma/d\Omega$. Thus, the counts observed in any experiment will be proportional to the number of incident beam particles, the number of scattering centers per unit area of target material, to the solid angle subtended by the detector, and to an effective cross section that each scattering center presents for bringing about the process of interest. (We are still assuming that corrections due to multiple collisions are small, that is, we are dealing with thin targets.) Geiger and Marsden performed very detailed measurements of dn as a function of θ, using different target material, different α-particle sources of different energy, different thicknesses of foil, and found their data to be in complete agreement with Rutherford's prediction, as given in Eq. 1.37. That is, knowing N_0, N/S, and $d\Omega$, they measured dn, and extracted a form for the differential cross section that agreed beautifully with Rutherford's prediction, and thereby verified the presence of nuclei within atoms. It should be recognized that, although Geiger and Marsden's measurements provided clear evidence for the existence of a nuclear center, these experiments shed very little light on the nature of the nuclear force. The low-energy α-particles never penetrated into the nuclear core of the atom because of the presence of the repulsive Coulomb barrier.

LABORATORY FRAME AND THE CENTER-OF-MASS FRAME

So far, we have discussed collisions of a particle with a fixed center. In reality, however, the target also moves (recoils) as a result of the scattering. In some experiments we may be interested in colliding two beams of particles of comparable energy with each other. Although such situations may appear to be extremely complicated at first glance, when the potential is central, the problem can be reduced to the one we have just studied; this can be achieved through the separation of the motion of the center of mass.

Let us assume that we have two particles with masses m_1 and m_2, at coordinates $\vec{r}_1$ and $\vec{r}_2$, interacting through a central potential. The equations for the motion can be written down as

$$\begin{aligned} m_1\ddot{\vec{r}}_1 &= -\vec{\nabla}_1 V(|\vec{r}_1 - \vec{r}_2|) \\ m_2\ddot{\vec{r}}_2 &= -\vec{\nabla}_2 V(|\vec{r}_1 - \vec{r}_2|) \end{aligned} \tag{1.42}$$

where $\vec{\nabla}$ is the gradient operator, which has the following form in spherical coordinates:

$$\vec{\nabla}_i = \hat{r}_i\frac{\partial}{\partial r_i} + \frac{\hat{\theta}_i}{r_i}\frac{\partial}{\partial\theta_i} + \frac{\hat{\phi}_i}{r_i\sin\theta_i}\frac{\partial}{\partial\phi_i} \qquad i = 1, 2 \tag{1.43}$$

Since the potential energy depends only on the relative separation of the two particles, let us define the variables:

$$\vec{r} = \vec{r}_1 - \vec{r}_2$$
$$\vec{R}_{CM} = \frac{m_1\vec{r}_1 + m_2\vec{r}_2}{m_1 + m_2} \tag{1.44}$$

where $\vec{r}$ denotes the coordinate of m_1 relative to m_2, and $\vec{R}_{CM}$ defines the coordinate of the center of mass of the system (see Fig. 1.5). From Eqs. 1.42 and 1.44 we can easily obtain the following:

$$\frac{m_1 m_2}{m_1 + m_2}\ddot{\vec{r}} \equiv \mu\ddot{\vec{r}} = -\vec{\nabla}V(|\vec{r}|) = -\frac{\partial V(|\vec{r}|)}{\partial r}\hat{r}$$
$$(m_1 + m_2)\ddot{\vec{R}}_{CM} = M\ddot{\vec{R}}_{CM} = 0 \quad \text{or} \quad \dot{\vec{R}}_{CM} = \text{constant} \cdot \hat{R} \tag{1.45}$$

where we have used the fact that $V(|\vec{r}|) = V(r)$ depends only on the radial coordinate r, and not on the angular variables associated with $\vec{r}$, and where we have defined

$$M = m_1 + m_2 = \text{total mass of the system}$$
$$\mu = \frac{m_1 m_2}{m_1 + m_2} = \text{"reduced" mass of the system} \tag{1.46}$$

It is clear from the preceding analysis that, when the potential is central, the motion of two particles can be decoupled when rewritten in terms of a relative coordinate and the coordinate of the center of mass.

We also note from Eq. 1.45 that the motion of the center of mass is trivial in the sense that it corresponds to that of a free, nonaccelerating particle. In other words, the center of mass moves in the laboratory with a constant velocity ($\dot{\vec{R}}_{CM}$ is fixed) independent of the specific form of the potential.

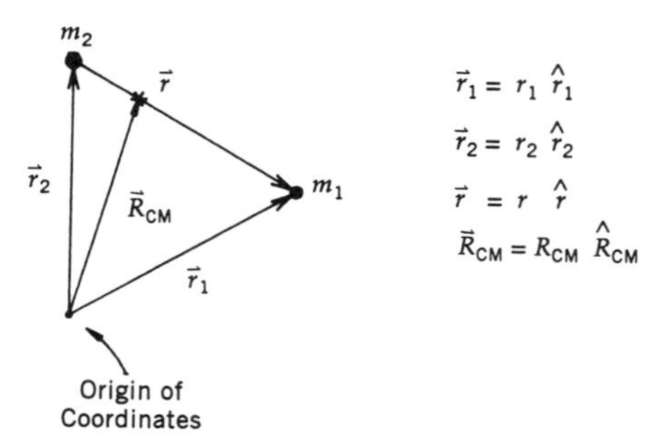

Figure 1.5 Position of the center of mass and the definition of the relative coordinate for the two particles with masses m_1 and m_2.

The dynamics is completely contained in the motion of a fictitious particle with the reduced mass μ and coordinate $\vec{r}$. In the frame in which the center of mass is at rest, the complete dynamics, then, becomes equivalent to the motion of a single particle, with mass μ, scattering from a fixed central potential, a situation that we have already analyzed in detail. A simplification that occurs in the center-of-mass frame is that the sum of the momenta of the interacting objects vanishes, which follows from Eqs. 1.44 and 1.45 when $\vec{R}_{CM}$ is set to zero. Because of this, it is more common to define the center-of-mass frame as the frame in which the total momentum vanishes, and we often refer to the center-of-mass frame equivalently as the center of momentum frame.

To understand how various quantities can be transformed between the laboratory frame and the center-of-mass frame, let us return to the scattering from a fixed target. In this case, in the laboratory frame, the particle with mass m_2 is initially at rest and the particle of mass m_1 is incident along the z-axis with a velocity v_1. Let the scattering angle of particle m_1 in the laboratory frame be given by θ_{Lab}, and its speed after scattering by v. The center of mass, in this case, moves along the z-axis with a speed (remember that particle m_2 is originally at rest).

$$v_{\text{CM}} = \dot{R}_{\text{CM}} = \frac{m_1 v_1}{m_1 + m_2} \tag{1.47}$$

In the center of mass, therefore, the two particles move toward each other (see Fig. 1.6) with speeds

$$\begin{aligned} \tilde{v}_1 &= v_1 - v_{\text{CM}} = \frac{m_2 v_1}{m_1 + m_2} \\ \tilde{v}_2 &= v_{\text{CM}} = \frac{m_1 v_1}{m_1 + m_2} \end{aligned} \tag{1.48}$$

where $\tilde{v}_1$ and $\tilde{v}_2$ are the speeds of beam and target particle, respectively, as viewed in the center-of-mass frame. Here we see explicitly that the momenta of the two particles in the center of mass are equal and opposite.

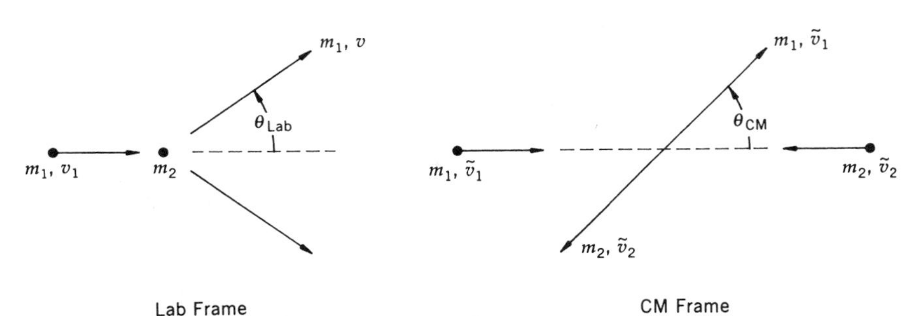

Figure 1.6 Collision of m_1 with m_2, as viewed in the Lab and in the CM frame.

For elastic scattering, the magnitudes of the velocities of the particles after the collision will remain unchanged, but the angles at which they emerge will depend on the dynamics. Let θ_{CM} denote the scattering angle as measured in the center-of-mass frame. Note that because θ_{CM} represents the change in the direction of the relative position vector ($\vec{r}$) before and after the collision, it must be identical to the scattering angle for the particle with reduced mass. To obtain a relation between θ_{Lab} and θ_{CM}, we note that the velocities in the laboratory frame and in the center-of-mass frame are related through the velocity of the center of mass. After scattering, we obtain for the z-component of the velocity of the particle of mass m_1

$$v\cos\theta_{Lab} - v_{CM} = \tilde{v}_1 \cos\theta_{CM}$$

or

$$v\cos\theta_{Lab} = \tilde{v}_1 \cos\theta_{CM} + v_{CM} \tag{1.49}$$

while, for the components of the velocity perpendicular to the z-axis (remember that the center of mass moves along the z-axis), we have

$$v\sin\theta_{Lab} = \tilde{v}_1 \sin\theta_{CM} \tag{1.50}$$

Dividing Eq. 1.50 by Eq. 1.49, we obtain the nonrelativistic result:

$$\tan\theta_{Lab} = \frac{\sin\theta_{CM}}{\cos\theta_{CM} + (v_{CM}/\tilde{v}_1)} = \frac{\sin\theta_{CM}}{\cos\theta_{CM} + \zeta} \tag{1.51}$$

where we have defined

$$\zeta = \frac{v_{CM}}{\tilde{v}_1} = \frac{m_1}{m_2} \tag{1.52}$$

The last equality in Eq. 1.52 holds only for the case of elastic scattering, which we are currently considering. For later use, we rewrite Eq. 1.51 in an alternate form:

$$\cos\theta_{Lab} = \frac{\cos\theta_{CM} + \zeta}{(1 + 2\zeta\cos\theta_{CM} + \zeta^2)^{1/2}} \tag{1.53}$$

Using the relationship given earlier between θ_{Lab} and θ_{CM}, we can also relate the differential cross sections in the two frames, arguing as follows. The particles that scatter through an angle θ_{Lab} into the solid angle $d\Omega_{Lab}$ in the laboratory frame, are the same ones that scatter by θ_{CM} into the corresponding solid angle $d\Omega_{CM}$ in the center of mass frame. (That is, these are two equivalent ways of looking at the same process.) Because ϕ is transverse to the boost direction between the two reference frames, it follows that $d\phi_{Lab} = d\phi_{CM}$. Thus, ignoring the azimuthal coordinate, we must have

$$\frac{d\sigma}{d\Omega_{\mathrm{Lab}}}(\theta_{\mathrm{Lab}}) \sin\theta_{\mathrm{Lab}} d\theta_{\mathrm{Lab}} = \frac{d\sigma}{d\Omega_{\mathrm{CM}}}(\theta_{\mathrm{CM}}) \sin\theta_{\mathrm{CM}} d\theta_{\mathrm{CM}}$$

or

$$\frac{d\sigma}{d\Omega_{\mathrm{Lab}}}(\theta_{\mathrm{Lab}}) = \frac{d\sigma}{d\Omega_{\mathrm{CM}}}(\theta_{\mathrm{CM}}) \frac{d(\cos\theta_{\mathrm{CM}})}{d(\cos\theta_{\mathrm{Lab}})} \tag{1.54}$$

The right-hand side of Eq. 1.54 can be evaluated using Eq. 1.53, leading to

$$\frac{d\sigma}{d\Omega_{\mathrm{Lab}}}(\theta_{\mathrm{Lab}}) = \frac{d\sigma}{d\Omega_{\mathrm{CM}}}(\theta_{\mathrm{CM}}) \frac{(1 + 2\zeta\cos\theta_{\mathrm{CM}} + \zeta^2)^{3/2}}{|1 + \zeta\cos\theta_{\mathrm{CM}}|} \tag{1.55}$$

RELATIVISTIC VARIABLES

In the appendix we review the basics of special relativity, and we use those results here to briefly discuss the kinematics in terms of relativistic variables. In the scattering of any two particles with rest masses m_1 and m_2, the velocity of the center of mass is obtained from the ratio of the total relativistic momentum and the total relativistic energy:

$$\frac{\vec{v}_{\mathrm{CM}}}{c} = \vec{\beta}_{\mathrm{CM}} = \frac{(\vec{P}_1 + \vec{P}_2)c}{E_1 + E_2} \tag{1.56}$$

If m_1 refers to the mass of the projectile and m_2 to that of a target particle, then using laboratory variables, we obtain (with the target initially at rest):

$$\vec{\beta}_{\mathrm{CM}} = \frac{\vec{P}_1 c}{E_1 + m_2c^2} = \frac{\vec{P}_1 c}{\sqrt{P_1^2c^2 + m_1^2c^4} + m_2c^2} \tag{1.57}$$

where our convention is to define $|\vec{P}_i| = P_i$ for $i = 1, 2$. At very low energies, namely when $m_1c^2 \gg P_1c$, this reduces to our nonrelativistic expressions (see Eq. 1.47):

$$\vec{\beta}_{\mathrm{CM}} = \frac{m_1\vec{v}_1 c}{m_1c^2 + m_2c^2} = \frac{m_1\vec{v}_1}{(m_1 + m_2)c} \tag{1.58}$$

At very high energies, when $m_1c^2 \ll P_1c$ and $m_2c^2 \ll P_1c$, we can write the following for the value of β_{CM}:

$$\beta_{\mathrm{CM}} = |\vec{\beta}_{\mathrm{CM}}| = \frac{1}{\sqrt{1 + ((m_1c^2)/(P_1c))^2} + (m_2c^2)/(P_1c)} \simeq 1 - \frac{m_2c}{P_1} - \frac{1}{2}\left(\frac{m_1c}{P_1}\right)^2 \tag{1.59}$$

When m_1 and m_2 are comparable, Eq. 1.59 simplifies to $\beta_{CM} \simeq (1 - m_2c/P_1)$, and, for this case, γ_{CM} becomes

$$\gamma_{CM} = \left(1 - \beta_{CM}^2\right)^{-1/2} \simeq \left[(1 + \beta_{CM})(1 - \beta_{CM})\right]^{-1/2}$$
$$\simeq \left[(2)\left(\frac{m_2c}{P_1}\right)\right]^{-1/2} = \sqrt{\frac{P_1}{2m_2c}} \tag{1.60}$$

In general, we can obtain an expression for γ_{CM} in the following way. We note from Eq. 1.57 that

$$\beta_{CM}^2 = \frac{P_1^2c^2}{(E_1 + m_2c^2)^2} \tag{1.61}$$

so that

$$1 - \beta_{CM}^2 = \frac{E_1^2 + 2E_1m_2c^2 + m_2^2c^4 - P_1^2c^2}{(E_1 + m_2c^2)^2}$$
$$= \frac{m_1^2c^4 + m_2^2c^4 + 2E_1m_2c^2}{(E_1 + m_2c^2)^2} \tag{1.62}$$

where we have substituted $m_1^2c^4$ for $E_1^2 - P_1^2c^2$. It follows, therefore, that

$$\gamma_{CM} = \left(1 - \beta_{CM}^2\right)^{-1/2} = \frac{E_1 + m_2c^2}{(m_1^2c^4 + m_2^2c^4 + 2E_1m_2c^2)^{1/2}} \tag{1.63}$$

which, in the high energy limit of $E_1 \simeq P_1c \gg m_1c^2$ and $P_1c \gg m_2c^2$, reduces to the result of Eq. 1.60.

The quantity in the denominator of Eq. 1.63, despite its appearance, is an invariant scalar. This can be deduced by evaluating the square of the following four-vector in the laboratory frame ($\vec{P}_2 = 0$):

$$s = (E_1 + E_2)^2 - (\vec{P}_1 + \vec{P}_2)^2c^2$$
$$= (E_1 + m_2c^2)^2 - P_1^2c^2 = E_1^2 + m_2^2c^4 + 2E_1m_2c^2 - P_1^2c^2$$
$$= m_1^2c^4 + m_2^2c^4 + 2E_1m_2c^2 \tag{1.64}$$

Because s is a scalar, it has the same value when calculated in any reference frame. In particular, it has a simple meaning in the center-of-mass frame, where the two particles have equal and opposite momenta (i.e., the total momentum vanishes in the center-of-mass frame):

$$s = m_1^2c^4 + m_2^2c^4 + 2E_1m_2c^2 = (E_{1CM} + E_{2CM})^2 - \left(\vec{P}_{1CM} + \vec{P}_{2CM}\right)^2 c^2$$
$$= (E_{1CM} + E_{2CM})^2 = \left(E_{CM}^{TOT}\right)^2 \tag{1.65}$$

Thus s is the square of the total energy available in the center of mass. Hence, in terms of the total center-of-mass energy, we can write

$$\gamma_{\text{CM}} = \frac{E_1 + m_2c^2}{E_{\text{CM}}^{\text{TOT}}} = \frac{E_{\text{Lab}}^{\text{TOT}}}{E_{\text{CM}}^{\text{TOT}}} \tag{1.66}$$

The variable s is used frequently in describing high-energy collisions, and $E_{\text{CM}}^{\text{TOT}}$ is often referred to as $\sqrt{s}$. Clearly, from its structure in Eq. 1.65, $\sqrt{s}/c^2$ can also be regarded as the rest mass or the invariant mass of the two colliding objects.

In discussing scattering problems, it is often convenient to define another invariant called t, the square of the four-momentum transfer in a collision. This variable is just the square of the difference in the energy-momentum four-vectors of the projectile before and after the scattering:

$$t = \left(E_1^f - E_1^i\right)^2 - \left(\vec{P}_1^f - \vec{P}_1^i\right)^2 c^2 \tag{1.67}$$

Because momentum and energy are conserved separately in all collisions, we can express t as well in terms of the target variables

$$t = \left(E_2^f - E_2^i\right)^2 - \left(\vec{P}_2^f - \vec{P}_2^i\right)^2 c^2 \tag{1.68}$$

Furthermore, since t is an invariant, just as s is, we can calculate its value in any reference frame. In particular, let us analyze this quantity in the center-of-mass frame. For simplicity, we restrict ourselves to the case of elastic scattering, for which $|\vec{P}_{\text{CM}}^i| = |\vec{P}_{\text{CM}}^f| = |\vec{P}_{\text{CM}}|$ and consequently $E_{\text{CM}}^i = E_{\text{CM}}^f$ for the two particles in the center-of-mass frame. In this case, we note from Eq. 1.67 that

$$t = -\left(P_{1\text{CM}}^{f2} + P_{1\text{CM}}^{i2} - 2\vec{P}_{1\text{CM}}^f \cdot \vec{P}_{1\text{CM}}^i\right)c^2 = -2P_{\text{CM}}^2c^2(1 - \cos\theta_{\text{CM}}) \tag{1.69}$$

where we have set $|\vec{P}_{1\text{CM}}^f| = |\vec{P}_{1\text{CM}}^i| = P_{\text{CM}}$, and where θ_{CM} denotes the scattering angle in the center-of-mass frame. Since $-1 \le \cos\theta_{\text{CM}} \le 1$, we conclude that for scattering through any finite angle, $t < 0$. On the other hand, from its definition in Eq. 1.67, we note that we can think of t as the square of the mass of an exchanged particle (with energy $E_1^f - E_1^i$ and momentum $\vec{P}_1^f - \vec{P}_1^i$) that mediates the scattering. Consequently, we must conclude that if such an exchange process can be used to describe scattering, then the object being exchanged cannot be physical since it has an imaginary rest mass. This means that although this "virtual" object cannot be detected, if the picture is correct, its consequences can be calculated and observed. Diagrams of the kind shown in Fig. 1.7 were pioneered by Richard Feynman in the calculation of scattering amplitudes in quantum electrodynamics (QED) and are referred to as Feynman graphs.

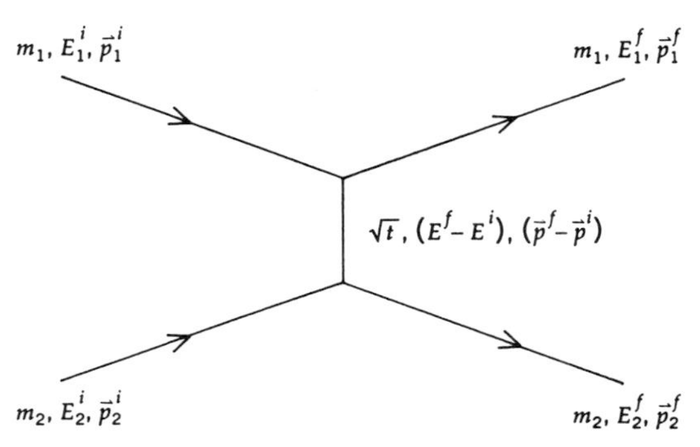

Figure 1.7 Exchange of a mediating object of mass $\sqrt{t}$ in the collision of masses m_1 and m_2.

For convenience, let us define a variable q^2 given by $q^2c^2 = -t$. In the laboratory frame, we have that $\vec{P}^i_{2\text{Lab}} = 0$ and, therefore, from Eq. 1.68 we obtain

$$\begin{aligned}
q^2c^2 &= -\left[\left(E^f_{2\text{Lab}} - m_2c^2\right)^2 - \left(P^f_{2\text{Lab}}c\right)^2\right] \\
&= -\left[\left(E^f_{2\text{Lab}}\right)^2 - \left(P^f_{2\text{Lab}}c\right)^2 - 2E^f_{2\text{Lab}}m_2c^2 + m_2^2c^4\right] \\
&= -\left[2m_2^2c^4 - 2E^f_{2\text{Lab}}m_2c^2\right] \\
&= 2m_2c^2\left(E^f_{2\text{Lab}} - m_2c^2\right) = 2m_2c^2T^f_{2\text{Lab}}
\end{aligned}$$

or

$$q^2 = 2m_2T^f_{2\text{Lab}} \tag{1.70}$$

where, in the last step, we replaced $E^f_{2\text{Lab}}$ by $T^f_{2\text{Lab}} + m_2c^2$. Thus, in the nonrelativistic limit, where $T_{2\text{Lab}} \sim \frac{1}{2}m_2v_2^2$, q^2 is just the square of the momentum transferred to the target, namely $q^2 \simeq (m_2v_2)^2$. We consequently expect q^2 to reflect the "hardness" of the collision, with small q^2 being characterized by long-range ($R \sim \hbar/q$) soft collisions. Referring back to Eq. 1.69, we see that small q^2 must correspond to small θ_{CM}. Also, we see that, for small θ_{CM}, $q^2 \simeq P^2_{\text{CM}}\theta^2_{\text{CM}} \simeq p^2_T$, or the square of the transverse momentum developed as a result of the collision.

We will leave it to the reader to show, with the help of Appendix A, that the relativistic equivalent of Eq. 1.51 is

$$\tan\theta_{\text{Lab}} = \frac{\beta^*\sin\theta_{\text{CM}}}{\gamma_{\text{CM}}(\beta^*\cos\theta_{\text{CM}} + \beta_{\text{CM}})} \tag{1.71}$$

where $\beta^* c$ is the velocity of the scattered particle in the center-of-mass frame. Note also that Eq. 1.71 reduces to Eq. 1.51 in the limit of low velocities.

Finally, let us rewrite the Rutherford cross section of Eq. 1.37 in terms of the momentum transferred between the two objects involved in the scattering process. From Eq. 1.69, we can deduce that

$$dq^2 = -2P^2 d(\cos\theta) = \frac{P^2 d\Omega}{\pi} \tag{1.72}$$

where we have ignored the small difference between CM and Lab variables ($P_{1\text{Lab}} \simeq P_{1\text{CM}} = P = m_1 v_0$). Specializing to the case of Rutherford scattering at low velocities and, for convenience, setting $m = m_1 \ll m_2$, $v = v_0$, in Eq. 1.37 we obtain

$$\frac{d\sigma}{(\pi/(mv)^2)dq^2} = \frac{(ZZ'e^2)^2}{(2mv^2)^2} \frac{1}{(1 - \cos\theta/2)^2}$$

or

$$\frac{d\sigma}{dq^2} = \frac{4\pi(ZZ'e^2)^2}{v^2} \frac{1}{q^4} \tag{1.73}$$

The q^{-4} divergence of the cross section is characteristic of Coulomb scattering, and reflects the r^{-1} dependence of the potential. It is important to recognize that there is a distribution in q^2, with different events having different momentum transfers. The rapid falloff with q^2 means that the typical value of momentum transfer will be small. The dispersion in this mean has important physical consequences that are brought up in Chapter VI. Although the minimum value of q^2 can be zero, this value corresponds to no scattering; the maximum value (a rare occurrence, indeed!) is $4P^2$. Although Eq. 1.73 was obtained using nonrelativistic kinematics, it also holds, in fact, as $v^2 \to c^2$ (see, however, our comments in the section on the Sizes of Nuclei in Chapter II).

QUANTUM TREATMENT OF RUTHERFORD SCATTERING

We arrived at Eq. 1.73 through a rather circuitous classical route. We now end this section by sketching how the Rutherford cross section can be calculated using quantum mechanics. This is done through an application of Fermi's Golden Rule,* from which the transition probability per unit

*A discussion of this famous result for transitions between states can be found in standard texts on quantum mechanics.

time in perturbation theory is given by

$$P = \frac{2\pi}{\hbar}|H_{fi}|^2 \rho(E_f) \tag{1.74}$$

where $\rho(E_f)$ is the density of final states and H_{fi} denotes the matrix element of the perturbation Hamiltonian between the initial and the final states

$$H_{fi} = \langle f|H|i\rangle = \int d^3r\, \psi_f^*(r) H(r) \psi_i(r) \tag{1.75}$$

For the case of elastic Rutherford scattering, the wave functions are plane waves, corresponding to free-particle states approaching (i) and leaving (f) the scattering center; and the perturbation Hamiltonian is the Coulomb potential energy given in Eq. 1.23. For the incident and outgoing momenta $\vec{p}$ and $\vec{p}\,'$, respectively, we define the wave vectors $\vec{k} = \vec{p}/\hbar$ and $\vec{k}\,' = \vec{p}\,'/\hbar$, and a momentum transfer that results from the scattering $\vec{q} = \hbar(\vec{k}\,' - \vec{k})$. Except for an overall normalization of the wave functions, our matrix element H_{fi} can now be written as follows:

$$H_{fi} \simeq \int_{\substack{\text{all}\\ \text{space}}} e^{i\vec{k}'\cdot\vec{r}}\, V(r) e^{-i\vec{k}\cdot\vec{r}}\, d^3r = \int_{\substack{\text{all}\\ \text{space}}} V(r) e^{(i/\hbar)\vec{q}\cdot\vec{r}}\, d^3r \tag{1.76}$$

The integral on the right is the Fourier transform of $V(r)$, and can be thought of as the potential energy in momentum space. Doing the integration,* we find that

$$V(q) = \int_{\substack{\text{all}\\ \text{space}}} V(r) e^{(i/\hbar)\vec{q}\cdot\vec{r}}\, d^3r = \frac{(ZZ'e^2)(4\pi\hbar^2)}{q^2} \tag{1.77}$$

Evaluating the density of final states,† substituting into Eq. 1.74, and relating the transition probability to the scattering cross section, leads to the same expression as obtained in Eq. 1.73. Thus Rutherford's result, without any apparent reference to $\hbar$, is also in agreement with quantum mechanics (when spin effects are ignored).

Problems

I.1 Using Eq. 1.38 calculate the total cross sections for Rutherford scattering of a 10-MeV α-particle from a lead nucleus for impact parameters

---

*The Fourier transform corresponds to a generalization of the Fourier decomposition of functions into series. Transforms of different functions can be found in mathematical tables and are useful for a variety of applications in physics. See, for example, L. Schiff, *Quantum Mechanics* (New York. McGraw-Hill, 1968), and A. Das and A. C. Melissinos, *Quantum Mechanics* (New York. Gordon & Breach, 1986).

†See the discussion of this issue, and matters pertaining to this entire section, in A. Das and A. C. Melissinos, *Quantum Mechanics*, pp. 199–204.

b less than 10^{-12}, 10^{-10}, and 10^{-8} cm. How well do these agree with the values of πb^2?

I.2 Prove that Eq. 1.55 follows from the relations in Eqs. 1.53 and 1.54.

I.3 Sketch $\cos\theta_{\text{Lab}}$ as a function of $\cos\theta_{\text{CM}}$ for the nonrelativistic elastic scattering of particles of unequal mass, for the cases when $\zeta = 0.05$ and $\zeta = 20$ in Eqs. 1.52 and 1.53.

I.4 What would be the approximate counting rate observed in the Rutherford scattering of 10-MeV α-particles off lead foil at an angle of $\theta = \pi/2$ in the laboratory? Assume an incident flux of 10^6 α-particles per second on the foil, a foil 0.1 cm thick, and a detector of transverse area 1 cm $\times$ 1 cm placed 100 cm from the interaction point, and density of lead of 11.3 g/cm^3. What would be the counting rate at $\theta = 5°$? By about how much would your answers change if the angles were specified instead in the center of mass–be quantitative, but use approximations where necessary. (Why don't you have to know the area of the foil?)

I.5 Sketch the cross section in the laboratory frame as a function of $\cos\theta_{\text{Lab}}$ for the elastic scattering of equal-mass particles when $d\sigma/d\Omega_{\text{CM}}$ is isotropic and equal to 100 mb/sr. What would be your result for $\zeta = 0.05$ in Eq. 1.52? (You may use approximations where necessary.)

I.6 Certain radioactive nuclei emit α particles. If the kinetic energy of these α particles is 4 MeV, what is their velocity if you assume them to be nonrelativistic? How large an error do you make in neglecting special relativity in the calculation of v? What is the closest that such an α-particle can get to the center of a Au nucleus?

I.7 An electron of momentum 0.5 MeV/c is observed in the laboratory. What are its $\beta = (v/c)$, $\gamma = (1 - \beta^2)^{-1/2}$, kinetic energy, and total energy?

I.8 What are the approximate values of kinetic energy for the recoiling lead nucleus and the momentum transfers (in eV units) at the cutoffs specified in Problem I.1?

I.9 Taking the ultrarelativistic limit of Eq. 1.71, find an approximate expression for θ_{Lab} at $\theta_{\text{CM}} = \pi/2$, and evaluate θ_{Lab} for $\gamma_{\text{CM}} = 10$ and $\gamma_{\text{CM}} = 100$. Does the approximation hold best for particles with small or large mass values?

Suggested Readings

Geiger, H. and E. Marsden, 1913. Philos. Mag. **25**: 604.

Rutherford, E. 1911. Philos. Mag. **21**: 669.

Thomson, J. J. 1910. Cambridge Lit. Philos. Soc. **15**: 465.

Chapter II

NUCLEAR PHENOMENOLOGY

INTRODUCTORY REMARKS

The Rutherford scattering experiments clearly demonstrated that each atom has a positively charged central core that we call its nucleus. Although before the discovery of the neutron by James Chadwick it was thought that the nucleus consisted of protons and electrons, it is now recognized that the constituents of the nucleus are protons and neutrons—collectively known as nucleons. In fact, we will argue later that the presence of electrons inside the nucleus is inconsistent with experimental observations. Let us emphasize once again that much of what we know about nuclei and the nuclear force has been obtained through decades of painstaking experimentation. In what follows, we merely summarize the main features, and only occasionally present some of the experimental underpinnings.

PROPERTIES OF NUCLEI

Labeling of Nuclei

The nucleus of any atom X can be labeled uniquely by its *charge*, or the atomic number Z, and its total number of nucleons A, and conventionally represented as ${}^{A}X^{Z}$. Alternatively, it can be specified by the number of protons (Z) and the number of neutrons ($N = A - Z$). Because the whole atom is electrically neutral, the nucleus must be surrounded by a cloud of Z electrons. A great many nuclei, with various Z and A values, have been found in nature or produced in the laboratory. Nuclei with the same number of protons but different number of neutrons are known as *isotopes*. All such atoms have similar chemical properties; thus, ${}^{A}X^{Z}$ and ${}^{A'}X^{Z}$ are isotopes. Nuclei that have the same total number of nucleons but different numbers of protons are called *isobars*. Thus ${}^{A}X^{Z}$ and ${}^{A}Y^{Z'}$ are isobars. Just as an atom can be found in an excited as well as in its ground state, so also can a nucleus be excited to higher levels, in which case such states are referred to as *resonances* or *isomers* of the ground state.

Masses of Nuclei

As we already mentioned, a nucleus, ${}^{A}X^{Z}$, contains Z protons and (A–Z) neutrons. Thus, naively, we would expect the mass of the nucleus to be

$$M(A, Z) = Z m_p + (A - Z) m_n \tag{2.1}$$

where m_p and m_n denote, respectively, the masses of the proton and the neutron, which we know experimentally to be

$$\begin{aligned} m_p &\simeq 938 \cdot 27\ \text{MeV}/c^2 \\ m_n &\simeq 939 \cdot 56\ \text{MeV}/c^2 \end{aligned} \tag{2.2}$$

The actual measurement of nuclear masses reveals, however, that the mass of a nucleus is smaller than the sum of the masses of its constituents.* Namely,

$$M(A, Z) < Z m_p + (A - Z) m_n \tag{2.3}$$

This explains why a nucleus, by itself, cannot break apart into its constituents, since that would violate conservation of energy. The mass deficit, defined as

$$\Delta M(A, Z) = M(A, Z) - Z m_p - (A - Z) m_n \tag{2.4}$$

---

*As an aside about masses, we should point out that isotope charts usually give masses of neutral atoms and not of the nuclei. To get the nuclear mass one must subtract the electron masses ($Z m_e$) from the atomic weights (ignoring the small differences in electron bindings). Unfortunately, chemists and physicists use different mass scales. Chemists assign 16.0 atomic mass units (amu) to the "natural" isotopic mixture of oxygen found on earth, while physicists assign 16.0 amu to the atom of ${}^{16}O^{8}$. One amu is the mass in grams of one fictitious atom that has an atomic weight of 1.0000 gm. Thus 1 amu $= (A_0^{-1})$ gm $= 1.6606 \times 10^{-24}$ gm. (The latest value for A_0 is $(6.022098 \pm 0.000006) \times 10^{23}$ mole^{-1}.) There is also the unified mass unit "u," defined as $\frac{1}{12}$ of the mass of the ^{12}C atom. We will use $m_p = 1.00728$ amu $= 938.27$ MeV/$c^2 = 1.6726 \times 10^{-24}$ gm, and $m_n = m_p + 1.29332$ MeV/c^2.

is negative, and can be thought of as being proportional to the nuclear binding energy (B.E.); the absolute value of ΔM is related to the minimum energy required to break up the nucleus into its components. Thus a negative B.E. assures that the nucleus holds together, and the more negative the value of ΔM, the more stable the nucleus. The mass deficit and the B.E. are simply related as

$$B.E. = \Delta M(A, Z)c^2 \tag{2.5}$$

where c is the speed of light. Thus, $-\Delta M c^2$ or $-$B.E. is the amount of energy required to release the nucleons from their captivity within the nucleus. It is instructive to define a binding energy per nucleon, or the average energy needed to release a nucleon from a nucleus. This is defined as

$$\frac{B}{A} = \frac{-\text{B.E.}}{A} = \frac{-\Delta M(A, Z)c^2}{A} = \frac{(Z m_p + (A - Z)m_n - M(A, Z))c^2}{A} \tag{2.6}$$

This quantity has been measured for a wide range of stable nuclei (see Fig. 2.1) and, except for some fine structures that we will discuss later, shows a remarkable feature.

For low-mass nuclei ($A \lesssim 20$), B/A oscillates somewhat and increases rapidly with A, and then saturates, reaching a peak value of about 9 MeV/nucleon near $A = 60$; for larger A, B/A drops very slowly. The approximate average value of B/A for a wide range of nuclei can therefore be taken as about 8 MeV/nucleon. As we will see, these characteristics have great implications for the nature of the nuclear force and the structure of the nucleus. One immediate deduction is that if we deposit about 8 MeV of kinetic energy inside the nucleus, and transfer it all to one

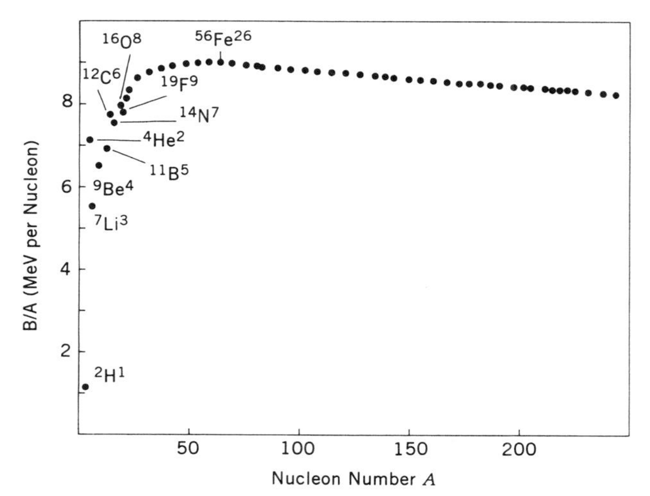

Figure 2.1 The binding energy per nucleon for the most stable nuclei.

nucleon, then we can, in principle, free that nucleon from the binding of the strong force, and it can escape the nucleus and materialize.

To appreciate the significance of the last observation, let us recall that quantum mechanically all particles display wave behavior. In fact, for any particle with momentum p, we have an associated wavelength (according to de Broglie's hypothesis)

$$\lambda\!\!\!^{-} = \frac{\hbar}{p} \tag{2.7}$$

where $\hbar$ and $\lambda\!\!\!^{-}$ are Planck's constant and the wavelength λ divided by 2π (referred to as the *reduced* wavelength). (The de Broglie bound-state requirement corresponds to $2\pi r = n\lambda$, and $\lambda\!\!\!^{-}$ consequently reflects a typical radial size.) Now, let us assume that we transfer about 8 MeV of kinetic energy to a nucleon within a nucleus. Being quite massive (m~940 MeV/c^2), the nucleon will be essentially nonrelativistic. Calculating its wavelength from nonrelativistic kinematics, we obtain

$$\begin{aligned}\lambda\!\!\!^{-} &= \frac{\hbar}{p} = \frac{\hbar}{\sqrt{2mT}} = \frac{\hbar c}{\sqrt{2mc^2T}} \\ &\simeq \frac{197 \text{ MeV-fm}}{\sqrt{2 \times 940 \times 8} \text{ MeV}} \simeq \frac{197}{120} \text{ fm} \simeq 1.6 \text{ fm}\end{aligned}$$

or

$$\lambda\!\!\!^{-} \simeq 1.6 \times 10^{-13} \text{ cm} \tag{2.8}$$

where 1 fm is a femto-meter (10^{-15} m) or one fermi (after Enrico Fermi). Thus, this wavelength is within the range of typical nuclear dimensions, and it is therefore reasonable to expect to localize nucleons of such energies within the nucleus. Consequently, a nucleon with about 8 MeV of kinetic energy (or 120 MeV/c momentum) could either be absorbed into or emitted from a nucleus. On the other hand, if electrons were present inside a nucleus, a kinetic energy of about 8 MeV would make them relativistic, and in this case $pc \simeq T \simeq 8$ MeV would yield a far larger de Broglie wavelength

$$\begin{aligned}\lambda\!\!\!^{-} &= \frac{\hbar}{p} \simeq \frac{\hbar c}{T} \simeq \frac{\hbar c}{8 \text{ MeV}} \simeq \frac{197 \text{ MeV-fm}}{8 \text{ MeV}} \\ &\simeq 25 \text{ fm} \simeq 2.5 \times 10^{-12} \text{ cm}\end{aligned} \tag{2.9}$$

With a de Broglie wavelength substantially larger than any nuclear radius, it would be unnatural to imagine electrons of ~8 MeV residing inside nuclei. Well, then what about an electron with momentum of 120 MeV/c^2? That kind of electron could, in principle, fit into the nucleus, but it would have 120 MeV of energy, and would therefore not be consistent with the energy scales of ~ 8 MeV characterizing nuclear binding. This is, of course, a rather heuristic argument against the presence of electrons inside a nucleus. However, more direct experimental observations also support this deduction. (We will return later to other implications of the B/A data.)

Sizes of Nuclei

The size of a subatomic object must be defined rather carefully. For a quantum mechanical system, the size normally refers to the expectation value of the coordinate operator in an appropriate state. For an atom, this would correspond to the average coordinate of the outermost electron, which can usually be calculated, at least perturbatively. In the nuclear domain, there is no simple expression for the force, and we have to rely on interpretation of experiments to determine the size.

There are several ways to go about this. First, in the low-energy Rutherford-scattering experiment, we note that when the impact parameter is zero, namely when the projectile collides head-on with the scattering center, the distance of closest approach is a minimum (see Eq. 1.25), given by

$$r_0^{\min} = \frac{ZZ'e^2}{E} \qquad (2.10)$$

Such particles will, of course, be scattered backwards ($\theta = \pi$), and this distance of closest approach will provide an upper bound on the size of the nucleus. The assumption, here, is that low-energy α-particles will not be able to overcome the repulsive Coulomb barrier of the nucleus, and will therefore not be able to penetrate into the nucleus. Such low-energy measurements yield relatively poor upper limits, typically,

$$R_{\mathrm{Au}} \lesssim 3.2 \times 10^{-12} \text{ cm} \qquad R_{\mathrm{Ag}} \lesssim 2 \times 10^{-12} \text{ cm} \qquad (2.11)$$

An alternate way to measure the sizes of nuclei is to scatter very high-energy charged particles such as electrons off nuclei. For head-on collisions (i.e., when the impact parameter vanishes) we see from Eq. 2.10 that when E is large, then

$$r_0^{\min} \longrightarrow 0 \qquad (2.12)$$

That is, such particles can probe deeper into the nucleus. Because electrons interact mainly through the electromagnetic force, however, and are not at all sensitive to the nuclear force, they will be influenced primarily by the charge structure of the nucleus. In other words, using electron scattering, one can deduce the distribution of charge (the "form factor") in a nucleus. One can, then, define the radius of the charge distribution as representing the size of the nucleus. At relativistic energies, the magnetic moment of the electron also contributes to the scattering cross section. Neville Mott was first to generalize Rutherford scattering to include these spin effects. Systematic studies of the scattering of high-energy electrons, initiated by Robert Hofstadter and his colleagues during the late 1950s, have revealed the effects of spin and the extended nature of the nuclear charge distribution.

One can use yet another method to study sizes of nuclei. Here the idea is to take advantage of the effects of the strong force to probe the nucleus. If one uses beams of sufficiently energetic strongly interacting particles (such as π mesons, protons, etc.) to scatter off nuclei, then the effect of the relatively weak Coulomb force can be neglected. These projectiles interact quite readily with nuclei and are thereby effectively "absorbed" out of the beam. The result of the absorption is a diffraction pattern (see Fig. 2.2)—very similar to that observed in optical phenomena in the scattering of light from a slit or grating. By studying this diffraction pattern, one can again infer the size of the nucleus, which acts in many ways like an absorbing disc.

All these phenomenological investigations have provided a remarkably simple relation for the radial size of the nucleus as a function of its nucleon number A:

$$\begin{aligned} R &= r_0 A^{1/3} \\ &\simeq 1.2 \times 10^{-13} A^{1/3}\ \text{cm} = 1.2 A^{1/3}\ \text{fm} \end{aligned} \tag{2.13}$$

From the preceding we can conclude that nuclei have enormous mass densities of $\sim 10^{14}$ gm/cm^3, and that nucleons are quite tightly packed inside the nucleus.

Nuclear Spins and Dipole Moments

Both the proton and the neutron have been found to have a spin angular momentum of $\frac{1}{2}\hbar$. Furthermore, just as electrons in an atom can have orbital angular momentum, so also can nucleons inside a nucleus. We know from quantum mechanics, that orbital angular momentum can only take on integral values. The total angular momentum of the consitituents—namely, the vector sum of the orbital and intrinsic spin angular momenta—defines the spin of the nucleus. Thus, it is not surprising that nuclei with even atomic numbers have integral nuclear spin, whereas nuclei with odd atomic numbers have half-integral nuclear spin. However, what is surprising is that all nuclei with an even number of protons and an even number of neutrons (even-even nuclei) have zero nuclear spin. It is equally surprising that large nuclei have very small nuclear spins in their ground states. These facts lend credence to the hypothesis that spins of nucleons inside a nucleus are very strongly paired so as to cancel their overall effect.

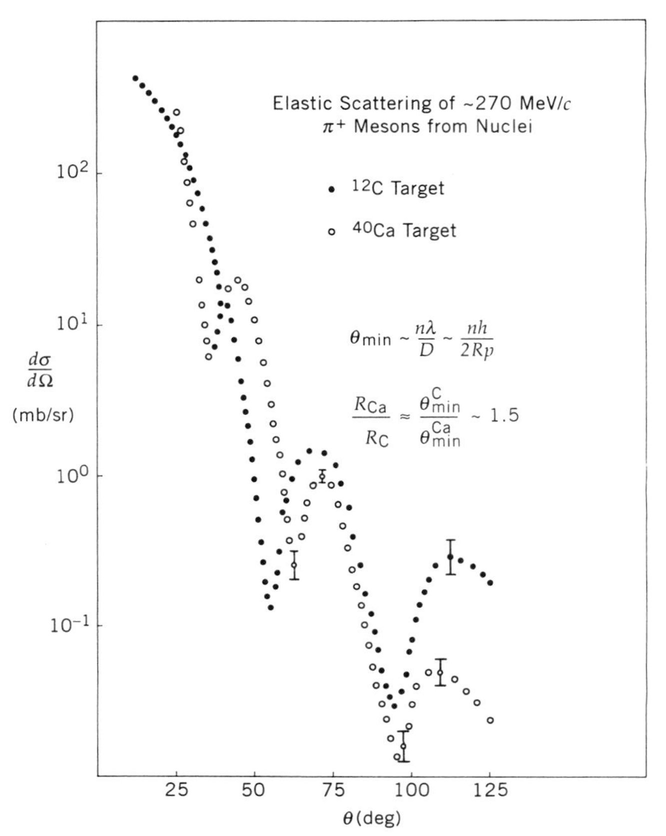

Figure 2.2 **Differential elastic cross sections for scattering of π^+ mesons of momentum of $\sim$ 270 MeV/*c* from carbon and calcium targets. The unit mb is 10^{-3} of a barn. Using the de Broglie wavelength for the π^+, and the optical analogy for the first minimum in the scattering, yields nuclear radii close to values expected from Eq. 2.13; also, the ratio of the angles at the minima for the two targets scale as the radii of these nuclei.**

Note: Data are based on C. H. Q. Ingram, *Meson-Nuclear Physics—1979*, AIP Conf. Proc. No. 54 (American Institute of Physics, New York).

We also know that a spinning charged particle has a magnetic dipole moment associated with it, given by

$$\vec{\mu} = g \frac{e}{2mc} \vec{S} \tag{2.14}$$

where e, m, and $\vec{S}$ are the charge, mass, and the intrinsic spin of the charged particle. The constant g is known as the Landé factor, which for a point particle, such as the electron, is expected to have the value $g = 2$. (In fact, small deviations at the level of 10^{-3} have been observed for the "pointlike" electron, which agrees with expectation from field-theoretical calculations of quantum electrodynamics (QED).) When $g \neq 2$, the particle is said to possess an anomalous magnetic moment, which is usually ascribed to the particle having a substructure. For the electron (with $|S_Z| = \frac{1}{2}\hbar$), the dipole moment $\mu_e \simeq \mu_B$, where μ_B is the Bohr magneton, defined as

$$\mu_B = \frac{e\hbar}{2m_e c} = 5.79 \times 10^{-11}\ \text{MeV/T} \tag{2.15}$$

where a magnetic field of 1 tesla (T) corresponds to 10^4 gauss (G). The magnetic dipole moment for nucleons is measured in terms of the nuclear magneton, defined using the proton mass

$$\mu_N = \frac{e\hbar}{2m_p c} \tag{2.16}$$

From the ratio of m_p/m_e, we deduce that the Bohr magneton is about 2000 times larger than the nuclear magneton.

The magnetic moments of the proton and the neutron have been measured to have the values

$$\begin{aligned} \mu_p &\simeq 2 \cdot 79\mu_N \\ \mu_n &\simeq -1 \cdot 91\mu_N \end{aligned} \tag{2.17}$$

It is clear, therefore, that both nucleons must have large anomalous contributions to their moments. This provides indirect evidence that these particles have additional structure. In fact, since the neutron is electrically neutral, its sizable magnetic moment is particularly dramatic and points to the fact that the neutron must have an extended charge distribution. The measurement of magnetic dipole moments for different nuclei have yielded the surprising result that all their values lie between $-3\mu_N$ and $10\mu_N$. This again is evidence for strong pairing inside the nucleus. Furthermore, this also shows that electrons cannot be present inside a nucleus because then it would be particularly hard to explain the small values of nuclear moments, since even one electron would produce a moment a thousand times that observed for nuclei.

Stability Curve

When we examine the characteristics of all stable nuclei, we find that for small $A (\lesssim 40)$ the number of protons equals the number of neutrons ($N = Z$). Beyond $A = 40$, the stable nuclei correspond to having $N \simeq 1.7Z$; namely, for these nuclei, the number of neutrons is larger than the number of protons (see Fig. 2.3). This can be understood from the fact

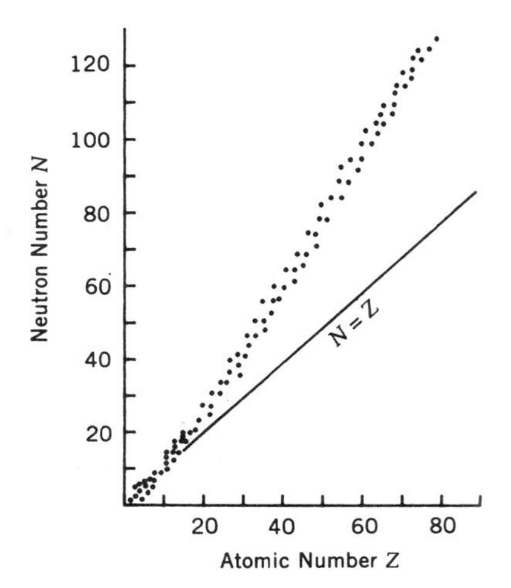

Figure 2.3 Neutron number as a function of atomic number for a representative sample of most stable nuclei.

that, for larger nuclei, the charge density, and therefore the destabilizing effect of Coulomb repulsion, will be smaller when there is a neutron excess.

Furthermore, a survey of the stable nuclei (see Table 2.1) reveals that even-even nuclei are the ones most abundant in nature. This again lends support to the strong pairing hypothesis, namely that pairing of nucleons leads to nuclear stability.

Instability of Nuclei

In 1896, through sheer accident, Henri Becquerel discovered natural radioactivity. He was studying fluorescent properties of uranium salts by exposing the material to the sun and then photographing the emission spectrum. Because the weather was cloudy, he stored the compound as well as some photographic plates inside a desk drawer. When he subsequently developed the plates, he noticed that they were overexposed, and surmised that the uranium compound must have emitted penetrating radiation, of a variety quite different from fluorescence. This was the first observation of natural nuclear radioactivity, and subsequent studies

Table 2.1 Number of stable nuclei as a function of even and odd number of nucleons

N	*Z*	*Number of Stable Nuclei*
Even	Even	156
Even	Odd	48
Odd	Even	50
Odd	Odd	5

have revealed that such spontaneous emission is a common phenomenon, especially characteristic of heavy nuclei.

Nuclear radioactivity involves the emission of essentially three kinds of radiation: α-radiation, β-radiation, and γ-radiation. Each of these emanations has distinct properties that were recognized in the following way. Consider a narrow and deep cavity within a piece of lead. Because lead easily absorbs nuclear radiation, the cavity will therefore function as a source of a well-collimated beam of radiation (see Fig. 2.4). If a magnetic field is applied perpendicular to the plane of the paper in Fig. 2.4, the beam will bend if it contains any charged components. The direction of bending will depend on the sign of the charge, whereas the amount of bending will be determined by the momentum of the particles in the beam. These simple experiments demonstrated that α-rays have positive charge, and because most of them arrived close to the same spot on the screen, they indicated that the α-particles in the beam were essentially monoenergetic, with typical velocities of about $0.1c$. Furthermore, the range of α-particles was found to be relatively short. (We discuss in Chapter VII how such measurements can be carried out.) In contrast, the most common forms of β-rays were found to bend in a direction opposite to that of α-rays, indicating that β-radiation consisted of negatively charged particles. The β-particles were observed to be well dispersed along the screen, which meant that, unlike the α-particles, the β-particles had a continuous spectrum of velocities, which were as high as $0.99\ c$. Other measurements revealed that β-particles had longer ranges and were less ionizing than α-particles. (We discuss ionization in more detail in Chapter VI.) It took about 3 mm of lead to stop typical β-particles, while a piece of paper sufficed to stop α-particles. Finally, a third form of emission, namely γ-radiation, was observed to arrive undeflected at the center of the screen, suggesting that such objects were charge–neutral. In fact, γ-rays behaved

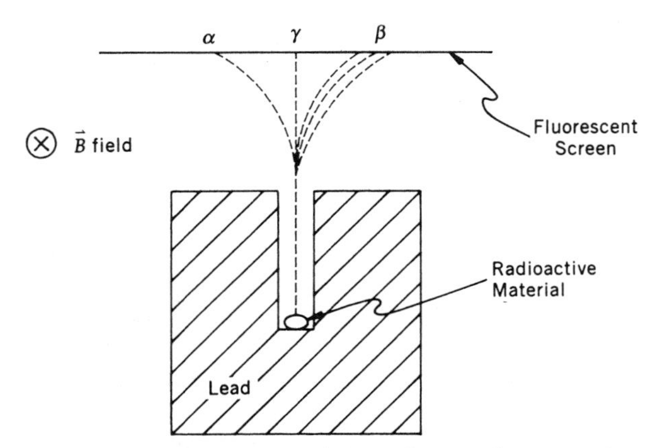

Figure 2.4 Separation of α- , β- , and γ-rays in a magnetic field.

in all respects like electromagnetic radiation, and therefore we can conclude that they are photons, which travel with the speed of light. Measurements reveal that the γ-rays have much longer ranges, and produce even less ionization per unit path, than the β-rays. Typically, it takes several centimeters of lead to completely stop (absorb) γ-rays.

It is, of course, also possible to deflect charged particles using an electric field. In fact, by applying an electric field in the plane of the paper in Fig. 2.4, perpendicular to both $\vec{B}$ and to the beam axis, and adjusting the magnitudes of the electric and the magnetic fields, the deflection of any charged particle can be varied, and even completely canceled. For a given electric and magnetic field, the deflection is a function of the charge and mass of the radioactive emission, and such measurements of deflections revealed that α-particles carried two units of positive charge and four units of atomic mass. In other words, α-particles were merely the very stable nuclei of helium atoms, namely $^4He^2$. Similarly, through such measurements, β-particles were identified as electrons. Thus, the most common forms of natural nuclear radiation, namely α-rays, β-rays, and γ-rays, correspond, respectively, to the spontaneous emission of helium nuclei, electrons, and energetic photons by heavy nuclei. It should be recognized, however, that any nuclear fragment can also be regarded as a form of radiation. More quantitative aspects of α, β, and γ emission are treated in Chapter IV.

NATURE OF THE NUCLEAR FORCE

In addition to determining the properties of nuclei, scattering experiments also provide more global information on the character of the nuclear force, as we summarize below.

First of all, it is clear that the nuclear force has no classical analogue. The gravitational attraction between nucleons is far too weak to bind them together. And the nuclear force cannot have an electromagnetic origin, since the nucleus of the deuteron contains only one proton and one neutron, and the neutron, being charge–neutral, has only very weak electromagnetic interactions (due to its magnetic dipole). In fact, as we have seen, the electromagnetic interaction (namely Coulomb repulsion) serves primarily to destabilize the nucleus.

It is also clear that the nuclear force must be extremely short-ranged. For example, the structure of the atom is explained exceedingly well by just using the electromagnetic interaction. Consequently, the range of the nuclear force cannot be much greater than the size of the nucleus, simply because otherwise it would affect the excellent agreement between theory and experimental observations in atomic physics. This argument would suggest that the range of the nuclear force is limited to about 10^{-13} cm–10^{-12} cm, which corresponds to the approximate size of nuclei.

Other important evidence for the short-ranged nature of the nuclear force comes from the fact that the binding energy per nucleon is a constant, essentially independent of the size of the nucleus. In fact, if the nuclear force had a long range like the Coulomb force, then given A nucleons, there would be $\frac{1}{2}A(A-1)$ pairwise interactions between them (that is, the total number of independent combinations for A nucleons, taken two at a time). Correspondingly, the binding energy, which basically measures the total potential energy of all possible interactions among the nucleons, would grow with the number of nucleons as

$$B \propto A(A-1) \tag{2.18}$$

Thus, for large values of A, we would have

$$\frac{B}{A} \propto A \tag{2.19}$$

In other words, if the force between any two nucleons were independent of the presence of other nucleons, the binding energy per nucleon would grow linearly with A. This is, in fact, what occurs for the Coulomb force, and it is primarily because a long-ranged force does not saturate, in the sense that any single particle can interact with as many other particles as are available. The net effect of this kind of force is that the binding becomes ever tighter as the number of interacting objects increases, and, as a result, the size of the interaction region remains fairly constant. This is the situation for the case of atomic binding, where atoms with a large number of electrons have sizes comparable to those with few electrons.

For the case of nuclei, however, we know from Fig. 2.1 that the binding energy per nucleon is essentially constant, and therefore we conclude that the nuclear force must saturate. Namely, a given nucleon can only interact with a finite number of nucleons in its neighborhood. Adding any more nucleons to a nucleus would therefore only increase the size of the nucleus, but not the binding energy per nucleon. As we have seen earlier (see Eq. 2.13), the size of a nucleus grows slowly with atomic number in such a way as to keep the nuclear density essentially fixed. These observations again lend support to the fact that the nuclear force is short-ranged.

We realize that, in general, the nuclear force must be attractive since it binds nucleons inside a nucleus. Experiments in which high-energy particles were scattered off nuclei have revealed, however, that the nuclear force has a repulsive core. Namely, we find that below a certain length scale, the nuclear force changes from being attractive to repulsive. (The presence of the repulsive core is best attributed to a quark substructure of the nucleon.) Conceptually, this result is appealing because, if the nuclear force were attractive at all distances, then the nucleus would collapse in on itself. Pictorially, we can represent the behavior of the nuclear force

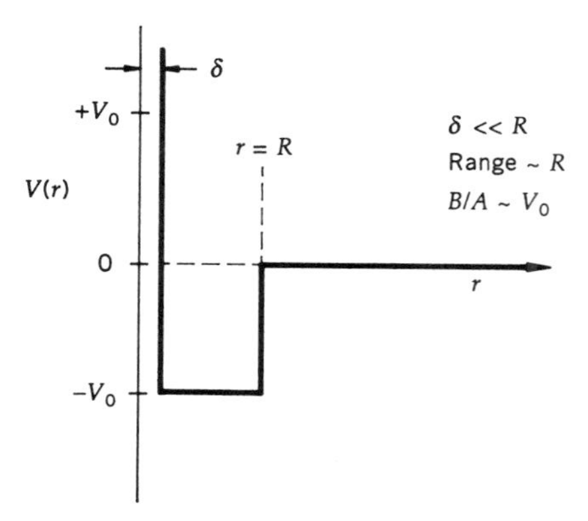

Figure 2.5 Approximate description of the nuclear potential as a function of distance to the center. The repulsive core is sensed only at small distances ($\delta \ll R$).

through a square-well potential that an incident nucleon can sense as it moves toward the nuclear center (see Fig. 2.5).

Because low-energy particles cannot probe short-distance behavior in the nucleus, to an excellent approximation the repulsive core can be ignored in problems pertaining to low-energy nuclear structure, and the nuclear force can be represented adequately through just a square-well potential.

We should point out that we do not expect the nuclear density or the nuclear force to cut off suddenly at some $r = R$, and so the square-well potential is meant to represent only the general effects of the nuclear force. It is more appropriate for incident neutrons than, for example, for protons, or for other incident nuclei, which, in addition, are subject to the repulsive Coulomb potential due to the positive charge of the nucleus (see Fig. 2.6). In the presence of Coulomb repulsion, an incident proton of total energy E_0 senses the Coulomb barrier as it approaches the nucleus. Classically, the proton cannot get closer than $r = r_0^{\min}$, because, for $R < r < r_0^{\min}$,

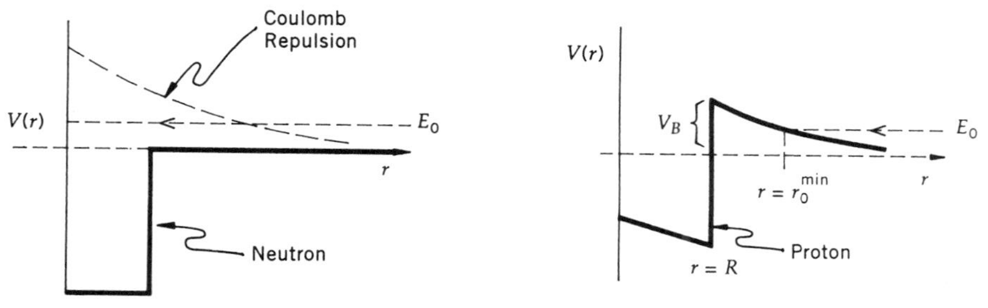

Figure 2.6 The potential energy of a proton and a neutron incident on a nuclear target. (Recall that, as a charged particle enters the nucleus, it sees less of the nuclear charge at the center, and the character of the potential changes from $1/r$ to $(3R^2 - r^2)$, and consequently the potential remains finite at $r = 0$.)

$V(r)$ would exceed E_0, and the kinetic energy would have to be negative, which is not physically possible. Ignoring the repulsive core for $r \lesssim \delta$ in Fig. 2.5, a neutron of same energy could, however, penetrate into the nuclear center.

It was once the hope that low-energy scattering experiments could be used to obtain the exact shape of the nuclear potential, but it turns out that the results of the scattering are not very sensitive to the details of the shape, but primarily to the range and the height of the potential. The square well is one of several forms of potentials that can provide a good phenomenological description of the nuclear force.

The fact that the nuclear force can be described through a potential energy function of the kind shown in Fig 2.5 suggests on the basis of quantum theory that nuclear systems can have discrete energy levels and corresponding bound states similar to the type found in atomic systems. The presence of such nuclear quantum states, and the transitions between them, have been confirmed in a variety of ways. They can be inferred from scattering experiments and through studies of the energies observed for emitted nuclear radiation. The modeling of both the ground levels and the excited nuclear states formed one of the early testing grounds for quantum mechanics. Some of the experimental evidence for nuclear levels, and several nuclear models, are described in the following chapters.

Studies of mirror nuclei,* and the scattering of protons and neutrons, demonstrate the interesting fact that, once we correct for known Coulomb effects, the force between two neutrons is the same as the force between two protons, which also coincides with the force between a proton and a neutron. This property of the nuclear force is referred to as *charge independence*. Namely, the strong nuclear force between two particles is independent of the electric charge carried by these particles. This is a remarkable result and, as we will see in Chapter IX, it leads to the concept of strong isotopic-spin symmetry. Very briefly, this symmetry implies that, just as the "spin-up" and the "spin-down" states of an electron correspond to two different spin states of the same particle, so the proton and the neutron correspond to two states of the same particle called the *nucleon*. If we could turn off the Coulomb field, the neutron and the proton would be indistinguishable in their nuclear interactions. This is analogous to the indistinguishability of spin-up and spin-down states in the absence of a magnetic field. We discuss this symmetry in more detail in Chapters IX and X.

Let us examine the question of range of the nuclear force from a somewhat different perspective, and note that the electromagnetic force between two interacting charged particles can be understood as a result of an exchange of a photon between them. The range of any force is related

---

*Mirror nuclei are isobars that have proton and neutron numbers interchanged as in ${}^{A}X^{Z}$ and ${}^{A}Y^{A-Z}$ (e.g., ${}^{15}O^{8}$ and ${}^{15}N^{7}$). Such pairs of nuclei have the same number of n–p interactions, but differ in their number of p–p and n–n interactions.

to the Compton wavelength ($\lambda\!\!\!^{-} = \hbar/mc$), or inverse of the mass of the particle (or quantum) being exchanged. Photon propagation is described by the Maxwell equations, which correspond to propagation at the speed of light (see Chapter XIII). Consequently, we presume that the photon is massless, and conclude that the range of the Coulomb force is infinite, which is consistent with what we know from experiment. Furthermore, we know that the Coulomb force is represented by the potential

$$V(r) \propto \frac{1}{r} \tag{2.20}$$

which, of course, shows explicitly that it is a long-ranged force.

In the case when the exchanged particle is massive, Hideki Yukawa showed in 1934 that the corresponding potential takes the form

$$V(r) \propto \frac{e^{-mcr/\hbar}}{r} \tag{2.21}$$

where m is the mass of the particle mediating the interaction.

In the limit that m vanishes, we recover the Coulomb potential of Eq. 2.20. This is, of course, what we would expect. From the form of the Yukawa potential, the range for the interaction is given by some characteristic value of r, which also corresponds to the Compton wavelength of the object of mass m:

$$\lambda\!\!\!^{-} = \frac{\hbar}{mc} \tag{2.22}$$

Therefore, once we know the mass of the exchanged particle, we can predict the range of the force. Conversely, if we know the range of the force, we can also predict the mass of the particle being exchanged. In the case of the nuclear force, a simple calculation shows that

$$m = \frac{\hbar}{\lambda\!\!\!^{-} c}$$

or

$$mc^2 = \frac{\hbar c}{\lambda\!\!\!^{-}} \simeq \frac{197 \text{ MeV-fm}}{1.2 \times 10^{-13} \text{ cm}} \simeq 164 \text{ MeV} \tag{2.23}$$

But this is approximately the mass of the well-known π meson (pion). There are, in fact, three pions, with masses

$$\begin{aligned} m_{\pi^+} &= m_{\pi^-} = 139.6 \, MeV/c^2 \\ m_{\pi^0} &= 135 \, MeV/c^2 \end{aligned} \tag{2.24}$$

This suggests that pions might be the mediators of the nuclear force. We will return later to a discussion of pions, other mesons, and their place in the development of the full story of charge independence of the strong force.

Problems

II.1 Calculate the approximate density of nuclear matter in gm/cm^3.

II.2 Calculate the difference between the binding energy of a nucleus of ^{12}C and the sum of the binding energies of three ^{4}He nuclei (α-particles). Assuming that ^{12}C is composed of three α-particles in a triangular structure, with three effective "α-bonds" between them, what would be the binding energy per α-bond? (See *CRC Handbook of Chemistry and Physics* for mass values.)

II.3 Calculate the binding energy of the last neutron in ^{4}He and the last proton in ^{16}O. How do these compare with B/A for these nuclei? What does this tell you about the stability of ^{4}He relative to ^{3}He, and of ^{16}O relative to ^{15}N? [*Hint:* The binding energy of the last neutron needed to form a nucleus (A, Z) is given by $[M(A-1, Z) + m_n - M(A, Z)]\,c^2$. An analogous expression holds for the last proton.]

II.4 Starting with cgs quantities, calculate the value of $\mu_B = e\hbar/2m_e c$, and convert it to MeV/ T units. (*Hint:* You can relate forces and magnetic fields through the Lorentz force equation $\vec{F} = q(\vec{v} \times \vec{B})/c$.)

II.5 Assume that the spin of a proton can be represented by a positive pion moving at a speed c in a circular orbit of radius 10^{-13} cm about a neutral center. Calculate the current and the magnetic moment associated with this motion. Compare this with the known magnetic moment of the proton. (*Hint:* Recall that using cgs units you can write a magnetic moment $\vec{\mu} = (I/c)\vec{A}$, where I is the current flowing around the area A.)

II.6 We argued previously that the π^+ mesons in Fig. 2.2 scattered from the entire nuclei. In fact, the first minima corresponded to $\theta \simeq h/2Rp$, with R being consistent with 1.2 $A^{1/3}$. At higher energies, when larger momenta can be transferred to nuclei, it is possible to dislodge a proton or neutron from the nucleus. When this can happen, the π^+ mesons can scatter elastically from a "free" nucleon. How would this affect the diffraction pattern in Fig. 2.2? What if you could scatter from very small pointlike constituents within nucleons? (Would the fact that a π^+ is not a point particle affect your answer?)

II.7 Normally, in optics, one looks at the diffraction pattern as a function of angle θ. In this case, the value of θ at the first minimum changes with wavelength or momentum. Can you see any advantage to using a variable such as $q^2 \simeq p_T^2 \simeq (p\theta)^2$ to examine diffraction patterns at different scattering energies? Sketch how the pattern might look for scattering of π^+ mesons of different energies from nuclear targets. Now, as energy increases, and larger q^2 become possible, what would be the effect of having nucleon substructure within the nucleus? What about point substructure within the nucleon? (Does your answer depend on whether the π^+ has such substructure?)

II.8 For magnetic fields of ~5 tesla, what are the frequencies that correspond to typical splitting of lines due to the presence of nuclear magnetic moments?

II.9 Show that when nonrelativistic neutrons of kinetic energy E_0 collide head-on with nuclei of mass number A at rest, the smallest energy that elastically scattered neutrons can have is given approximately by

$$E_{\text{min}} = E_0 \left(\frac{A-1}{A+1} \right)^2$$

What will be the approximate energies of the neutrons after one, two, and any number j of such consecutive collisions, if the target nucleus is hydrogen, carbon, and iron?

II.10 Using the results of Problem II.9, calculate the number of collisions needed to reduce the energy of a 2-MeV neutron to 0.1 MeV through elastic collisions between the neutron and carbon nuclei.

Suggested Readings

Chadwick, J. 1932. Proc. R. Soc. **A136**: 692.

Evans, R. D. 1955. *The Atomic Nucleus.* New York: McGraw-Hill.

Hofstadter, R., et al. 1960. Phys. Rev. Lett. **5**: 263.

——. 1956. Phys. Rev. **101**: 1131.

Yukawa, H. 1935. Proc. Phys. Math. Soc., Japan **17**: 48.

Chapter III

NUCLEAR MODELS

INTRODUCTORY REMARKS

A variety of early experiments clearly demonstrated that the character of the nuclear force differed markedly from any previously encountered in classical physics. However, a quantitative description of the nuclear force has turned out to be more elusive. As we know from atomic physics, where the correct level structure was found only after the classical Coulomb interaction between the nucleus and the electrons was extended to the atomic domain through quantum mechanics, knowing the properties of a force is only the first step in developing a theory of structure. Although neutrons and protons were known to be the nuclear constituents, the absence of a fundamental understanding of the nuclear force made it difficult indeed to determine the structure of the nucleus. It is not surprising therefore that, instead of a theory, phenomenological models of the nucleus were constructed to explain the many remarkable experimental findings. In the following, we describe only several such models. We should also keep in mind that, unlike the case of atomic physics, most of these nuclear models were proposed to explain only limited aspects of the experimental findings, which is precisely what they do.

LIQUID DROP MODEL

The liquid drop model of the nucleus was one of the earliest phenomenological successes constructed to account for the binding energy of a nucleus. As we have already discussed, experiments revealed that nuclei were essentially spherical objects, with sizes that could be characterized by radii proportional to $A^{1/3}$, which suggested that nuclear densities were almost independent of nucleon number. This leads quite naturally to a model that envisioned the nucleus as an incompressible liquid droplet, with nucleons playing the role analogous to molecules in a drop of normal liquid. In this picture, known as the *liquid drop model*, the individual quantum properties of nucleons are completely ignored.

As in the case of a liquid drop, here one imagines the nucleus as composed of a stable central core of nucleons for which the nuclear force is completely saturated, and a surface layer of nucleons, not as tightly bound (forces not entirely saturated). This leads to a decrease in the binding energy per nucleon (B/A) due to the weaker binding of the surface layer, and provides a "surface tension," or an effective attraction of the surface nucleons toward the center (see Fig. 3.1). If we attribute a constant binding energy (B.E.) per nucleon to the saturation of the nuclear force, as experiments appear to suggest, then, on the basis of these considerations, we can write a general form for the binding energy of a nucleus as follows:

$$\text{B.E.} = -a_1 A + a_2 A^{2/3} \tag{3.1}$$

where the first term represents a volume energy for the case of uniform saturated binding (remember that volume $\sim R^3 \sim A$), and the second term corrects for any overestimation due to the surface tension. It is clear that the correction to the binding energy per nucleon in Eq. 3.1 is higher for lighter nuclei because these have a larger surface-to-volume ratio of nucleons. That is, small nuclei have relatively more nucleons on the surface

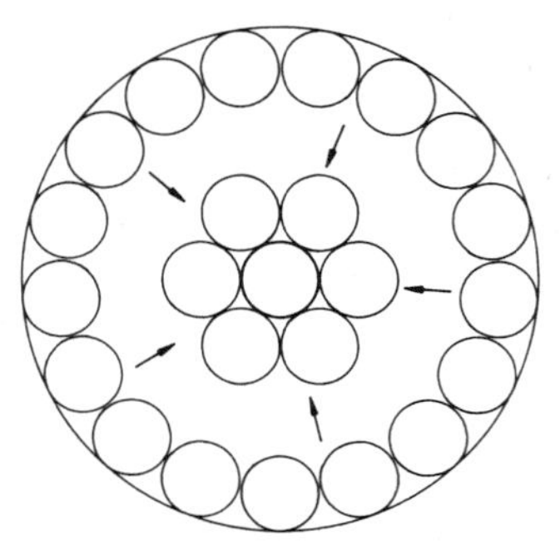

Figure 3.1 Surface layer and core of nucleus in the liquid drop model.

than in the core. This explains, therefore, why the binding energy per nucleon is smaller for lighter nuclei.

In this same model, the small decrease in the binding energy per nucleon for very heavy nuclei can be understood as due to Coulomb repulsion. Namely, if the nucleus has Z protons, then the Coulomb self-energy of these protons, which has a destabilizing effect, has the form $\sim Z^2/R$. Thus, adding such a positive term to reduce the binding strength, we can write

$$\text{B.E.} = -a_1 A + a_2 A^{2/3} + a_3 \frac{Z^2}{A^{1/3}} \tag{3.2}$$

The three terms in Eq. 3.2 arise from purely classical considerations. Unfortunately, they do not explain why lighter nuclei with an equal number of protons and neutrons are particularly stable. In other words, Eq. 3.2 does not yield the known fact that in lighter nuclei the binding is stronger (that is, the binding energy is more negative) when $N = Z$, and that these nuclei are more stable. Similarly, Eq. 3.2 does not explain the natural abundance of even-even nuclei nor the paucity of odd-odd nuclei. Such observations can be understood mainly as due to quantum effects (spin, statistics, etc.). Within the framework of the liquid drop model, however, they can be included by generalizing the empirical formula for the binding energy to contain additional phenomenological terms:

$$\text{B.E.} = -a_1 A + a_2 A^{2/3} + a_3 \frac{Z^2}{A^{1/3}} + a_4 \frac{(N-Z)^2}{A} \pm a_5 A^{-3/4} \tag{3.3}$$

where all the coefficients a_1, a_2, a_3, a_4, a_5 are assumed to be positive. Note that the fourth term implies that, unless $N = Z$, the binding energy will contain a positive contribution that will destabilize the nucleus. For small Z, where destabilization from the a_3 term is not very important, the a_4 term reflects the stability of $N = Z$ nuclei. In the last term, the positive sign is chosen for odd-odd nuclei, implying that such nuclei are relatively unstable. On the other hand, for even-even nuclei, the sign of this term is taken to be negative, implying greater stability and, therefore, abundance of such nuclei in nature. For odd A nuclei, the value of a_5 is chosen to be zero, primarily because the binding energy for such nuclei can be described quite well without the last term in Eq. 3.3.

The coefficients a_1, a_2, a_3, a_4, and a_5 can be determined by fitting the empirical formula to experimentally observed binding energies of a wide range of nuclei. The following set of values provides a rather good fit:

$$\begin{gathered} a_1 \simeq 15.6\text{MeV} \qquad a_2 \simeq 16.8\text{ MeV} \qquad a_3 \simeq 0.72\text{ MeV} \\ a_4 \simeq 23.3\text{ MeV} \qquad a_5 \simeq 34\text{ MeV} \end{gathered} \tag{3.4}$$

Given the phenomenological formula for the binding energy, we can also write an equivalent empirical relation for masses of nuclei as follows (see Eqs. 2.4 and 2.5):

$$\begin{aligned} M(A,Z) &= (A-Z)m_n + Zm_p + \frac{\text{B.E.}}{c^2} \\ &= (A-Z)m_n + Zm_p - \frac{a_1}{c^2}A \\ &\quad + \frac{a_2}{c^2}A^{2/3} + \frac{a_3}{c^2}\frac{Z^2}{A^{1/3}} + \frac{a_4}{c^2}\frac{(A-2Z)^2}{A} \pm \frac{a_5}{c^2}A^{-3/4} \end{aligned} \tag{3.5}$$

This expression, known as the Bethe–Weizsäcker semiempirical mass formula, can be used to predict stability and masses of unknown nuclei of arbitrary A and Z. It also plays a crucial role in a quantitative understanding of the theory of fission, as we will see in Chapter V.

THE FERMI-GAS MODEL

The *Fermi-gas model* was one of the earliest attempts to incorporate quantum mechanical effects into the discussion of nuclear structure. Here one assumes that a nucleus consists of a gas of free protons and neutrons confined to a very small region of space, namely to the nuclear volume. Under such conditions, the nucleons would be expected to populate discrete (quantized) energy levels within the nucleus. We can think of the protons and neutrons as moving inside a spherically symmetric well, whose range is given by the radius of the nucleus, and whose depth can be adjusted to obtain the correct binding energy. Because protons carry electric charge, as discussed in Chapter II, they will sense a potential that differs from that sensed by neutrons. The observed energy levels for neutrons and protons will therefore differ somewhat, depending on the specific range and depth of the individual potentials. We see in Chapter IX that all elementary particles can be classified as either bosons or fermions, and that protons and neutrons being fermions obey Fermi–Dirac statistics. This implies that any given energy level can be filled by no more than two identical nucleons (i.e., of same energy and charge) of opposite spin projection.

Since the lowest levels in a well have strongest binding, we expect that, in order for the ground state of any nucleus to have greatest stability, the energy levels will fill up from the bottom. The highest level that is completely filled defines what is referred to as the *fermi level* of energy E_F. If there is no fermion beyond the fermi level, the binding energy of the last nucleon is obtained simply from E_F. Otherwise, the energy of the fermion in the next level reflects the binding energy of the last nucleon.

If the depths of the wells for neutrons and protons were the same, then, in heavier nuclei, where the number of neutrons exceeds the number of protons, the fermi level for neutrons would lie higher than for protons.

If this were the case, then the binding energy of the last nucleon would be charge dependent, namely, different for protons and neutrons. This is inconsistent with experimental observations, and leads us to conclude that protons in such nuclei must move typically in a shallower potential well, so as to provide fermi levels for neutrons and protons of the same energy (see Fig. 3.2). In fact, if this were not the case, all such nuclei would be unstable, and neutrons would drop down to lower proton levels through β^- emission (such β^- decay processes are discussed in Chapter IV).

Let us next relate the energy of the fermi level to the number of fermions. We define the momentum associated with the fermi level through

$$E_F = \frac{p_F^2}{2m} \tag{3.6}$$

where m is the mass of a nucleon. Ignoring the presence of fermions beyond the fermi level, we can write the volume for states in momentum space as

$$V_{p_F} = \frac{4\pi}{3} p_F^3 \tag{3.7}$$

If V denotes the physical nuclear volume, then the total volume for states in what we may call "phase space" will be given by the product:

$$\begin{aligned} V_{\text{TOT}} = V V_{p_F} &= \frac{4\pi}{3} r_0^3 A \cdot \frac{4\pi}{3} p_F^3 \\ &= \left(\frac{4\pi}{3}\right)^2 A (r_0 p_F)^3 \end{aligned} \tag{3.8}$$

which is proportional to the total number of quantum states of the system. From the Heisenberg uncertainty principle, we know that for any quantum

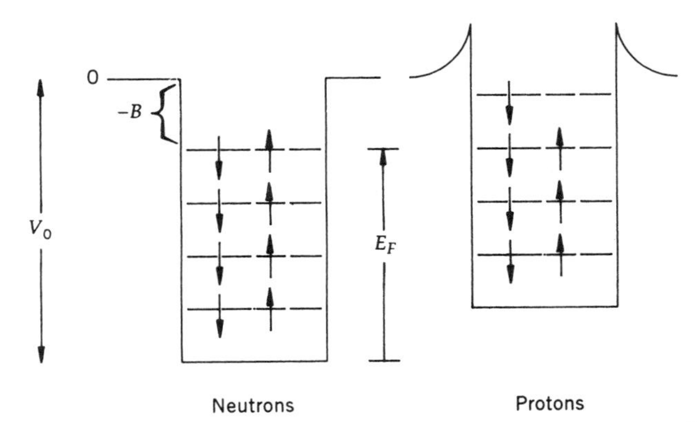

Figure 3.2 Ground-state energy levels for neutrons and for protons in nuclei.

mechanical state the same components of momentum and position obey the inequality:

$$\Delta x \Delta p_x \geq \frac{\hbar}{2} \tag{3.9}$$

This relation can be used to provide a restriction on the minimum volume that can be associated with any physical state of the system, which can be shown to be

$$V_{\text{state}} = (2\pi\hbar)^3 = h^3 \tag{3.10}$$

It follows, therefore, that the number of fermions that can fill states up to and including the fermi level is

$$n_F = 2\frac{V_{\text{TOT}}}{(2\pi\hbar)^3} = \frac{2}{(2\pi\hbar)^3}\left(\frac{4\pi}{3}\right)^2 A\,(r_0 p_F)^3 = \frac{4}{9\pi}A\left(\frac{r_0 p_F}{\hbar}\right)^3 \tag{3.11}$$

where the factor of 2 arises because each state can be occupied by two fermions with opposite spins.

Let us now consider, for simplicity, a nucleus with $N = Z = A/2$, and assume that all the states up to and including the fermi level are filled. In this case we have

$$N = Z = \frac{A}{2} = \frac{4}{9\pi}A\left(\frac{r_0 p_F}{\hbar}\right)^3$$

or

$$p_F = \frac{\hbar}{r_0}\left(\frac{9\pi}{8}\right)^{1/3} \tag{3.12}$$

In other words, the fermi momentum for this case is a constant independent of the nucleon number. It follows that

$$E_F = \frac{p_F^2}{2m} = \frac{1}{2m}\left(\frac{\hbar}{r_0}\right)^2\left(\frac{9\pi}{8}\right)^{2/3} \simeq \frac{2.32}{2mc^2}\left(\frac{\hbar c}{r_0}\right)^2 \simeq \frac{2.32}{2\times 940}\left(\frac{197}{1.2}\right)^2 \simeq 33\text{MeV} \tag{3.13}$$

Taking the average binding energy per nucleon of about -8 MeV to represent the binding of the last nucleon, it follows from our simple approximation that the depth of the potential well is about 40 MeV:

$$V_0 = E_F + B \simeq 40 \text{ MeV} \tag{3.14}$$

This result is consistent with the value of V_0 obtained through other considerations. The Fermi-gas model has been used to study excited states of complex nuclei, which can be accessed by "raising the temperature" (adding kinetic energy) to the nucleon gas. The model can also be shown to account in a natural way for the presence of the a_4 term in the Bethe–Weizsäcker mass formula of Eq. 3.5.

SHELL MODEL

The shell model of the nucleus is based on the analogous orbital structure of electrons in complex atoms. The model successfully accommodates many nuclear properties, and it is therefore useful to review briefly some of the relevant features of atomic structure before discussing the specifics of the nuclear shell model.

The binding of electrons to a nucleus in a complex atom is due to the Coulomb potential, which we know to be central. The electron orbits and the energy levels for such a quantum system can be obtained by solving the appropriate Schrödinger equation. In general, the solutions are quite complicated because they involve the Coulomb field of the nucleus as well as that of the other electrons, and cannot be obtained in closed analytic form. Certain characteristic features of the motion of the electron in a hydrogen atom have general relevance, however, and we discuss these first. Thus, for example, the orbits and the atomic energy levels that the electrons can occupy are labeled by a principal quantum number n (this determines the value of the energy for the case of hydrogen), which can take integral values

$$n = 1, 2, 3, \ldots \tag{3.15}$$

Furthermore, for any given value of the principal quantum number, there are energy-degenerate levels with orbital angular momentum given by

$$\ell = 0, 1, 2, \ldots, (n-1) \tag{3.16}$$

For any value of the orbital angular momentum, there are $(2\ell+1)$ substates with different values of the projection of orbital angular momentum along any chosen axis:

$$m_\ell = -\ell, -\ell+1, \ldots, 0, 1, \ldots, \ell-1, \ell \tag{3.17}$$

Due to the rotational symmetry of the Coulomb potential, all such substates will be degenerate in energy. Furthermore, since electrons have a spin angular momentum of $\frac{1}{2}\hbar$, each of the preceding states can be occupied by an electron with spin "up" or "down," corresponding to the spin-projection quantum number

$$m_s = \pm\frac{1}{2} \tag{3.18}$$

and, again, the energy corresponding to either of these spin configurations will be the same.

Thus, any energy eigenstate in the hydrogen atom is labeled by four quantum numbers, namely (n, ℓ, m_ℓ, m_s). For a given value of n, it follows that the number of such states that will be degenerate in energy is given by

$$\begin{aligned} n_d &= 2\sum_{\ell=0}^{n-1}(2\ell+1) \\ &= 2\left(2\sum_{\ell=0}^{n-1}\ell + n\right) \\ &= 2\left(2\cdot\tfrac{1}{2}\cdot(n-1)n + n\right) \\ &= 2(n^2 - n + n) = 2n^2 \end{aligned} \tag{3.19}$$

All of these states are degenerate in energy, however, only if there is no preferred direction in space that can break the rotational symmetry of the Coulomb interaction. That is, if there is a preferred direction given, for example, by some magnetic field, then the energy of the system could depend on the m_ℓ and m_s quantum numbers. Consequently, an interaction term such as $-\vec{\mu}\cdot\vec{B}$ added to the Coulomb potential could split previously degenerate energy levels. We know of interactions in an atom (see Fig. 3.3) such as the spin–orbit coupling between the spin angular momentum of the electron ($\vec{S} \sim \vec{\mu}$) and the magnetic field ($\vec{B} \sim \vec{L}$) due to the motion of the nucleus (as observed in the electron's rest frame), which change the energies of levels and thereby remove some of the degeneracies. In particular, spin–orbit interactions in atoms lead to a fine structure in the energy

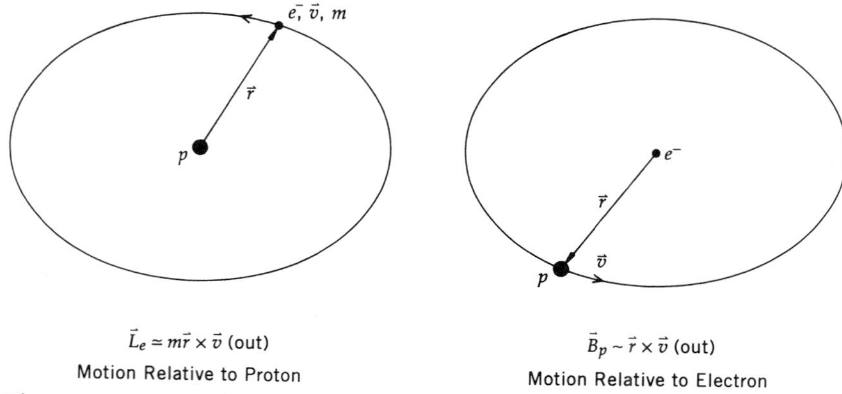

Figure 3.3 Spin–orbit coupling of the electron and proton in a hydrogen atom. Classically, the orbital motion of an electron is equivalent to a magnetic field due to the circulating proton. Thus a $\vec{\mu}_e \cdot \vec{B}_p$ term is equivalent to an $\vec{L}\cdot\vec{S}$ operator for the electron.

levels that has been well-studied. Because the effects of such interactions are usually quite small, they are often neglected in elementary discussions of atomic structure; however, as we shall see, they provide a key element in determining the nature of nuclear structure.

Consequently, ignoring fine structure, we can view the hydrogen atom as consisting of allowed electron orbits corresponding to shells of a given value of n, with each shell having degenerate subshells specified by the value of the orbital angular momentum. Going beyond hydrogen, and introducing the electron–electron Coulomb interactions, leads to a splitting in any energy level n according to the ℓ-value of the state. The larger the ℓ, the more aspherical is the orbit (and the larger is the average orbit radius), which yields less average binding, and produces a greater shift up in energy. The degeneracy in m_ℓ and m_s is not affected for more complex atoms. Any shell can accommodate $2n^2$ electrons, in consistency with the Pauli principle. It is quite straightforward to see that if a shell or a subshell is filled, then we have

$$\begin{aligned} \sum m_s &= 0 \\ \sum m_\ell &= 0 \end{aligned} \tag{3.20}$$

In other words, there is a strong pairing effect for closed shells. From the antisymmetry of the fermionic wave function (see Chapter IX), one can show that, in such cases, in general

$$\begin{aligned} \vec{L} &= 0 = \vec{S} \\ \vec{J} &= \vec{L} + \vec{S} = 0 \end{aligned} \tag{3.21}$$

For any atom with a closed shell or a closed-subshell structure, all electrons are paired off, and consequently no valence electrons are available. As a result, such atoms will be chemically inert. In fact, if we examine the inert elements, we find that they have just such a structure. For example, both electrons in the helium atom, He ($Z = 2$), fill up the shell corresponding to $n = 1$. Similarly, Ne ($Z = 10$) has closed shells corresponding to $n = 1$ and $n = 2$; Ar ($Z = 18$), on the other hand, has closed shells corresponding to $n = 1, 2$ and closed subshells corresponding to $n = 3$, $\ell = 0, 1$. The electrons in Kr ($Z = 36$) fill up the shells corresponding to $n = 1, 2, 3$, as well as the subshells corresponding to $n = 4$, $\ell = 0, 1$. Finally, Xe ($Z = 54$) has closed shells corresponding to $n = 1, 2, 3$ and closed subshells corresponding to $n = 4$, $\ell = 0, 1, 2$, as well as $n = 5$, $\ell = 0, 1$. (The energies of the oblong $n = 4$, $\ell = 3$ levels are above the more spherical $n = 5$, $\ell = 0, 1$ levels; the latter therefore get filled first.) These inert elements are exceedingly stable. In fact, their ionization energies are particularly large, as is consistent with their greater stability. The atomic numbers just given, namely,

$$Z = 2, 10, 18, 36, 54 \tag{3.22}$$

are called the *magic* numbers of atomic physics, and correspond to closed-shell structures.

In nuclei, as well, there is evidence for magic numbers. In fact, even though the binding energy per nucleon varies smoothly on a broad scale, a close examination shows peaks corresponding to specific values of nucleon numbers:

$$\begin{aligned} N &= 2, 8, 20, 28, 50, 82, 126 \\ Z &= 2, 8, 20, 28, 50, 82 \end{aligned} \tag{3.23}$$

Nuclei with either proton or neutron number corresponding to any of these magic values appear to be particularly stable, and are referred to as *magic nuclei*. Nuclei where both the proton and the neutron numbers are magic numbers (e.g., $^4He^2$, $^{16}O^8$, $^{208}Pb^{82}$) are known as *doubly magic*, and have even greater stability.

In addition to stronger binding of magic nuclei, other interesting nuclear features also suggest that nuclei possess shell structure. For example, magic nuclei have many more stable isotopes and isotones than their neighbors do. (Isotones are nuclei with the same number of neutrons but a different number of protons.) Thus, Sn ($Z = 50$) has ten stable isotopes, whereas both In ($Z = 49$) and Sb ($Z = 51$) have only two each. Similarly, for $N = 20$, there are five stable isotones, whereas $N = 19$ has none and $N = 21$ has only one, $^{40}K^{19}$, which is not very stable (has a mean life of about 10^9 years). Also, we know that a departure from a spherical charge distribution inside a nucleus can give rise to an electric quadrupole moment. Such moments are known to vanish for magic nuclei, whereas neighboring nuclei display large values. Again, this is reminiscent of behavior expected from shell structure. Similarly, neutrons capture cross sections—measured by scattering neutrons from nuclei of different neutron number—show a sharp drop for magic nuclei relative to their neighbors. This again suggests a shell structure for neutrons within nuclei.

Although there are many suggestive indications for shell structure in nuclei, in trying to set up and solve an appropriate Schrödinger equation, we face two essential differences from the situation in the case of atoms. First, there is no apparent central core that can provide the binding potential. Consequently, for the nuclear analog, we must picture the nucleons as moving in some effective mean potential within the nucleus. Second, whereas the well-understood Coulomb potential provides binding in the atomic case, the exact form of the nuclear potential is unknown. Nevertheless, since we are interested in obtaining shell structure, it is not unreasonable to at first assume that the mean potential in which the nucleons move is central. The Schrödinger equation for a central potential $V(r)$ has the form:

$$\left(-\frac{\hbar^2}{2m}\vec{\nabla}^2 + V(r)\right)\psi(\vec{r}) = E\psi(\vec{r})$$

or

$$\left(\vec{\nabla}^2 + \frac{2m}{\hbar^2}(E - V(r))\right)\psi(\vec{r}) = 0 \tag{3.24}$$

where E is the energy eigenvalue. Because we assume that the potential is spherically symmetric, the energy eigenstates will also be eigenstates of the angular momentum operator. (In other words, the system has rotational invariance, as a result of which, its angular momentum will be conserved. The angular momentum operator will therefore commute with the Hamiltonian of the system, and will have simultaneous eigenstates.) The energy eigenstates can therefore be labeled by the angular momentum quantum numbers. Under these circumstances it is convenient to use spherical coordinates, where we can write

$$\vec{\nabla}^2 = \frac{1}{r^2}\frac{\partial}{\partial r}r^2\frac{\partial}{\partial r} - \frac{1}{\hbar^2 r^2}\vec{L}^2 \tag{3.25}$$

where $\vec{L}^2$ is the angular momentum operator in coordinate space, whose eigenstates are the spherical harmonics $Y_{\ell,m_\ell}(\theta,\phi)$, satisfying

$$\begin{aligned}\vec{L}^2 Y_{\ell,m_\ell}(\theta,\phi) &= -\hbar^2\left[\frac{1}{\sin\theta}\frac{\partial}{\partial\theta}\sin\theta\frac{\partial}{\partial\theta} + \frac{1}{\sin^2\theta}\frac{\partial^2}{\partial\phi^2}\right]Y_{\ell,m_\ell}(\theta,\phi)\\ &= \hbar^2\ell(\ell+1)Y_{\ell,m_\ell}(\theta,\phi) \qquad (3.26)\\ L_z Y_{\ell,m_\ell}(\theta,\phi) &= -i\hbar\frac{\partial}{\partial\phi}Y_{\ell,m_\ell}(\theta,\phi) = \hbar m_\ell Y_{\ell,m_\ell}(\theta,\phi)\end{aligned}$$

Now, writing the Schrödinger wave function in a separable form, namely*

$$\psi_{n\ell m_\ell}(\vec{r}) = \frac{u_{n\ell}(r)}{r}Y_{\ell,m_\ell}(\theta,\phi) \tag{3.27}$$

where n, ℓ, and m_ℓ are, respectively, the radial, orbital and projection quantum numbers. Substituting Eq. 3.27 back into Eq. 3.24, we obtain the

*The symmetry of a wave function, that is, its response to some particular transformation, has important consequences. We discuss these issues in greater detail in Chapters X and XI, when we get to particle physics. Here we only wish to point out that, under inversion of coordinates, namely $\vec{r} \to -\vec{r}$, the length r does not change, $\theta \to \pi - \theta$, and $\phi \to \phi + \pi$. The net effect of this transformation is that the $Y_{\ell,m_\ell}(\theta,\phi)$, and therefore the total wave function, picks up a phase of $(-1)^\ell$. This defines the "parity" of a state. Thus when ℓ is even, there is no change in sign of the wave function, and the parity of the state is termed even. When the sign changes (for odd ℓ), the parity of the level is termed odd. Atomic and nuclear states have unique parity—they are either even or odd, but not mixtures of the two. (See Appendix B for a description of the properties of the $Y_{\ell,m_\ell}(\theta,\phi)$.)

radial equation:

$$\left(\frac{d^2}{dr^2} + \frac{2m}{\hbar^2}\left(E_{n\ell} - V(r) - \frac{\hbar^2 \ell(\ell+1)}{2mr^2}\right)\right)u_{n\ell}(r) = 0 \tag{3.28}$$

The radial equation has the form of a one-dimensional Schrödinger equation, but with two differences. First, for $\ell \neq 0$, there is an additional potential term due to a centrifugal barrier resulting from the orbital motion. Second, the boundary condition for the radial wave function, $u_{n\ell}(r)$, is that it must vanish at infinite separation, and also at the origin when $\ell \neq 0$. (This is essential for having a normalizable wave function.) Note that n is the radial quantum number and measures directly the number of nodes in the radial solution, and determines the energies of the states. It is related to the usual principal (or total) quantum number "n" of the hydrogen atom through $n = (\text{"}n\text{"} - \ell - 1)$. In the general case, therefore, n and ℓ are not correlated, and can take on any integral values.

It is impossible to obtain any further information about the energy levels of a nucleus without assuming a specific form for the potential. Two simple potentials that are commonly used in solving Eq. 3.28 are the infinite square well and the harmonic oscillator potentials. Although these potentials yield exact solutions for the system, they are not realistic because, among other things, they do not provide the possibility of barrier penetration through tunneling. A more realistic potential, such as a finite square well, can only yield numerical solutions, and is therefore not very useful for gaining qualitative insights. Fortunately, the qualitative features of the solutions are not too sensitive to the specific form of the potential, so in what follows we will restrict ourselves to the simpler potentials.

INFINITE SQUARE WELL

In this case, the potential is given by

$$V(r) = \begin{cases} \infty, & r \geq R \\ 0, & R \geq r \geq 0 \end{cases} \tag{3.29}$$

where R denotes the nuclear radius. The radial equation for $R \geq r \geq 0$ takes the form

$$\left(\frac{d^2}{dr^2} + \frac{2m}{\hbar^2}\left(E_{n\ell} - \frac{\hbar^2 \ell(\ell+1)}{2mr^2}\right)\right)u_{n\ell}(r) = 0 \tag{3.30}$$

The solutions that are regular at the origin are given by the oscillatory "spherical Bessel" functions (see Appendix C), namely

$$u_{n\ell}(r) = j_\ell(k_{n\ell} r) \tag{3.31}$$

where

$$k_{n\ell} = \sqrt{\frac{2mE_{n\ell}}{\hbar^2}} \tag{3.32}$$

Since the height of the well is infinite, the nucleons cannot escape from the well, and consequently the radial wave function must vanish at the boundary. In other words, we must have

$$u_{n\ell}(R) = j_\ell(k_{n\ell}R) = 0 \qquad \ell = 0, 1, 2, 3, \ldots \quad n = 1, 2, 3, \ldots \quad \text{for any } \ell \tag{3.33}$$

This boundary condition leads to the quantization of energy levels. In fact, the energy eigenvalue corresponding to any $k_{n\ell}$ is given by the nth zero of the ℓth spherical Bessel function. Since the zeros of Bessel functions are all distinct (nondegenerate), it follows that in the present case, there will be no degeneracy in energy corresponding to different combinations of n and ℓ values. Rotational invariance, however, would still provide a $(2\ell + 1)$ fold degeneracy in the energy levels corresponding to different m_ℓ values for a given ℓ. Also, since nucleons have a spin angular momentum of $\frac{1}{2}$, as usual each state can accommodate two neutrons or two protons, in consistency with the Pauli principle. Thus, we conclude that, for the case of an infinite square well, each shell can contain $2(2\ell + 1)$ protons or neutrons. It now follows that, for $n = 1$, closed shells can occur for the following proton or neutron numbers:

$$\mathbf{2},\ 2 + 6 = \mathbf{8},\ 8 + 10 = \mathbf{18},\ 18 + 14 = \mathbf{32},\ 32 + 18 = \mathbf{50}, \ldots \tag{3.34}$$

It is heartening to see that we can obtain several of the known magic numbers. This simple analysis, however, clearly does not yield the desired magic numbers 20, 82, 126. (We should add that we were somewhat sloppy in presenting the preceding results, in that we ignored all but the $n = 1$ solutions. The specific order in which energy levels are filled depends on the exact values of the zeros of the different Bessel functions. Taking other n values into account does not greatly affect our overall conclusions, namely, that the infinite square well potential does not reproduce all the nuclear magic numbers.)

HARMONIC OSCILLATOR

The radial equation for the three-dimensional harmonic oscillator potential

$$V(r) = \tfrac{1}{2}m\omega^2 r^2 \tag{3.35}$$

takes the form

$$\left(\frac{d^2}{dr^2} + \frac{2m}{\hbar^2}\left(E_{n\ell} - \frac{1}{2}m\omega^2 r^2 - \frac{\hbar^2 \ell(\ell+1)}{2mr^2}\right)\right)u_{n\ell}(r) = 0 \qquad (3.36)$$

In this case, the radial solutions are related to the associated Laguerre polynomials, as follows:

$$u_{n\ell}(r) \sim e^{-(m\omega r^2)/2\hbar} r^{\ell+1} L_{n+\ell-1/2}^{\ell+1/2}\left(\sqrt{\frac{m\omega}{\hbar}}r\right) \qquad (3.37)$$

and the energy eigenvalues of the bound states are given by:

$$E_{n\ell} = \hbar\omega\left(2n + \ell - \frac{1}{2}\right) \qquad n = 1, 2, 3, \ldots \quad \ell = 0, 1, 2, \ldots \quad \text{for any } n \qquad (3.38)$$

This can be rewritten in the more familiar form obtained from an analysis using Cartesian coordinates by defining the quantum number Λ:

$$\Lambda = 2n + \ell - 2 \qquad (3.39)$$

so that

$$E_{n\ell} = \hbar\omega\left(\Lambda + \frac{3}{2}\right) \qquad \Lambda = 0, 1, 2, \ldots \qquad (3.40)$$

where the ground state $\Lambda = 0$ has the characteristic nonvanishing zero-point energy.

Rotational invariance, as in the case of the infinite square well, implies a $(2\ell + 1)$-fold degeneracy for every value of ℓ, corresponding to different m_ℓ values. In the present case, however, there is a further degeneracy in the energy eigenvalues corresponding to different ℓ and n combinations that yield the same Λ. In fact, we note from Eq. 3.39 that when Λ is an even integer, then all the states with the following (ℓ, n) values

$$(\ell, n) = \left(0, \frac{\Lambda+2}{2}\right), \left(2, \frac{\Lambda}{2}\right), \left(4, \frac{\Lambda-2}{2}\right), \ldots, (\Lambda, 1) \qquad (3.41)$$

will be degenerate in energy. Similarly, if Λ is an odd integer, the states with (ℓ, n) values

$$(\ell, n) = \left(1, \frac{\Lambda+1}{2}\right), \left(3, \frac{\Lambda-1}{2}\right), \left(5, \frac{\Lambda-3}{2}\right), \ldots, (\Lambda, 1) \qquad (3.42)$$

will have the same energy. Thus, the total number of degenerate states corresponding to some even value of Λ is

$$
\begin{aligned}
n_\Lambda &= \sum_{\ell=0,2,4,\ldots}^{\Lambda} 2(2\ell + 1) \\
&= \sum_{k=0}^{\Lambda/2} 2(4k + 1) \\
&= 2\left(4 \cdot \frac{1}{2}\frac{\Lambda}{2}\left(\frac{\Lambda}{2} + 1\right) + \left(\frac{\Lambda}{2} + 1\right)\right) \\
&= 2\left(\frac{\Lambda}{2} + 1\right)(\Lambda + 1) = (\Lambda + 1)(\Lambda + 2) \qquad (3.43)
\end{aligned}
$$

Similarly, for a given odd value of Λ, the total number of degenerate states is

$$
\begin{aligned}
n_\Lambda &= \sum_{\ell=1,3,5,\ldots}^{\Lambda} 2(2\ell + 1) \\
&= \sum_{k=0}^{(\Lambda-1)/2} 2(2(2k + 1) + 1) \\
&= 2 \sum_{k=0}^{(\Lambda-1)/2} 4(4k + 3) \\
&= 2\left(4 \cdot \frac{1}{2}\frac{\Lambda - 1}{2} \cdot \left(\frac{\Lambda - 1}{2} + 1\right) + 3\left(\frac{\Lambda - 1}{2} + 1\right)\right) \\
&= 2\left(\frac{\Lambda + 1}{2}\right)(\Lambda - 1 + 3) = (\Lambda + 1)(\Lambda + 2) \qquad (3.44)
\end{aligned}
$$

Thus, we see that for any value of Λ, the total degeneracy of states is given by

$$
n_\Lambda = (\Lambda + 1)(\Lambda + 2) \qquad (3.45)
$$

It now follows that for the three-dimensional harmonic oscillator potential, closed shells can occur for proton or neutron numbers of 2, 8, 20, 40, 70, ... Once again, we see that this model predicts some of the magic numbers, but not all of them.

SPIN–ORBIT POTENTIAL

It was fairly clear by the 1940s that a central potential could not reproduce all the magic numbers. The crucial breakthrough came in 1949 when Maria Goeppert Mayer and Hans Jensen suggested—once again following the lead from atomic physics—that inside the nucleus, in addition to the central potential, there is a strong spin–orbit interaction, and therefore the total potential sensed by a nucleon has the form

$$V_{\mathrm{TOT}} = V(r) - f(r)\vec{L}\cdot\vec{S} \tag{3.46}$$

where $\vec{L}$ and $\vec{S}$ are the orbital and the spin angular momentum operators for a nucleon, and $f(r)$ is an arbitrary function of the radial coordinates. As we know from atomic physics, a spin–orbit interaction splits the $j = \ell \pm \frac{1}{2}$ energy levels and produces a fine structure. The spin–orbit interaction in Eq. 3.46 has precisely the same form as in atomic physics, except for the presence of the function $f(r)$. Secondly, the sign of this interaction is chosen to be consistent with the data, so that the state with $j = \ell + \frac{1}{2}$ will have a lower energy than the state with $j = \ell - \frac{1}{2}$. This is, in fact, opposite to what we have in atoms.

Now, the total angular momentum operator is given by

$$\vec{J} = \vec{L} + \vec{S} \tag{3.47}$$

so that we have

$$\vec{J}^2 = \vec{L}^2 + \vec{S}^2 + 2\vec{L}\cdot\vec{S}$$

or

$$\vec{L}\cdot\vec{S} = \tfrac{1}{2}(\vec{J}^2 - \vec{L}^2 - \vec{S}^2) \tag{3.48}$$

Here we have used the fact that orbital and spin angular momentum operators commute, and therefore their order in a product does not matter. Thus, in a state with definite ℓ, s, and j values (we note that a quantum state can be labeled either by the eigenvalues ℓ, m_ℓ, s, m_s or ℓ, s, j, m_j, and it is the second basis that is more useful for this calculation), we have

$$\begin{aligned}
\langle \vec{L}\cdot\vec{S}\rangle &= \left\langle \frac{1}{2}(\vec{J}^2 - \vec{L}^2 - \vec{S}^2)\right\rangle \\
&= \frac{\hbar^2}{2}[j(j+1) - \ell(\ell+1) - s(s+1)] \\
&= \frac{\hbar^2}{2}\left[j(j+1) - \ell(\ell+1) - \frac{3}{4}\right] \\
&= \begin{cases} \frac{\hbar^2}{2}\ell & \text{for } j = \ell + \frac{1}{2} \\ -\frac{\hbar^2}{2}(\ell+1) & \text{for } j = \ell - \frac{1}{2} \end{cases}
\end{aligned} \tag{3.49}$$

where we have substituted $s = \frac{1}{2}$ for the spin of a nucleon.

The shifts in the energies from their degenerate central values can now be written:

$$\Delta E_{n\ell}\left(j = \ell + \tfrac{1}{2}\right) = -\frac{\hbar^2 \ell}{2} \int d^3r |\psi_{n\ell}(\vec{r})|^2 f(r)$$

$$\Delta E_{n\ell}\left(j = \ell - \tfrac{1}{2}\right) = \frac{\hbar^2(\ell + 1)}{2} \int d^3r |\psi_{n\ell}(\vec{r})|^2 f(r) \tag{3.50}$$

so that the total splitting between the two levels becomes

$$\begin{aligned} \Delta &= \Delta E_{n\ell}\left(j = \ell - \tfrac{1}{2}\right) - \Delta E_{n\ell}\left(j = \ell + \tfrac{1}{2}\right) \\ &= \hbar^2\left(\ell + \tfrac{1}{2}\right) \int d^3r |\psi_{n\ell}(\vec{r})|^2 f(r) \end{aligned} \tag{3.51}$$

We see that the splitting in the energy levels due to the spin–orbit interaction is larger for higher values of the orbital angular momentum and can consequently produce level crossing. Namely, for large ℓ, the splitting of any two neighboring degenerate levels can shift the $j = \ell - \frac{1}{2}$ state of the initially lower level to lie above the $j = \ell + \frac{1}{2}$ of the previously higher level. For an appropriately chosen $f(r)$, the energy levels for a finite square well can split upon the addition of a spin–orbit interaction, as shown in Fig. 3.4. Thus, including the spin–orbit interaction, we can

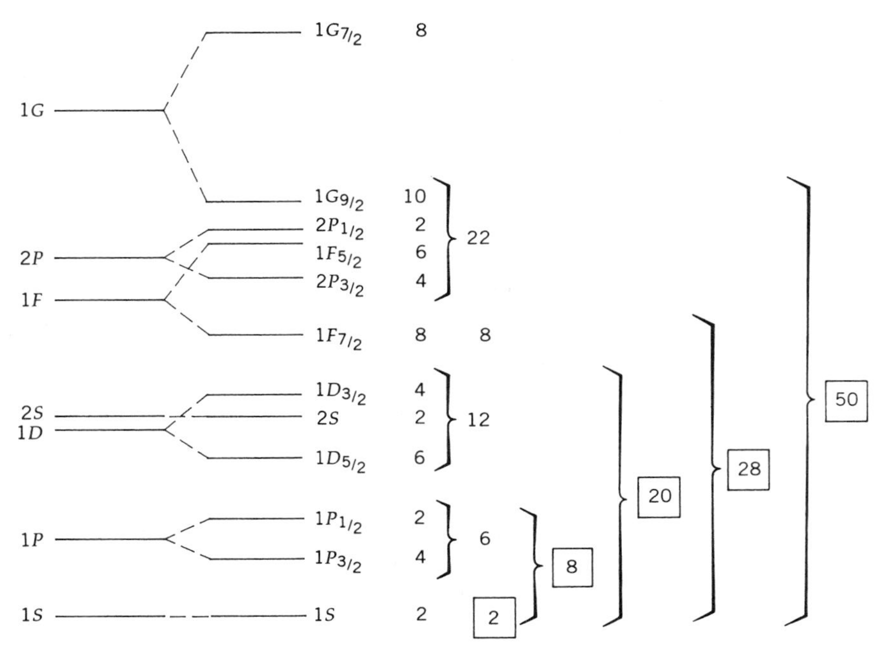

Figure 3.4 Energy levels in a single-particle shell model. The boxed integers correspond to the magic nuclear numbers.

reproduce all the desired magic numbers, and thereby confirm a shell-like structure in nuclei. The energy level diagram of Fig. 3.4 is labeled according to the spectroscopic notation of atomic physics, namely as (nL_j), that is, the split (n, ℓ, j) values. The multiplicity of any final level is given, as usual, by $(2j + 1)$. We have not shown the levels beyond $1G_{7/2}$; these are $2D_{5/2}$, $2D_{3/2}$, $3S_{1/2}$, $1H_{11/2}$, and so forth.

It is worth pointing out that in our discussion of the energy spectrum, we have treated protons and neutrons on an equal footing. It is clear, however, that the effect of the Coulomb potential must be to shift the energy levels for protons to somewhat higher values. Upon applying such corrections, it is found that the qualitative features of the spectrum remain essentially unchanged.

PREDICTIONS OF THE SHELL MODEL

The shell model accounts for a wide variety of properties of complex nuclei. For example, it provides the correct spin-parity assignment for the ground states of a large number of odd-A nuclei. According to the shell model, the proton and the neutron levels fill up independently and, due to the exclusion principle, only two neutrons or two protons can occupy any given level (with their spins being antiparallel). If we assume that nucleons pair off in every filled level, yielding zero total angular momentum, then it is the last unpaired nucleon that will determine the spin-parity of the ground state. An immediate consequence of this picture is that the ground states of all even-even nuclei will have zero spin, which is experimentally correct. The single-particle shell model cannot predict the ground-state spins of odd-odd nuclei because there is no a priori constraint on how unpaired protons and neutrons should couple.

Let us next examine the spin-parity assignments of several odd-A nuclei in greater detail. Consider the isobars $^{13}C^6$ and $^{13}N^7$. (Note that these are, in fact, mirror nuclei.) The six protons in ^{12}C and the six neutrons in ^{13}N will be completely paired off, while the remaining seven nucleons in both cases will fill up the shells as follows:

$$(1S_{1/2})^2\,(1P_{3/2})^4\,(1P_{1/2})^1 \tag{3.52}$$

Thus, the last unpaired nucleon—a neutron for $^{13}C^6$ and a proton for $^{13}N^7$—will have total angular momentum $\frac{1}{2}$ and orbital angular momentum $\ell = 1$. (Recall, from our previous comments in the footnote pertaining to Eq. 3.27, that $\ell = 1$ corresponds to odd parity for the state.) Thus, according to the shell model, the ground-state spin-parity for these nuclei is $\left(\frac{1}{2}\right)^-$, which is the experimentally observed value. Similarly, for the isobars $^{17}O^8$ and $^{17}F^9$, the nine neutrons for $^{17}O^8$ and nine protons for $^{17}F^9$ will fill up the levels as follows:

$$(1S_{1/2})^2\,(1P_{3/2})^4\,(1P_{1/2})^2\,(1D_{5/2})^1 \tag{3.53}$$

The total angular momentum of the last unpaired nucleon in the $\ell = 2$ state is $\frac{5}{2}$. Thus, the spin-parity of these nuclei would be expected to be $\left(\frac{5}{2}\right)^+$, which is consistent with experimental measurements. For $^{33}S^{16}$, the measured value of the ground-state spin-parity is $\left(\frac{3}{2}\right)^+$. According to the shell model, the seventeen neutrons will fill up the levels as:

$$(1S_{1/2})^2\,(1P_{3/2})^4\,(1P_{1/2})^2\,(1D_{5/2})^6\,(2S_{1/2})^2\,(1D_{3/2})^1 \tag{3.54}$$

again leading to a prediction consistent with experiment. Some spin-parity assignments of the shell model, however, do not agree with observations. For example, the neutrons in $^{47}Ti^{22}$ would be expected to fill up the levels as follows:

$$(1S_{1/2})^2\,(1P_{3/2})^4\,(1P_{1/2})^2\,(1D_{5/2})^6\,(2S_{1/2})^2\,(1D_{3/2})^4\,(1F_{7/2})^5 \tag{3.55}$$

leading to a ground-state spin-parity of $\left(\frac{7}{2}\right)^-$, whereas the experimental value is $\left(\frac{5}{2}\right)^-$. Such discrepancies can be remedied by slightly modifying the assumptions of the single-particle shell model to allow pairing between all "valence" nucleons, namely between any nucleons that occupy unfilled levels.

The shell model can also be used to calculate magnetic moments of various nuclei. As measurements show, the proton and the neutron have intrinsic dipole moments of 2.79 μ_N and -1.91 μ_N, respectively. Thus, we would expect the intrinsic magnetic moment of any unpaired nucleon to contribute to the total magnetic moment of the nucleus. In addition, since protons are charged, the orbital motion of any unpaired proton will also contribute to the magnetic moment of the nucleus. For the deuteron, for example, if we assume that the proton and the neutron are in $1S_{1/2}$ states, then, without orbital angular momentum for the proton ($\ell = 0$), we get for the magnetic moment of the deuteron the sum of the intrinsic dipole moments of the proton and the neutron:

$$\mu_d = 2.79\mu_N - 1.91\mu_N = 0.88\mu_N \tag{3.56}$$

The experimentally observed magnetic moment of the deuteron is $0.86\mu_N$, in good agreement with expectation. The nucleus of tritium $\left(^3H^1\right)$ has two neutrons and one proton, all in the $1S_{1/2}$ state. Since the neutrons are paired, they should not contribute to the magnetic moment. The unpaired proton, having $\ell = 0$, will have no contribution from its orbital motion. Consequently, the total magnetic moment of $^3H^1$ will be the same as that of the unpaired proton, namely $2.79\mu_N$, which is in good agreement with the measured value of $2.98\mu_N$. For $^3He^2$, the unpaired nucleon is a neutron in a $1S_{1/2}$ state. Consequently, the total magnetic moment should be the same as that of the neutron, which is $-1.91\mu_N$, again close to the observed value of $-2.13\mu_N$. $^4He^2$ (α-particle) has a closed shell structure (in fact, it is doubly magic), and the shell model would therefore predict no spin and no magnetic moment, which is indeed experimentally correct. In $^{10}B^5$, the five protons and the five neutrons have the same level structure, namely,

$$(1S_{1/2})^2\,(1P_{3/2})^3 \tag{3.57}$$

Thus, there is one unpaired proton and one unpaired neutron. The unpaired proton will be in an $\ell = 1$ state, and therefore the orbital motion will contribute $\mu = (e\hbar/2m_N c)\ell = \mu_N$ to the total magnetic moment, which will have a value

$$2.79\mu_N - 1.91\mu_N + \mu_N = 1.88\mu_N \tag{3.58}$$

This compares quite well with the measured value of $1.80\mu_N$.

Thus, we see that the shell model, in addition to providing the known magic numbers, also describes various important properties of light nuclei. For heavy nuclei, however, there is a marked difference between the predictions of the shell model and the measured quantities.

COLLECTIVE MODEL

In heavy nuclei, many predictions of the single-particle shell model do not agree quantitatively with experimentally measured values. Such discrepancies are particularly problematic for values of magnetic dipole moments. Also, the shell model predicts vanishingly small quadrupole moments for closed shells, and quadrupole moments of opposite sign for neighboring nuclei with atomic numbers $Z \pm 1$. Although this agrees qualitatively with experiment, the measured values of quadrupole moments are very different from the predictions of the model. In fact, some heavy nuclei appear to have large permanent electric quadrupole moments, suggesting a nonsphericity in the shape of these nuclei. This is certainly not consistent with the assumptions of the shell model, where rotational symmetry plays a crucial role.*

In a revival of the liquid drop model, Aage Bohr noted that many properties of heavy nuclei could be attributed to a surface motion of the nuclear liquid drop. Furthermore, James Rainwater showed that excellent agreement between the expected and the measured values of magnetic dipole and electric quadrupole moments could be obtained under the assumption that the liquid drop had an aspherical shape. These successes presented somewhat of a dilemma, because the liquid drop model and the single-particle shell model had fundamentally opposite viewpoints about the nature of nuclear structure. Individual particle characteristics, such as

*Finite quadrupole moments of charge distributions arise when the second moments $\langle x^2\rangle$, $\langle y^2\rangle$, and $\langle z^2\rangle$ differ from each other, namely when the distribution of charge is not spherical.

intrinsic spin and orbital angular momentum, play no role in a liquid drop picture, where collective motion that involves the entire nucleus has prime importance. On the other hand, individual nucleon properties, especially of the valence nucleons, are crucial to the success of the independent-particle shell model. The shell model had yielded too many important nuclear features to be abandoned outright, and a reconciliation between the two extreme views was needed.

The reconciliation was brought about by Aage Bohr, Ben Mottelson, and James Rainwater who proposed a collective model for the nucleus that provided many features that were not present in either the shell or the liquid drop model. In what follows, we describe this model only qualitatively. Its basic assumption is that a nucleus consists of a hard core of nucleons in the filled shells of the shell model, and outer valence nucleons that behave like the surface molecules in a liquid drop. The surface motion (rotation) of the valence nucleons introduces a nonsphericity in the central core, which in turn affects the quantum states of the valence nucleons. In other words, one can think of the surface motion as a perturbation that causes the quantum states of the valence nucleons to change from the unperturbed states of the shell model. This adjustment accounts for the difference in the predictions for dipole and quadrupole moments from those given by the shell model.

Physically, one can view the collective model as a shell model with a potential that is not spherically symmetric. That is, the nucleus can be regarded as aspherical. Spherically symmetric nuclei are, of course, insensitive to rotations, and consequently rotational motion cannot produce additional (rotational) energy levels in such nuclei. Aspherical nuclei, on the other hand, can have additional energy levels because of the presence of rotational and vibrational degrees of freedom. Such effects would, clearly, modify the predictions of the simple shell model. In particular, large nonsphericity in nuclei could provide large permanent dipole and quadrupole moments. Mathematically, one can incorporate these ideas as follows. If, for simplicity, we assume the nucleus to be an ellipsoid defined by

$$ax^2 + by^2 + \frac{z^2}{ab} = R^2 \tag{3.59}$$

where a and b are parameters related to the deformation from a spherical shape, then the mean potential for nuclear motion can be chosen as follows:

$$V(x, y, z) = \begin{cases} 0 & \text{for } ax^2 + by^2 + \frac{z^2}{ab} \leq R^2 \\ \infty & \text{otherwise} \end{cases} \tag{3.60}$$

Needless to say, more realistic calculations in the collective model provide even better descriptions of nuclear properties, but they also become far more complicated.

One of the important predictions of the collective model is the existence of rotational and vibrational levels in a nucleus. These levels can be derived much the same way as is done for the case of molecules. Thus, we can choose the Hamiltonian for rotations to be

$$H = \frac{\vec{L}^2}{2I} \tag{3.61}$$

with eigenvalues $(\ell(\ell+1)/2I)\,\hbar^2$, where the effective moment of inertia I is a function of the nuclear shape. If there is rotation along an axis perpendicular to the symmetry axis of the ellipsoid, then it can be shown that the angular momentum of the rotational levels can only be even. Thus, we see that rotational and vibrational levels in a nucleus are predicted with specific angular momentum and parity values. Such excitations have, indeed, been found experimentally through the observation of photon quadrupole transitions ($\Delta\ell = 2$) between these levels.

Finally, the collective model explains quite naturally why the level spacing of the first excited state in even-even nuclei decreases with increasing A, and becomes largest for nuclei with closed shells. This can be understood as follows. We would expect the energy eigenvalue of any first excited state corresponding to a rotational level to decrease with increasing A, simply because the moment of inertia grows with increasing A. The spacing to the ground level would therefore decrease with A. For a closed-shell nucleus, however, there should be no rotational level since such a nucleus would tend to be spherical. On the other hand, such a nucleus can have vibrational excitations. Such vibrational excitations, however, involve the entire core and not just the surface. This core being much more massive implies that the energy level for vibration will lie far higher, and the spacing between the ground state and the first excited state will be much larger.

SUPERDEFORMED NUCLEI

Throughout our discussion of nuclear phenomena we have emphasized that nuclei tend to have relatively small intrinsic spins. We can imagine that under certain circumstances, however, nuclei could be greatly deformed and yet not fission (see Chapter V). In fact, for some time, superdeformed nuclei were predicted to be particularly stable for values of A between 150 and 190. Such nuclei were expected to be spheroidal in character, with semimajor and semiminor axes differing by about a factor of 2. During the late 1980s, a series of experiments was carried out in the scattering of heavy ions on heavy ions. When such collisions take place, superdeformed nuclei are produced with remarkably large angular momenta of about $60\hbar$. These nuclei deexcite through a series of emissions of $\sim$50 keV γ-rays down to lower levels with more symmetric nuclear shapes. However, the observed level spacings (photon energies) are essentially all the same. This result poses a problem from the point of view of the collective model, where we would expect the moment of inertia to decrease and therefore the level spacings to change, with a decrease in deformation. In fact, different nuclei appear to have essentially identical emissions as they "spin down." This is an even greater puzzle, because of the known effects of nucleon pairing on binding energy and level spacing. This is currently a very active area of study in nuclear physics, that may yet offer additional surprises.

Problems

III.1 The Bethe–Weizsäcker formula of Eq. 3.5 provides an excellent representation of the mass systematics of nuclei. Show explicitly that for fixed A, $M(A, Z)$ has a minimum value. Is there evidence for the "valley of stability" observed in Fig. 2.3? What is the stablest nucleus with $A = 16$? What about $A = 208$? (You can differentiate Eq. 3.5, or simply plot M as a function of Z.)

III.2 Using Eq. 3.3, compute the total binding energy and the value of B/A for $^8\text{Be}^4$, $^{12}\text{C}^6$, $^{56}\text{Fe}^{26}$, and $^{208}\text{Pb}^{82}$. How do these values compare with the experimental ones? (See *CRC Handbook of Chemistry and Physics* for data.)

III.3 You might conclude from Problem III.2 that $^8\text{Be}^4$ is stable. This is, in fact, not the case. Can you provide a model to explain this result? (*Hint:* See Problem II.2.)

III.4 Calculate the binding energy of the last neutron in $^{15}\text{N}^7$ and of the last proton in $^{15}\text{O}^8$, and contrast with the last neutron in $^{16}\text{N}^7$ and in $^{16}\text{O}^8$.

III.5 What would you expect for the spin and parity of the ground states of $^{23}\text{Na}^{11}$, $^{35}\text{Cl}^{17}$, and $^{41}\text{Ca}^{20}$ on the basis of the single-particle shell model? Do these predictions agree with experimental values? What about the magnetic moments of these nuclei? (See *CRC Handbook* for data.)

III.6 Consider a somewhat more sophisticated model for the anomalous contribution to the magnetic moment of a nucleon. Assume that the proton can be regarded as a fixed neutral center with a π^+ meson circling about in an $\ell = 1$ orbit. Similarly, take a neutron as an effective proton center with a π^- meson in an $\ell = 1$ orbit around it. Using $m_\pi = 140$ MeV/c^2, calculate $\mu = (e\hbar/2m_\pi c)\ell$, and compare your results with those of Problem II.5.

III.7 The ground state of $^{137}Ba^{56}$ has spin-parity $\frac{3}{2}^{+}$. That is, its spin is $\frac{3}{2}$ and parity +. The first two excited states have spin parity $\frac{1}{2}^{+}$ and $\frac{11}{2}^{-}$. According to the shell model, what levels would be expected for these excited states? (*Hint:* The surprise has to do with "pairing energy.")

Suggested Readings

Frauenfelder, H., and E. M. Henley. 1991. *Subatomic Physics*. Englewood Cliffs, N.J.: Prentice-Hall.

Krane, K. S. 1987. *Introductory Nuclear Physics*. New York: Wiley.

Williams, W. S. C. 1991. *Nuclear and Particle Physics*. London/New York: Oxford Univ. Press.

Chapter IV

NUCLEAR RADIATION

INTRODUCTORY REMARKS

In earlier chapters we indicated that many nuclei are unstable and often emit α, β, and γ particles. Here we discuss some of the more quantitative aspects of nuclear radioactivity and its historical impact on our understanding of nuclear structure and nuclear transmutation.

ALPHA DECAY

As we have seen before, α-decay represents the disintegration of a parent nucleus to a daughter through the emission of a nucleus of a helium atom, and the transition can be characterized as follows:

$$ {}^{A}X^{Z} \longrightarrow {}^{A-4}Y^{Z-2} + {}^{4}\mathrm{He}^{2} \tag{4.1} $$

As we will see in Chapter V, one can regard α-decay as the spontaneous fission of the parent nucleus into two daughter nuclei with highly asymmetric masses. If we assume that the parent nucleus is initially at rest, then conservation of energy will require

$$ M_P c^2 = M_D c^2 + T_D + M_\alpha c^2 + T_\alpha \tag{4.2} $$

where M_P, M_D and M_α are the masses of the parent, daughter, and the α-particle, respectively. Similarly, T_D and T_α represent, respectively, the kinetic energies of the daughter and of the α-particle. It follows, therefore, that

$$T_D + T_\alpha = (M_P - M_D - M_\alpha)\,c^2 = \Delta M c^2 \tag{4.3}$$

We note that, although the right-hand side of Eq. 4.3 involves nuclear masses, we can, in fact, use atomic masses in the expression, since the masses of the electrons will cancel. Thus, we can write

$$T_D + T_\alpha = (M(A,Z) - M(A-4, Z-2) - M(4,2))\,c^2 \equiv Q \tag{4.4}$$

where we have defined the disintegration energy, or Q-value, as the difference in the rest masses of the initial and final states. It is clear that Q also equals the sum of the kinetic energies of the final-state particles. For nonrelativistic particles, the kinetic energies can be written as

$$\begin{aligned} T_D &= \tfrac{1}{2} M_D v_D^2 \\ T_\alpha &= \tfrac{1}{2} M_\alpha v_\alpha^2 \end{aligned} \tag{4.5}$$

with v_D and v_α representing the magnitude of the velocities of the daughter and the α-particle.

Since the parent nucleus decays from rest, the daughter nucleus and the α-particle must necessarily move in opposite directions to conserve momentum, satisfying

$$M_D v_D = M_\alpha v_\alpha$$

or

$$v_D = \frac{M_\alpha}{M_D} v_\alpha \tag{4.6}$$

Since the mass of the daughter nucleus is in general much greater than that of the α-particle, it follows that $v_D \ll v_\alpha$. Consequently, the kinetic energy of the daughter nucleus is relatively small.

Let us eliminate v_D and write expressions for T_D and T_α in terms of the Q-value:

$$\begin{aligned} T_D + T_\alpha &= \tfrac{1}{2} M_D v_D^2 + \tfrac{1}{2} M_\alpha v_\alpha^2 \\ &= \tfrac{1}{2} M_D \left(\frac{M_\alpha}{M_D} v_\alpha\right)^2 + \tfrac{1}{2} M_\alpha v_\alpha^2 \end{aligned}$$

$$= \frac{1}{2}M_\alpha v_\alpha^2 \left(\frac{M_\alpha}{M_D} + 1\right)$$

$$= T_\alpha \frac{M_\alpha + M_D}{M_D} \tag{4.7}$$

Using Eq. 4.4, this result can be rewritten as

$$T_\alpha = \frac{M_D}{M_\alpha + M_D} Q = \frac{1}{1 + (M_\alpha/M_D)} Q \tag{4.8}$$

The kinetic energy of the emitted α-particle cannot be negative, that is, $T_\alpha \geq 0$. Consequently, for α-decay to occur, we must have an exothermic process

$$\begin{aligned} \Delta M &\geq 0 \\ Q &\geq 0 \end{aligned} \tag{4.9}$$

As we can see, most of the energy is carried by the α-particle. The kinetic energy of the daughter nucleus is obtained from Eqs. 4.4 and 4.8:

$$T_D = Q - T_\alpha = \frac{M_\alpha}{M_\alpha + M_D} Q = \frac{M_\alpha}{M_D} T_\alpha \ll T_\alpha \tag{4.10}$$

If we use the approximation $M_\alpha/M_D \simeq 4/(A-4)$, then we can write approximately

$$\begin{aligned} T_\alpha &\simeq \frac{A-4}{A} Q \\ T_D &\simeq \frac{4}{A} Q \end{aligned} \tag{4.11}$$

which can be used to estimate the energy released in a decay.

We note from Eq. 4.8 that the kinetic energy (and therefore the magnitude of the velocity) of the α-particle in the decay is unique, which is consistent with our earlier discussion. This result is a direct consequence of the fact that the process is a two-body decay of a parent initially at rest. Careful measurements, however, have revealed a fine splitting in the energies of α-particles emitted from any radioactive material, corresponding to possibly different Q-values. Experimentally, the most energetic α-particles are observed to be produced alone. The less energetic α-decays, however, are always accompanied by the emission of photons. This is highly suggestive of the existence of energy levels and of an underlying quantum structure of discrete states in nuclei. If this were the case, a parent nucleus could decay to a ground state of the daughter nucleus, emitting an α-particle with energy corresponding to the entire Q-value. However, it could instead also decay to an excited state of the daughter nucleus, in which case the effective Q-value would be lower. Subsequently, the daughter nucleus could deexcite to its ground state by emitting a photon, as happens in the case of atomic transitions. Namely, in this case the decay chain would proceed as follows:

$$^{A}X^{Z} \longrightarrow {}^{A-4}Y^{*Z-2} + {}^{4}He^{2}$$

with

$$^{A-4}Y^{*Z-2} \longrightarrow {}^{A-4}Y^{Z-2} + \gamma \qquad (4.12)$$

The difference in the two Q-values would then correspond to the energy of the emitted photon. For example, the spectrum of observed α-particle energies in the decay of ^{228}Th to ^{224}Ra can be schematically associated with the level structure shown in Fig. 4.1.

The underlying level structure shown in Fig. 4.1 can be determined experimentally by measuring the kinetic energies of the different α-particles observed in these decays, which in turn yield the Q-values for the transitions through Eq. 4.8. Based on the assumption of discrete nuclear levels, the difference in the Q-values will then yield the expected energies of the emitted photons. The measured energies of such accompanying (coincident) photons have, in fact, confirmed the overall picture, and therefore the existence of discrete nuclear levels.

Example 1

Consider the α-decay of $^{240}Pu^{94}$:

$$^{240}Pu^{94} \longrightarrow {}^{236}U^{92} + {}^{4}He^{2}$$

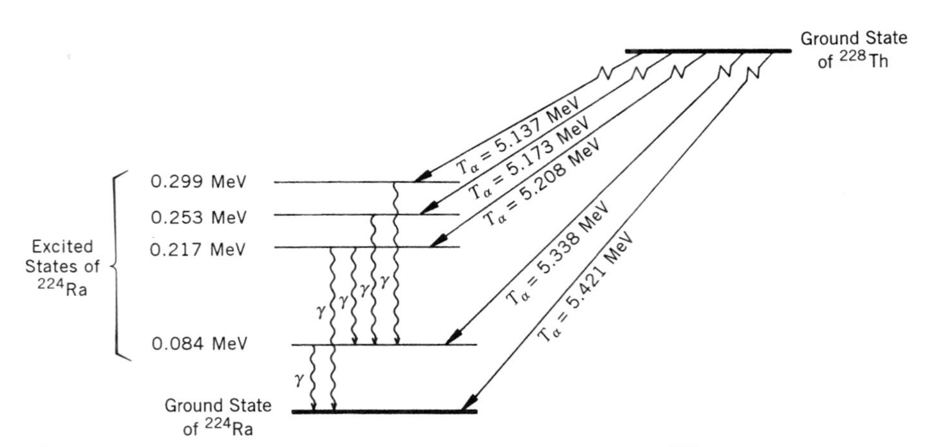

Figure 4.1 α-particle transitions observed in the decay of ^{228}Th.

Experimentally, the α-particles are observed to have energies 5.17 MeV and 5.12 MeV. Substituting the two values of the kinetic energy into Eq. 4.11

$$Q \simeq \frac{A}{A-4} T_\alpha$$

leads to the two Q-values

$$Q_1 \simeq \frac{240}{236} \times 5.17 \text{ MeV} \simeq 1.017 \times 5.17 \text{ MeV} \simeq 5.26 \text{ MeV}$$

$$Q_2 \simeq \frac{240}{236} \times 5.12 \text{ MeV} \simeq 1.017 \times 5.12 \text{ MeV} \simeq 5.21 \text{ MeV}$$

Thus, when $^{240}\mathrm{Pu}$ decays with disintegration energy $Q_2 \simeq 5.21$ MeV, the daughter nucleus $^{236}\mathrm{U}^{92}$ is left in an excited state, and will transform to the ground state by emitting a photon of energy

$$Q_1 - Q_2 \simeq 5.26 \text{ MeV} - 5.21 \text{ MeV} = 0.05 \text{ MeV}$$

This is, indeed, consistent with the experimentally observed energy of 0.045 MeV for the photon. Thus, we can conclude from such experimental studies of α-decays that there are discrete energy levels in the nucleus, very much like the energy levels in an atom, and that the minimum spacing between the nuclear levels is about 100 keV, whereas the corresponding spacing in atomic levels is of the order of 1 eV.

BARRIER PENETRATION

The α-particles emitted in nuclear decays have typical energies of about 5 MeV. When such low-energy particles are scattered from a heavy nucleus, they cannot penetrate closely enough to interact through the strong force because of the presence of the Coulomb barrier. The height of the Coulomb barrier for $A \simeq 200$ is about 20–25 MeV, and therefore a 5-MeV α-particle cannot overcome this barrier and be absorbed. On the other hand, a low-energy α-particle, bound in a nuclear potential well, sees the same barrier, and yet is able to escape. This constituted a great puzzle, until it was recognized that the emission of the α-particle was a quantum mechanical phenomenon.

The first quantitative understanding of α-decay came in 1929 from the work of George Gamow and of Ronald Gurney and Edward Condon. Assuming that the α-particle and the daughter nucleus can exist prior to dissociation within the parent nucleus, we can treat the problem as an α-particle moving in the potential of the daughter nucleus, with the

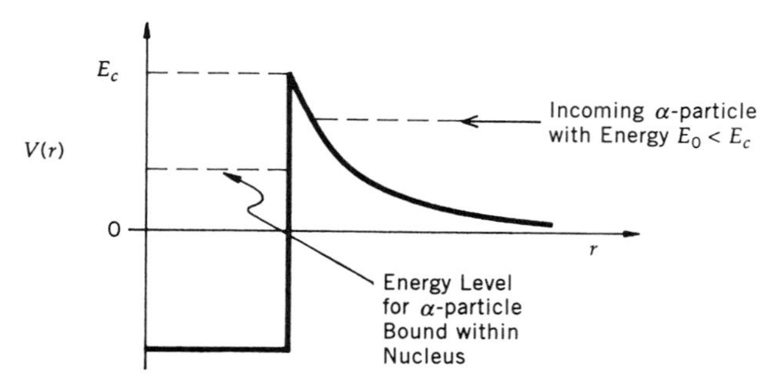

Figure 4.2 Potential energy function for an α-particle interacting with a nucleus.

Coulomb potential preventing their separation (see Fig. 4.2). For concreteness, consider the decay

$$^{232}\text{Th} \longrightarrow {}^{228}\text{Ra} + {}^{4}\text{He} \tag{4.13}$$

Experimentally, the energy of the α-particle is known to be $E = 4.05$ MeV, and the lifetime of ^{232}Th is $\tau = 1.39 \times 10^{10}$ years. Also, the radius of the thorium nucleus is about $R = 1.2 \times 10^{-13} A^{1/3}$ cm $\simeq 7.4 \times 10^{-13}$ cm. The α-particle must penetrate the Coulomb barrier in order for the decay to occur. The calculation of barrier penetration for the three-dimensional Coulomb potential is rather complicated. However, since we are interested only in order-of-magnitude estimates, we can ignore the angular dependence of the Schrödinger equation and consider the potential as effectively one-dimensional. Furthermore, we can replace the Coulomb potential by a square barrier of equal area, which approximates the effect of the Coulomb repulsion, and is calculationally much simpler (see Fig. 4.3). As long as V_0 is chosen so that it is larger than E, the transmission through the barrier is sensitive primarily to $(\sqrt{V_0 - E})a$ and not so much to the value of V_0.

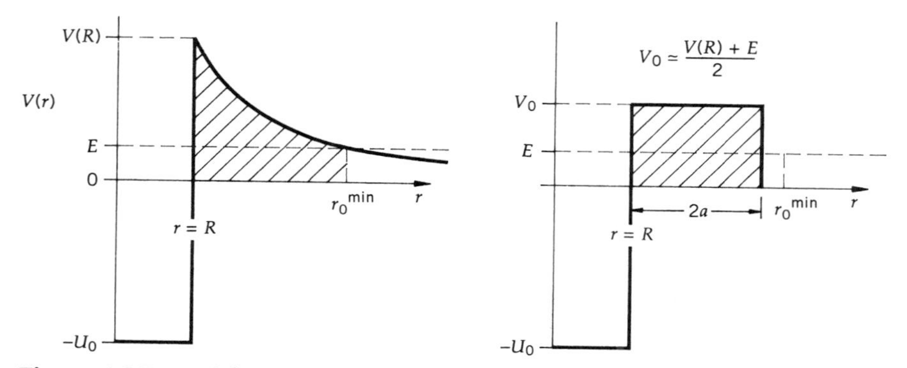

Figure 4.3 Potential energy for scattering of 4-MeV α-particle from ^{228}Ra, and the equivalent one-dimensional square-well potential.

For $Z \simeq 90$, we can choose

$$\begin{aligned} V_0 &= 14 \text{ MeV} \\ 2a &= 33 \text{ fm} = 33 \times 10^{-13} \text{ cm} \end{aligned} \tag{4.14}$$

A straightforward quantum mechanical calculation of the transmission through the square barrier shown in Fig. 4.3, yields the following transmission coefficient:

$$T = \frac{[(4k_1k)/(k_1+k)^2]}{1+\left[1+((\kappa^2-k_1k)/\kappa(k_1+k))^2\right]\sinh^2 2\kappa a} \tag{4.15}$$

with

$$\begin{aligned} k_1 &= \left[\frac{2M_\alpha}{\hbar^2}(E+U_0)\right]^{1/2} \\ k &= \left[\frac{2M_\alpha}{\hbar^2}E\right]^{1/2} \\ \kappa &= \left[\frac{2M_\alpha}{\hbar^2}(V_0-E)\right]^{1/2} \end{aligned} \tag{4.16}$$

where M_α is the rest mass and E is the kinetic energy of the emitted α particle (outside of the barrier). For $M_\alpha c^2 \simeq 4000$ MeV, $E = 4.05$ MeV, $V_0 = 14$ MeV, and $U_0 \simeq 40$ MeV (the calculation is not very sensitive to the depth of the nuclear potential), we have

$$\begin{aligned} \kappa &= \frac{1}{\hbar c}\left[2M_\alpha c^2(V_0-E)\right]^{1/2} \\ &\simeq \frac{1}{197 \text{ MeV-fm}}[2\times 4000 \text{ MeV}(14-4)\text{ MeV}]^{1/2} \sim 1.4 \text{ fm}^{-1} \\ k &\simeq 0.9 \text{ fm}^{-1} \\ k_1 &\simeq 3.0 \text{ fm}^{-1} \end{aligned} \tag{4.17}$$

Now, $2\kappa a \simeq 33 \text{ fm} \times 1.4 \text{ fm}^{-1} \simeq 46$, which means that $2\kappa a \gg 1$, and allows us to write

$$\sinh^2 2\kappa a \simeq \left(\frac{e^{2\kappa a}}{2}\right)^2 = \frac{1}{4}e^{4\kappa a} \simeq \frac{1}{4}e^{92} \gg 1 \tag{4.18}$$

We see that the transmission coefficient T is determined essentially by this exponent, and is not very sensitive to the choice of k_1 and k. Because we are only interested in an estimate for T, let us therefore simplify Eq. 4.15

by taking the limit of large $k_1 (k_1^2 \gg \kappa^2$ and $k_1^2 \gg k^2)$. In this limit, the transmission coefficient of Eq. 4.15 becomes

$$\begin{aligned} T &\simeq \frac{4\kappa^2}{\kappa^2 + k^2} \frac{k}{k_1} (\sinh^2 2\kappa a)^{-1} \\ &\simeq \frac{4(V_0 - E)}{V_0} \left(\frac{E}{E + U_0}\right)^{1/2} \left[4e^{-(4a/\hbar)[2M_\alpha(V_0 - E)]^{1/2}}\right] \\ &\simeq \frac{4(10)}{14} \left(\frac{4}{44}\right)^{1/2} \left(4 \cdot e^{-92}\right) \\ &\simeq 3.5 \times e^{-92} \simeq 4 \times 10^{-40} \end{aligned} \tag{4.19}$$

Thus, the α-particle has an exceedingly small probability for penetrating the barrier. This explains why low-energy α-particles cannot be absorbed by heavy nuclei. For an α-particle bound in a nucleus, however, the situation is quite different. The kinetic energy of the α-particle within the well is

$$T_\alpha \simeq U_0 + E \simeq 44 \text{ MeV} \tag{4.20}$$

and the corresponding velocity is

$$\begin{aligned} v_\alpha &= \sqrt{\frac{2T_\alpha}{M_\alpha}} = c\sqrt{\frac{2T_\alpha}{M_\alpha c^2}} \\ &\simeq c\sqrt{\frac{2 \times 44 \text{ MeV}}{4000 \text{ MeV}}} \simeq 0.15c \end{aligned} \tag{4.21}$$

Being confined to a small region of dimension $\simeq 10^{-12}$ cm, the α-particle will bounce against the barrier with a frequency given approximately by

$$\begin{aligned} \frac{v_\alpha}{R} &\simeq \frac{0.15 \times 3 \times 10^{10} \text{ cm/sec}}{7.4 \times 10^{-13} \text{ cm}} \\ &\simeq 6.0 \times 10^{21}/\text{sec} \end{aligned} \tag{4.22}$$

Every time the α-particle hits the barrier, the probability of escape is given by Eq. 4.19. We conclude, therefore, that the probability for the α-particle to escape per second is simply

$$\begin{aligned} P(\alpha\text{−emission}) &\simeq \frac{v_\alpha}{R} \cdot T \simeq 6.0 \times 10^{21}/\text{sec} \times 4 \times 10^{-40} \\ &\simeq 2.4 \times 10^{-18}/\text{sec} \end{aligned} \tag{4.23}$$

This is what is called the decay constant (denoted by λ), and is the probability of decay per unit time. The mean life time for the decay process (discussed in the next chapter) is the inverse of the decay constant

$$\begin{aligned}\tau &= \frac{1}{P(\alpha-\text{emission})}\\ &\simeq \frac{1}{2.4\times 10^{-18}/\text{sec}}\\ &\simeq 0.4\times 10^{18}\ \text{sec}\\ &\simeq 1.3\times 10^{10}\ \text{yr}\end{aligned} \tag{4.24}$$

This lifetime is remarkably close to the observed value.

We have presented an oversimplified calculation of α-decay. The quantitative result for the coefficient, therefore, cannot be trusted in detail. In general, however, the decay constant, for $V_0 \gg E$, can be represented as

$$P(\alpha-\text{emission}) \propto E^{1/2} e^{-(4a/\hbar)[2M_\alpha(V_0-E)]^{1/2}} \tag{4.25}$$

This result shows that the probability for decay is quite sensitive to the mass and the energy of the α-particle. In particular, it shows why spontaneous fission through barrier penetration into heavier daughter nuclei (of large M), a subject that we discuss more fully in the next chapter, is a slow process. It also connects the decay constant and the lifetime with the energy of the α-particle. We note that $P(\alpha$-emission) is proportional to $E^{1/2}$, and, as a result, the larger the E, the smaller the lifetime. That is, for large E, the decay is fast, which is consistent with our naive expectation. We also note from Eq. 4.25 that, for $V_0 \gg E$, and $(V_0-E)^{1/2}$ slowly varying with E, we can write approximately

$$\log P(\alpha-\text{emission}) \propto (\log E + \text{constant}) \tag{4.26}$$

This result provides a quantitative relation between the decay constant and the energy of the decaying particle, and is known as the Geiger-Nuttal rule. This result was first noted on the basis of experimental evidence.

BETA DECAY

A nucleus with an overabundance of neutrons (i.e., with a value of N/Z greater than that for stable nuclei) can transform to a stabler nucleus by emitting an electron. Such a process is known as *β-decay*, and the change observed in the decay can be denoted by

$$^{A}X^{Z} \longrightarrow {}^{A}Y^{Z+1} + e^- \tag{4.27}$$

From electric-charge conservation, it follows that the proton number of the daughter nucleus increases by one unit in such decays. However, the nucleon number remains unchanged. There are two other processes that are also referred to as β-decays. In one case, a proton-rich nucleus emits a positron (positrons are antiparticles of electrons; they have the same mass as electrons, but opposite electric charges), and thereby reduces the nuclear charge by one unit. In this case, the process can be represented by

$$^{A}X^{Z} \longrightarrow {}^{A}Y^{Z-1} + e^{+} \tag{4.28}$$

In addition, a proton-rich nucleus can also reduce its nuclear charge by one unit through absorbing an atomic electron. This process is referred to as *electron capture*, and can be represented as

$$^{A}X^{Z} + e^{-} \longrightarrow {}^{A}Y^{Z-1} \tag{4.29}$$

The electron is normally captured from an inner K-shell of an atom. As a result, the outer electrons of the atom cascade down to fill the lower atomic levels, and one or more X-rays are usually emitted. In all of these three β-decay processes, the nuclear transformation can be characterized by $\Delta A = 0$ and $|\Delta Z| = 1$.

Because in β-decay only the electron and the recoiling daughter nucleus were observable, the process was initially assumed to be a two-body disintegration, very much like that of α-decay. Thus, for the decay in Eq. 4.27, conservation of energy would yield

$$E_X = E_Y + E_{e^-} = E_Y + T_{e^-} + m_e c^2$$

or

$$\begin{aligned} T_{e^-} &= (E_X - E_Y - m_e c^2) = (M_X - M_Y - m_e)c^2 - T_Y \\ &= Q - T_Y \simeq Q \end{aligned} \tag{4.30}$$

In other words, for a two-body process, just as in α-decay, the lighter particle emitted, namely the electron, would be expected to carry away most of the released energy, which would have a unique value given by Eq. 4.30. Experimentally, however, as we discussed earlier in Chapter II, the electrons are emitted with a continuous spectrum of energies. In fact, the observed distribution in the number of electrons emitted with a given energy has the shape given in Fig. 4.4. The energy spectrum is continuous, and, within experimental accuracy, has an endpoint (the maximum energy of the emitted electron) given by the value in Eq. 4.30. In fact, most of the electrons have energy below that predicted by energy conservation in two-body decays. When this was first observed, it appeared to threaten the survival of one of the most cherished conservation laws in physics, namely energy conservation! In addition, a consideration of the change in angular momentum in β-decay processes reveals that angular momentum

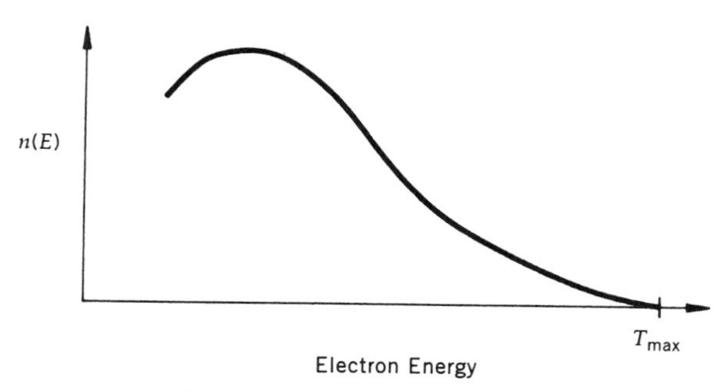

Figure 4.4 The energy spectrum of electrons emitted in **β**-decay.

could not be conserved if the decays produced only two particles in the final state. Examining the decay in Eq. 4.27, we note that the number of nucleons does not change in the transition. However, an electron, which is a fermion, is emitted in the process. The electron, as well as each of the nucleons, have spin angular momentum of $s = \frac{1}{2}$. Consequently, independent of any possible change in the value of the orbital angular momentum, which must always have integral value, it is clear that angular momentum cannot be conserved in such a process.

For a while it seemed as if physicists would have to give up the conservation of momentum, energy, and angular momentum in order to account for β-decay. This would have implied, through Emmy Noether's theorem (discussed in Chapter X), that the universe is not isotropic, and that there is an absolute coordinate system and an absolute time scale, all of which could severly impact physical behavior. Physics, as we know it, would have to have been abandoned. To extricate science from such an abyss, Wolfgang Pauli proposed that an additional particle, one that was difficult to detect, was emitted in β-decay. Conservation of charge required such a particle to be charge–neutral, just like the neutron and the photon. This suggestion could partially explain why it was so hard to detect this particle. We know now that this neutral particle, the neutrino, does not interact readily with matter, and this is the main reason why it is so difficult to observe. Because the maximum energies for electrons emitted in β-decay corresponded to the disintegration energy of the nucleus, it meant that this new particle had to be essentially massless. Furthermore, if the postulated neutrino were to restore the conservation of angular momentum, then it would have to be a fermion with spin angular momentum $s = \frac{1}{2}$. In some ways, such a particle would resemble a neutron, except that it would be much lighter, and Fermi therefore coined for it the name *neutrino* (diminutive for neutron), and it is denoted by the Greek letter ν.

Every elementary particle has an antiparticle, and the neutrino is no exception; its antiparticle is known as the *antineutrino* ($\overline{\nu}$). Since both the neutrino and the antineutrino are charge–neutral, an interesting question is what specific property distinguishes them. (Note that the neutron and the antineutron are also charge–neutral, but they have magnetic dipole moments of opposite sign, and, as we discuss in Chapters IX and XI, opposite nucleon or "baryon" numbers that can distinguish them. The neutrino, however, is a massless point particle, without structure, and has neither nucleon number nor a magnetic dipole moment.) Experimental studies of β-decays indicate that the neutrino that accompanies positrons ("ν_{e^+}") is always left-handed, whereas the one that accompanies electrons ("ν_{e^-}") is always right-handed, where by left-handed we mean that the particle has its spin pointing opposite to its momentum, and by right-handed that it has its spin pointing parallel to its line of flight. (This convention is opposite to the left-handed and right-handed polarizations for photons used in optics.) If we define e^- as the particle and e^+ as its antiparticle, then it is tempting to call the ν_{e^-} the antineutrino ($\overline{\nu}$) and the ν_{e^+} the neutrino (ν_e). (This assignment will, in fact, be justified shortly.) For this choice, we see that the handedness is therefore one of the distinguishing characteristics between a ν and a $\overline{\nu}$, and has far-reaching consequences, as we will see later. Using our new nomenclature, we can rewrite our three β-decay processes as follows:

$$\begin{aligned} {}^A X^Z &\longrightarrow {}^A Y^{Z+1} + e^- + \overline{\nu} \\ {}^A X^Z &\longrightarrow {}^A Y^{Z-1} + e^+ + \nu \\ {}^A X^Z + e^- &\longrightarrow {}^A Y^{Z-1} + \nu \end{aligned} \tag{4.31}$$

If the parent nucleus decays from rest, then conservation of energy for the case of electron emission will yield

$$M_P c^2 = T_D + M_D c^2 + T_{e^-} + m_e c^2 + T_{\overline{\nu}} + m_\nu c^2$$

or

$$T_D + T_{e^-} + T_{\overline{\nu}} = (M_P - M_D - m_e - m_\nu)\, c^2 = \Delta M c^2 = Q \tag{4.32}$$

where M_P, M_D, m_e, and m_ν are, respectively, the masses of the parent nucleus, the daughter nucleus, the electron, and the antineutrino. Similarly, T_D, T_{e^-}, and $T_{\overline{\nu}}$ represent the kinetic energies of the decay particles, namely, the daughter nucleus, the electron, and the antineutrino. We see from Eq. 4.32 that electron emission can take place only if the disintegration energy, Q, is positive, that is, when the mass of the parent nucleus is greater than the sum of the masses of the decay products. In fact, neglecting small differences in atomic binding energies, we conclude that electron emission will take place if

$$\begin{aligned} Q &= (M(A,Z) - M(A,Z+1) - m_\nu)c^2 \\ &\simeq (M(A,Z) - M(A,Z+1))c^2 \geq 0 \end{aligned} \tag{4.33}$$

where the $M(A,Z)$ values in Eq. 4.33 are atomic weights that include the atomic electrons. Furthermore, because the daughter nucleus is much heavier than both the electron and the antineutrino, the recoil energy of the daughter is negligible, and we can write for any β-decay

$$T_{e^-} + T_{\overline{\nu}} \simeq Q \tag{4.34}$$

It is now clear that, with a $\overline{\nu}$ in the final state, the energy of the electron is no longer unique. In fact, any continuous value $0 \leq T_{e^-} \leq Q$ is kinematically allowed, and the maximum electron energy, corresponding to $T_{\overline{\nu}} = 0$, is given by endpoint value of Eq. 4.32:

$$(T_{e^-})_{\text{max}} = Q \tag{4.35}$$

The postulate of Pauli therefore accommodates the continuous spectrum in β-decay, and simultaneously restores all the accepted conservation laws.

For completeness, let us note that the disintegration energy for positron emission is given by

$$\begin{aligned} Q &= (M_P - M_D - m_e - m_\nu)c^2 \\ &= (M(A,Z) - M(A,Z-1) - 2m_e - m_\nu)c^2 \\ &\simeq (M(A,Z) - M(A,Z-1) - 2m_e)c^2 \end{aligned} \tag{4.36}$$

where again all the $M(A,Z)$ values in the last line of Eq. 4.36 refer to full atomic weights, and Q must be positive for the decay to occur. Similarly, electron capture can take place only if

$$\begin{aligned} Q &= (M_P + m_e - M_D - m_\nu)c^2 \\ &= (M(A,Z) - M(A,Z-1) - m_\nu)c^2 \\ &\simeq (M(A,Z) - M(A,Z-1))c^2 \geq 0 \end{aligned} \tag{4.37}$$

As stated before, all of these relations neglect the difference in binding energies of electrons in atoms.

Just as a proton or a neutron is defined to be a nucleon with nucleon or baryon number +1, so is an electron defined to be a "lepton" with a lepton number +1. A positron, being the antiparticle of an electron, has lepton number −1, just as an antiproton or an antineutron has nucleon number −1. We will see in Chapter IX that both lepton number and nucleon number appear to be conserved in all interactions; we can therefore conclude from the three processes in Eq. 4.31 that the neutrino must also be a lepton with lepton number +1, while the lepton number for the antineutrino is −1.

We know from experiment that there are three charged leptons in nature, all with their own associated neutrinos, namely (e^-, ν_e), (μ^-, ν_μ), and (τ^-, ν_τ). The muon and the tau lepton appear to have properties similar to that of the electron, but are far more massive. The three types of neutrinos are also known to be distinct from one another (firmly established for ν_e and ν_μ). For example, when neutrinos produced in a decay such as $\pi^+ \to \mu^+ + \nu_\mu$ are allowed to interact with matter, they never produce charged leptons other than μ^-. That is,

$$\begin{aligned} \nu_\mu + {}^A X^Z &\to {}^A Y^{Z+1} + \mu^- \\ \nu_\mu + {}^A X^Z &\not\to {}^A Y^{Z+1} + e^- \end{aligned} \tag{4.38}$$

Similarly, a ν_e interacting with matter only produces an electron:

$$\begin{aligned} \nu_e + {}^A X^Z &\to {}^A Y^{Z+1} + e^- \\ \nu_e + {}^A X^Z &\not\to {}^A Y^{Z+1} + \mu^- \end{aligned} \tag{4.39}$$

This family structure for leptons and their antiparticles is quite important in constructing theories of fundamental interactions.

The issue of whether neutrinos have mass also has important ramifications. As is clear from Eqs. 4.33 and 4.35, the mass of the neutrino can be determined from the endpoint of the β-spectrum. If $m_\nu = 0$, then the endpoint of the spectrum is tangential to the abscissa, whereas for the case when $m_\nu \neq 0$, the endpoint is tangential to the ordinate (see Fig. 4.5). Thus, by studying the β-spectrum one can, in principle, determine the mass of the neutrino. However, in practice, the shape at the endpoint of the spectrum is very sensitive to measurement resolution. There are other methods for determining the mass of the neutrino, and at present the most stringent limit on the mass of the electron neutrino is $m_{\nu_e} \lesssim 10\ \text{eV}/c^2$.

Theoretically, a small but finite neutrino mass is attractive from a cosmological point of view. This is because standard cosmology suggests a closed universe, but the present estimate of all the luminous matter is too small to provide sufficient gravitational attraction to eventually halt the current expansion. A massive neutrino, which is hard to detect, can add to the mass of the universe as dark matter, and thereby provide the required closure. A finite mass for the neutrino leads naturally to concepts such as mixing between the different neutrino states, and to "neutrino

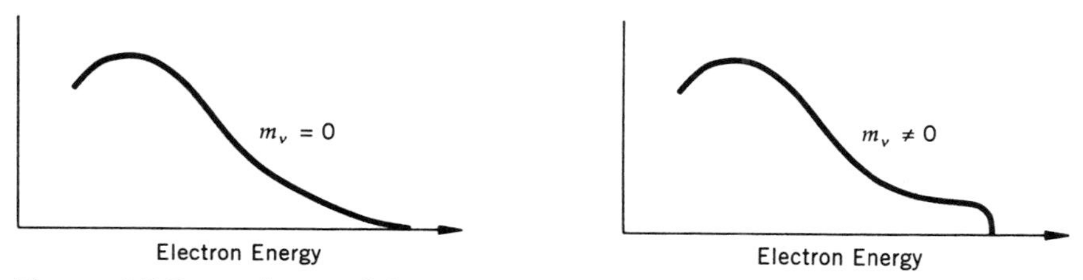

Figure 4.5 Dependence of the endpoint of β spectrum on the mass of the neutrino.

oscillations" from one species to another, much in the spirit of beat phenomena and energy exchange in the case of two weakly coupled pendulums or oscillators. This property provides one of the experimental methods for detecting a nonvanishing neutrino mass. For example, starting out with ν_μ we can look for the evolution of e^- production in matter as a function of distance traveled by the neutrino.

The β-decay processes of Eq. 4.31 can be written equivalently as

$$\begin{aligned} n &\longrightarrow p + e^- + \overline{\nu}_e \\ p &\longrightarrow n + e^+ + \nu_e \\ p + e^- &\longrightarrow n + \nu_e \end{aligned} \tag{4.40}$$

Because a neutron is more massive than a proton, a free neutron can decay as in Eq. 4.40. By the same token, however, since the proton is lighter than a neutron, a proton cannot β-decay in free space. That is, a proton can undergo β-decay only inside a nucleus. Free neutrons, on the other hand, do, in fact, decay in the laboratory, with lifetimes of about 900 sec. This lifetime is much longer than the time scales involved in nuclear and electromagnetic reactions. (We note that the typical time scale for a nuclear reaction is about 10^{-23} sec, while the corresponding quantity for an electromagnetic process is about 10^{-16} sec.) Thus, we conclude that although β-decay is a nuclear phenomenon, it does not involve the strong nuclear force. (Its origin also cannot be electromagnetic.) This result led Fermi to postulate the existence of a new force that is responsible for β-decay. It is called the *weak force,* and is short-ranged, since it is effective only within the nuclear domain. The weakness of the strength of this force is responsible for the long lifetimes observed in β-decays. In terms of relative strengths, the nuclear, electromagnetic, weak, and the gravitational forces can be characterized by the ratios $1 : 10^{-2} : 10^{-5} : 10^{-39}$. As in the case of electromagnetism, the weak coupling strength of this force also allows us to calculate any of its effects through perturbative techniques.

As we have noted earlier, nuclei do not contain electrons. Consequently, electrons produced in β-decay cannot originate from within the nucleus. Rather, they must be produced at the time of the decay. This is quite analogous to the situation in atomic transitions, where we know that photons do not exist within atoms, but are produced during the transitions. Just as a transition in an atom can be understood as being induced, for example, by a dipole interaction, and can be calculated using perturbation theory, in a similar way, β-decay can be understood as being induced by the weak force of the weak interaction Hamiltonian. The transition probability per unit time, or the "width," for the process can also be calculated from perturbation theory using Fermi's Golden Rule:

$$P = \frac{2\pi}{\hbar}|H_{fi}|^2\rho(E_f) \tag{4.41}$$

where $\rho(E_f)$ is the density of states for the decay products and H_{fi} denotes the matrix element of the weak interaction Hamiltonian, H_{wk}, between the initial and the final states:

$$H_{fi} = \langle f|H_{\text{wk}}|i\rangle = \int d^3x\,\psi_f^*(x)H_{\text{wk}}\psi_i(x) \tag{4.42}$$

From Eq. 4.40 we see that the weak interaction Hamiltonian must connect four fermionic states; otherwise, the matrix element in Eq. 4.42 would not describe β-decay. The Hamiltonian for the theory of β-decay as proposed by Fermi—also known as the *four fermion theory*—was relativistic, and based on the properties of the Dirac equation for fermions. Over the years, experimental studies have greatly restricted the structure of the four fermion theory to a form that is in excellent agreement with all experimental measurements of low-energy β-decay processes.

Among the predictions of the current theory is that there are only left-handed neutrinos, which accounts naturally for parity violation in weak interactions. We discuss parity in detail in Chapter XI. Let us simply note here that a system is parity invariant if it does not change under reflection of spatial coordinates; that is, if it is indistinguishable from its mirror image. For the left-handed neutrinos emitted in β-decay, however, the mirror reflected process, that is, the mirror image, will involve right-handed neutrinos. This follows because under reflection $\vec{r} \to -\vec{r}$, $\vec{p} \to -\vec{p}$, $\vec{L} = \vec{r} \times \vec{p} \to (-\vec{r}) \times (-\vec{p}) = \vec{L}$. Therefore, as shown in Fig. 4.6, the direction of motion changes under reflection, but spin, being an angular momentum, does not, leading to a change in the handedness. (Note that handedness is given by $(\vec{p}\cdot\vec{s})/|\vec{p}|$.) Thus, we see that the process of β-decay is distinguishable from its mirror image. However, since right-handed neutrinos (and left-handed antineutrinos) do not exist in nature, the parity transformed process, in fact, does not occur, and consequently parity must be violated in weak interactions. This has, of course, been proved experimentally. (We urge the reader to look at a rotating screw and its image in a mirror to be convinced that the sense of rotation is preserved under reflection.)

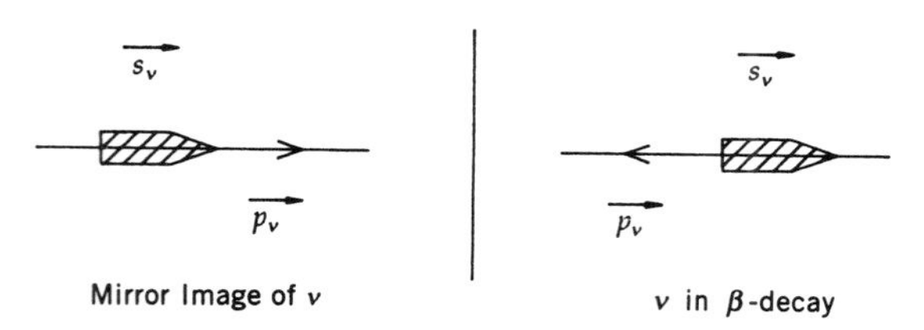

Figure 4.6 Sketch of the momentum and spin vector of a neutrino and its mirror (inverted) image.

GAMMA DECAY

As we have already seen, when a heavy nucleus disintegrates by emitting an α-particle or a β-particle, the daughter nucleus may be left in an excited state. If the excited nucleus does not break apart or emit another particle, it can deexcite to the ground state by emitting a high-energy photon or gamma (γ) ray. As we have seen, the minimum spacing of nuclear energy levels is about 50 keV, and therefore the typical energies of nuclear γ-rays range from a fraction to several MeV. Because this kind of deexcitation is electromagnetic, we expect typical lifetimes for such processes to be about 10^{-16} sec.* As in atomic transitions, the photon carries away at least one unit of angular momentum (the photon, being described by the vector electromagnetic field, has spin angular momentum of $s = 1$), and the process conserves parity.

The study of the emission and absorption of nuclear γ-rays forms an essential part of the development of nuclear spectroscopy. The subject has a direct parallel in the study of atomic spectroscopy, however, there are important differences. For example, consider a system initially in a state of energy E_i making a transition to a state with energy E_f through the absorption or emission of a photon of frequency ν. In such a process, we have what is known as a resonant or recoilless transition, for which we have the relationship

$$h\nu = \mp(E_i - E_f) \tag{4.43}$$

where "$-$" corresponds to absorption and "$+$" to emission. Thus, measuring ν determines the level spacings. In absorbing or emitting a photon, however, any system must, in fact, recoil to conserve momentum. If M denotes the mass of the system and v the magnitude of its recoil velocity, then from conservation of momentum it follows that

$$\frac{h\nu}{c} = Mv \tag{4.44}$$

Consequently, conservation of energy will yield

$$\begin{aligned} E_i - E_f &= \mp h\nu + \frac{1}{2}Mv^2 \\ &= \mp h\nu + \frac{1}{2M}\left(\frac{h\nu}{c}\right)^2 \end{aligned}$$

*When we speak about "typical" lifetimes for different interactions, it should be recognized that these can vary substantially for any specific process, because of differences in the phase space and transition operators under different circumstances. Thus, for example, electromagnetic reactions have "typical" lifetimes ranging between $\sim 10^{-19}$ sec in certain particle decays to $\sim 10^{-8}$ sec for atomic dipole transitions.

or

$$h\nu = \mp\left(E_i - E_f - \frac{h^2\nu^2}{2Mc^2}\right) = \mp(E_i - E_f - \Delta E_R) \tag{4.45}$$

where ΔE_R denotes the recoil energy of the system.

We also know that with every unstable energy level there is associated a width ΔE and a natural life time τ, related through the uncertainty principle as

$$\tau \Delta E \simeq \hbar$$

or

$$\Delta E \simeq \frac{\hbar}{\tau} \sim \text{uncertainty in } (E_i - E_f) \tag{4.46}$$

In other words, the exact value of each energy level is uncertain, and cannot be defined to better than ΔE. Consequently, if the recoil energy is such that $\Delta E_R \ll \Delta E$, then Eq. 4.45 is essentially equivalent to Eq. 4.43, and resonant absorption can take place. On the other hand, if $\Delta E_R \gg \Delta E$, it is then impossible to excite the system to a higher level through resonant absorption within the bounds (or "umbrella") provided by the uncertainty relation.

To appreciate this more fully, consider an atom with $A = 50$. The typical spacing of atomic levels is of the order of 1 eV; let us therefore consider absorption of a photon of energy of $h\nu = 1$ eV. For this case, $Mc^2 \simeq 50 \times 10^3$ MeV $= 5 \times 10^{10}$ eV, and, consequently,

$$\Delta E_R = \frac{(h\nu)^2}{2Mc^2} \simeq \frac{1(\text{eV})^2}{2 \times 5 \times 10^{10}\ \text{eV}} = 10^{-11}\ \text{eV} \tag{4.47}$$

Because typical lifetimes associated with excited atomic levels are about 10^{-8} sec, we see that

$$\Delta E \simeq \frac{\hbar}{\tau} \simeq \frac{6.6 \times 10^{-22}}{10^{-8}\ \text{sec}}\ \text{MeV-sec}$$
$$= 6.6 \times 10^{-14}\ \text{MeV} = 6.6 \times 10^{-8}\ \text{eV} \tag{4.48}$$

Consequently, $\Delta E_R \ll \Delta E$, and for atomic transitions, resonant absorption can take place.

In contrast, typical nuclear spacings have $h\nu \gtrsim 100$ keV $= 10^5$ eV. If we consider a nucleus with $A = 50$, we again have $Mc^2 \simeq 5 \times 10^{10}$ eV, and therefore, in this case, the recoil energy is given by

$$\Delta E_R = \frac{(h\nu)^2}{2Mc^2} \simeq \frac{(10^5 \text{ eV})^2}{10^{11} \text{ eV}} = 10^{-1} \text{ eV} \tag{4.49}$$

A typical lifetime for nuclear levels can be taken as about 10^{-12} sec , so that

$$\begin{aligned}\Delta E &\simeq \frac{\hbar}{\tau} \simeq \frac{6.6 \times 10^{-22} \text{ MeV-sec}}{10^{-12} \text{ sec}} \\ &= 6.6 \times 10^{-10} \text{ MeV} = 6.6 \times 10^{-4} \text{ eV}\end{aligned} \tag{4.50}$$

It is clear, therefore, that for these nuclear transitions $\Delta E_R \gg \Delta E$, and resonant absorption is not possible.

For resonant absorption to take place in nuclei, the recoil energy must somehow be reduced, and this is done beautifully through what is known as the Mössbauer effect (named after its discoverer Rudolf Mössbauer). The basic idea rests on the fact that, the heavier the recoiling system, the smaller is the recoil energy (see Eq. 4.49). An enormous increase in the mass of the recoil can be achieved by freezing the nucleus into a rigid crystal lattice, which, of course, has a much larger mass than a single nucleus. As a result, the mass of the recoiling system becomes the mass of the macroscopic crystal, thereby increasing the effective mass of the recoil by many orders of magnitude, and consequently making the energy loss ΔE_R negligible relative to ΔE. Because of this feature, the Mössbauer technique can provide exceedingly precise estimates of widths of levels. For example, level widths in iron have been measured to an accuracy of about 10^{-7} eV, which leads to an accuracy of about 1 part in 10^{12} in level spacing. The technique is therefore extremely useful in the study of hyperfine splittings of energy levels.

Problems

IV.1 Calculate the Q-values for the following α-decays between ground-state levels of the nuclei: (a) $^{208}\text{Po} \to {}^{204}\text{Pb} + \alpha$; (b) $^{230}\text{Th} \to^{226} \text{Ra} + \alpha$. What are the kinetic energies of the α-particles and of the nuclei in the final state if the decays proceed from rest?

IV.2 Estimate the relative contribution of the centrifugal barrier and the Coulomb barrier in the scattering of a 4-MeV α-particle from ^{236}U. In particular, consider impact parameters of $b = 1$ fm and $b = 7$ fm. What are the orbital quantum numbers in such collisions? (*Hint:* $|\vec{L}| \sim |\vec{r} \times \vec{p}| \sim \hbar k b \sim \hbar \ell$.)

IV.3 Free neutrons decay into protons, electrons, and antineutrinos, with a mean life of 889 sec. If the neutron–proton mass difference is taken as 1.3 MeV/c^2, calculate to at least 10 percent accuracy the maximum kinetic energies that electrons and protons can have. What would be the maximum energy that the antineutrinos can have? (Assume decay from rest and that the antineutrino is massless.)

IV.4 If the stable isotope of sodium is ^{23}Na, what kind of radioactivity would you expect from (a) ^{22}Na and (b) ^{24}Na?

IV.5 Specify any additional particles needed in the following weak reactions to assure the conservation of lepton number:

(a) $\mu^- \to e^- + ?$

(b) $\tau^+ \to e^+ \; ?$

(c) $e^- + {}^AX^Z \to ?$

(d) $\nu_\mu + n \to ?$

(e) ${}^AX^Z \to {}^AY^{Z-1} + ?$

(f) $\overline{\nu}_e + p \to ?$

IV.6 Calculate the typical kinetic energy expected of an α-particle confined within a nucleus if its emitted energy is 10 MeV. What is the momentum of such an α-particle inside the nucleus and after it is emitted. Is the wavelength of such an α-particle reasonable for it to be contained within a nucleus of ^{12}C? What about ^{238}U?

IV.7 When you examine the dependence of Z on N for stable nuclei, you find that β^+ emitters lie above the region of stability (have proton excess) and β^- emitters lie below that region (have neutron excess). For example, ^{8}B emits β^+, while ^{12}B emits β^-. Stable nuclei are those that do not seem to have sufficient mass for either emission to occur; namely, they are the nuclei with greatest binding or smallest mass. As discussed in Problem III.1, this suggests that the stable nuclei should correspond to a "valley" in the M–Z space, that is, specified by $\partial M/\partial Z = 0$. Using the semiempirical mass formula for M, show that the relationship between Z and A for this valley of stability is $Z \simeq A/(2 + 0.008A^{2/3})$. Nuclei with Z beyond 105 have never been observed. Is it possible that there could be "islands" of stability for $Z > 110$? Consider, for specifics, the possibility of binding of $Z = 125$ and $Z = 126$.

Suggested Readings

Frauenfelder, H. and E. M. Henley. 1991. *Subatomic Physics.* Englewood Cliffs, N.J.: Prentice-Hall.

Krane, K. S. 1987. *Introductory Nuclear Physics.* New York: Wiley.

Williams, W. S. C. 1991. *Nuclear and Particle Physics.* London/New York: Oxford Univ. Press.

Chapter V

APPLICATIONS OF NUCLEAR PHYSICS

INTRODUCTORY REMARKS

Studies of properties of nuclei and of the nuclear force have contributed significantly to the development of an understanding of the fundamental laws of nature. The understanding of physical laws has, in the past, led to applications that have benefited mankind. For example, the principles of electromagnetism led to the commercialization of electricity, which has proved indispensable in our daily life. Similarly, the understanding of the principles of atomic physics has given us the laser, the transistor, and a host of amazing devices. Needless to say, many applications have also been developed based on nuclear phenomena. However, because these applications have been put to both constructive as well as destructive use, they have consequently led to controversy. In this chapter, we describe only a few of these applications and the principles behind them.

FISSION

Neutrons, being electrically neutral, do not sense the direct Coulomb force. As a consequence, unlike protons that are repulsed by the nuclear charge, low-energy neutrons can get quite close to the nuclei and interact with them through the attractive nuclear potential to form bound states. In the early days of nuclear physics, the capture of low-energy neutrons within nuclei was promoted as a technique for producing new nuclei of higher A values. In experiments designed to make transuranic elements through neutron capture, it was often observed that scattering of low-energy thermal neutrons (at room temperature $T \simeq 300\,\mathrm{K}$, $kT \sim \frac{1}{40}$ eV) from odd-A nuclei such as ^{235}U did not produce heavier nuclei, but instead the parent nucleus fragmented into two smaller mass daughter nuclei. Such a fragmentation of a heavy nucleus into two medium-size nuclei and any other remnants is known as *nuclear fission.* Some heavy nuclei can undergo spontaneous fission with only minimal external perturbation. A typical example of induced fission of an odd-A nucleus is given by the absorption of thermal neutrons by ^{235}U:

$$^{235}\mathrm{U} + n \longrightarrow {}^{148}\mathrm{La} + {}^{87}\mathrm{Br} + n \tag{5.1}$$

On the other hand, the scattering of thermal neutrons from even-A nuclei, such as ^{238}U, does not produce fragmentation. However, fission does take place in such nuclei when the neutrons have energies of the order of 2 MeV.

Fission, therefore, appears to be an inherent characteristic of large nuclei, and it has come to play an important role in our lives, because the process can release a large amount of energy. An estimate of the energy released in the fission of a heavy nucleus can be obtained from the graph of the binding energy per nucleon (see Fig. 2.1). The binding energy per nucleon is smaller for very large A values than for medium nuclei, where it attains a maximum value. The process of fission therefore involves the breakup of a comparatively lightly bound heavy nucleus into two tightly bound medium-A nuclei, and as a result must lead to a release of energy. Thus, if we use -7.5 MeV as the approximate binding energy per nucleon for ^{235}U (remember that B/A is negative of the binding energy per nucleon) and about -8.4 MeV for the fission products, then we obtain an energy release of about 0.9 MeV per nucleon in a typical fission. Consequently, the total energy released per fission of a ^{235}U nucleus, and shared among the end products, is estimated to be:

$$235 \times 0.9\,\mathrm{MeV} = 211.5\,\mathrm{MeV} \simeq 200\,\mathrm{MeV} \tag{5.2}$$

This is, indeed, a lot of kinetic energy, and consequently the harnessing of nuclear fission can, in principle, provide a substantial source of power.

The phenomenon of nuclear fission can be understood both qualitatively and quantitatively on the basis of the liquid drop model. Qualitatively, let us recall that the model assumes nuclei to be spherical. This is, indeed, consistent with much of the data. However, for very large nuclei, a spherical shape need not necessarily be stable. Furthermore, an external perturbation, such as an incident neutron, could create surface waves that could lead to a change in the shape of a liquid drop. The liquid drop could, for example, elongate as a result of the perturbation. If the produced deformation is large enough, Coulomb repulsion between the elongated portions of the drop can produce a two-lobe structure that would push the lobes further apart, and cause a complete split or fission of the initial nuclear drop into two or more droplets. On the other hand, if the initial deformation is not very large, then the deformed liquid drop could form an excited state of the compound nucleus (consisting of the incident neutron and the parent nucleus of nucleon number A), which could eventually deexcite to a lower energy state of the nucleus of nucleon number $(A + 1)$ through the emission of a photon. This second scenario is commonly referred to as radiative capture of a neutron. These processes can be represented pictorially, as shown in Fig. 5.1.

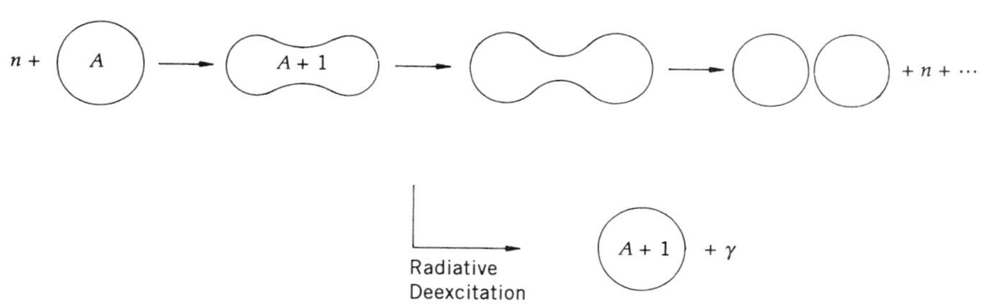

Figure 5.1 Neutron absorption leading either to fission or to radiative capture.

The liquid drop model also provides an excellent quantitative description of nuclear fission. As we have already seen, the model provides a natural and successful parameterization of the binding energy of nuclei. The empirical formula for the binding energy (see Eq. 3.3) has three classical terms that depend explicitly on the shape of the drop, namely, the volume energy, the surface energy, and the Coulomb energy. We can therefore perform a simple classical calculation to analyze the stability of a liquid drop under any external perturbation. Thus, if we assume that a spherical liquid drop, of radius R, deforms very slightly under some external perturbation to an ellipsoid of the same volume (recall that nuclear matter behaves like an incompressible liquid), with semimajor and semiminor axes a and b, respectively, we can parameterize a and b as follows (see Fig. 5.2):

$$a = R(1 + \epsilon)$$

$$b = \frac{R}{(1 + \epsilon)^{1/2}} \tag{5.3}$$

(See Problem V.13 for the connection between the infinitesimal parameter of deformation ϵ and the eccentricity of the ellipsoid.) This choice of parametrization guarantees that the volume remains unchanged:

$$V = \frac{4}{3}\pi R^3 = \frac{4}{3}\pi a b^2 \tag{5.4}$$

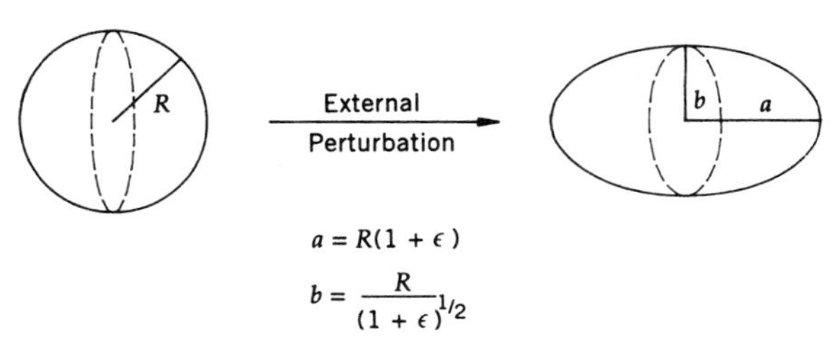

Figure 5.2 Deformation of a sphere into an ellipsoid of same volume.

Since the volume is identical for the sphere and for the ellipsoid, the volume energy will be the same for both the original and the deformed liquid drops. The surface energy and the Coulomb energy will, however, be different for the two cases. In fact, it can be shown (by comparing the surface areas of the ellipsoid to that of the sphere) that the surface energy for the ellipsoid assumes the form

$$a_2 A^{2/3} \longrightarrow a_2 A^{2/3}\left(1 + \frac{2}{5}\epsilon^2\right) \tag{5.5}$$

while the Coulomb energy changes to

$$a_3 \frac{Z^2}{A^{1/3}} \longrightarrow a_3 \frac{Z^2}{A^{1/3}}\left(1 - \frac{1}{5}\epsilon^2\right) \tag{5.6}$$

(A simple way to remember the correction terms is to note that, for small ϵ, we can think of the transformation as $A \to A\left(1 + \frac{3}{5}\epsilon^2\right)$.) Note that the chosen deformation increases the surface energy while decreasing the Coulomb term. The stability of the original droplet would therefore depend on how these two terms compete with each other. The total change in the binding energy (B.E.) due to the deformation can now be written as

$$\begin{aligned} \Delta &= \text{B.E. (ellipsoid)} - \text{B.E. (sphere)} \\ &= \frac{2}{5}\epsilon^2 a_2 A^{2/3} - \frac{1}{5}\epsilon^2 a_3 \frac{Z^2}{A^{1/3}} \\ &= \frac{1}{5}\epsilon^2 A^{2/3}\left(2a_2 - a_3 \frac{Z^2}{A}\right) \end{aligned} \tag{5.7}$$

Clearly, if this energy difference is positive, then the spherical drop will be more tightly bound, and consequently will be stable under a small external perturbation. In fact, from the values of a_2 and a_3 given in Eq. 3.4, namely, $a_2 \simeq 16.8$ MeV and $a_3 \simeq 0.72$ MeV, we find that $\Delta > 0$ if

$$2a_2 - a_3 \frac{Z^2}{A} > 0$$

or

$$\frac{Z^2}{A} < 47 \tag{5.8}$$

This simple classical analysis shows, therefore, that a spherical nucleus is stable under infinitesimal perturbations only if $Z^2 < 47A$. There are, of course, quantum mechanical corrections that have to be considered, but they do not affect the qualitative features of the result. Spherical nuclei with $Z^2 > 47A$ would be expected to be highly unstable and subject to spontaneous fission. Because large nuclei have $Z < \frac{1}{2}A$ (see Fig. 2.3), it follows, in fact, that all known nuclei have $Z^2 < 47A$, and, for these, a spherical shape appears to provide maximal binding. However, even for $Z^2 < 47A$, the binding energy of two daughter nuclei can be smaller than the binding energy of the spherical parent nucleus. In such cases, the spherical parent nucleus will have a tendency to fission, and thereby transform to a lower energy state.

Let us consider the simple example of a parent nucleus fragmenting into two identical daughter nuclei. (We are therefore assuming that both the A and Z values for the parent nucleus are even.) Neglecting the quantum mechanical terms, namely, the terms with a_4 and a_5 in Eq. 3.2, we can calculate the difference in the binding energy of the initial nucleus and the fission products when they are far apart from each other. Since the volume energy cancels out, we obtain

$$\begin{aligned}
\Delta(\text{B.E.}) &= \text{B.E.}(A, Z) - 2\text{B.E.}\left(\frac{A}{2}, \frac{Z}{2}\right) \\
&= a_2 A^{2/3}\left(1 - 2\left(\frac{1}{2}\right)^{2/3}\right) + a_3\,\frac{Z^2}{A^{1/3}}\left(1 - 2\,\frac{(1/2)^2}{(1/2)^{1/3}}\right) \\
&= a_2 A^{2/3}\left(1 - 2^{1/3}\right) + a_3\,\frac{Z^2}{A^{1/3}}\left(1 - 2^{-2/3}\right)
\end{aligned} \tag{5.9}$$

Using the values of a_2 and a_3 from Eq. 3.4, we obtain

$$\begin{aligned}
\Delta(\text{B.E.}) &\simeq A^{2/3}\left(-0.27\,a_2 + 0.38\,a_3\,\frac{Z^2}{A}\right) \\
&= A^{2/3}\left(-0.27 \times 16.8\,\text{MeV} + 0.38 \times 0.72\,\text{MeV}\,\frac{Z^2}{A}\right) \\
&\simeq 0.27\,A^{2/3}\left(-16.5 + \frac{Z^2}{A}\right)\text{MeV}
\end{aligned} \tag{5.10}$$

This calculation shows, therefore, that for $Z^2 > 16.5\,A$, we will have $\Delta(\text{B.E.}) > 0$, which corresponds to the condition that the two daughter nuclei will be more tightly bound than the parent nucleus. It follows, therefore, that for $16.5A < Z^2 < 47A$, when the spherical shape of the parent nucleus is stable under small perturbations, it is, nevertheless, energetically favorable for the parent to fragment into two lighter nuclei.

Our previous discussion can now be incorporated more quantitatively into a graph of the potential energy of the two fission fragments as a function of their separation distance (see Fig. 5.3). When the two daughter nuclei are far apart, their potential energy relative to the parent is given by Eq. 5.10. For $A \sim 240$ and $Z \sim 92$, this corresponds to ~200 MeV for two smaller nuclei of comparable size. As the fragments are brought closer together, they sense the repulsive Coulomb potential, which increases as the separation between them decreases. For $r = r_0$, approximately when the daughter nuclei start touching, the Coulomb potential is at its maximum, and is of the order of ~250 MeV. (This reduces by ~10–15 percent for daughter nuclei with asymmetric Z-values of ~2:1.) For $r < r_0$, the two nuclei begin to fuse into a single deformed nucleus and, as we have discussed, there are two possibilities for the evolution of the system. (Note that when $r < r_0$, the value of r provides an effective measure of the elongation of the deformed nucleus and is therefore proportional to the deformation parameter ϵ as given in Eq. 5.3.) First, when $Z^2 > 47A$, we know that the spherical shape is unstable, and that the energy decreases quadratically with deformation (see Eq. 5.7). This corresponds to Branch-I in the potential energy. In this situation, at the slightest perturbation, a spherical parent nucleus will evolve into two separate nuclei, because continued separation of the original object is energetically favorable for all r values. The nucleus will therefore "roll downhill" very rapidly, and spontaneously fission. When $Z^2 < 47A$, the spherical parent nucleus corresponds to a stable bound state whose energy increases quadratically with the deformation. This is represented in Branch-II of the potential energy graph. In this case, classically, the parent nucleus will be at the bottom of the potential well, but, due to quantum corrections, the ground state

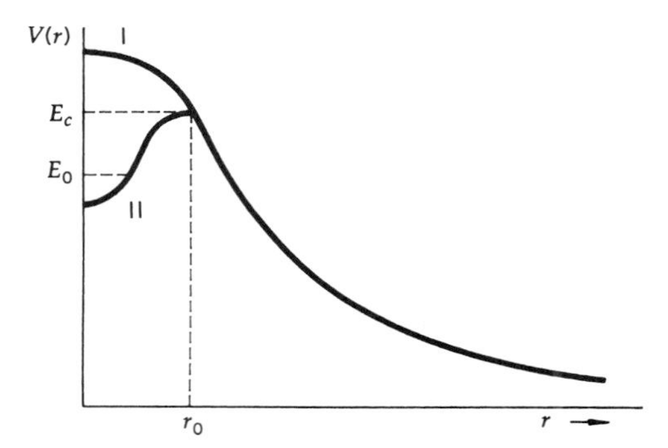

Figure 5.3 Potential energy for the interaction of two medium-size nuclei as a function of their separation distance. The nuclei just touch at $r = r_0$, and coalesce at $r = 0$.

acquires a zero-point energy given by some E_0. If E_c denotes the peak of the Coulomb barrier, then this is the classical amount of energy a nucleus must have in order to undergo fission. In other words, a nucleus must acquire an amount of energy $E_c - E_0$ in order to split apart. This is known as the activation energy, and its value is typically between 6 and 8 MeV for nuclei with $A \simeq 240$. For Branch-II, the parent nucleus can also fission through quantum mechanical tunneling through the barrier. However, the probability for this, as mentioned in our treatment of barrier penetration in Chapter IV, will be exceedingly small, since the fragment masses are large and, correspondingly, the lifetime for such a process will be quite long. Because the evolution for nuclei into two daughter nuclei on Branch-I is always energetically favorable, such fissions will be very fast.

This simple theory of fission, based on the liquid drop model, is due to Niels Bohr and John Wheeler, and although classical in its conception, it leads to a surprisingly good understanding of both natural and induced fission. It explains, in particular, why thermal neutrons induce fission in ^{235}U, whereas only higher energy neutrons can produce fission of ^{238}U. This difference can be argued in two ways as follows. First, from a qualitative viewpoint, because ^{235}U is an odd-even nucleus and ^{238}U is even-even, it follows that the ground state of ^{235}U will lie higher (less tightly bound) in the potential well of its fragments than that of ^{238}U. Hence, to induce fission, a smaller activation energy will be needed for ^{235}U than for ^{238}U. More quantitively, we can estimate the activation energy that is required for the fissioning of ^{236}U and ^{239}U; this can be calculated as about 5 MeV for the former and over 6 MeV for the latter. Now, in capturing another neutron, ^{235}U becomes an even-even excited compound nucleus. The process therefore changes an odd-even nucleus to a more tightly bound even-even nucleus; consequently, this kind of transformation releases energy (the binding energy of the last neutron is -6.5 MeV), which is sufficient for providing the activation energy needed for the compound nucleus to fission into its fragments. The kinetic energy of the incident neutron is irrelevant in this process, and hence even thermal neutrons can induce fission in ^{235}U. In contrast, the capture of a neutron by ^{238}U changes it from an even-even to an odd-even nucleus. In other words, neutron capture in this case changes a tightly bound nucleus to a less tightly bound one, which is a less exothermic process (the binding energy of the last neutron is -4.8 MeV in ^{239}U, short of the more than 6 MeV required for fission). It is for this reason that higher energy neutrons, of energy greater than 1.2 MeV, are needed to provide the additional activation energy required for ^{238}U to fission. We wish to note that, although the pairing term (i.e., the last term in Eq. 3.2) is negative for even-even nuclei and zero for odd-even nuclei, and reflects the qualitative behavior of the two systems, this term alone does not account for the entire difference observed in the neutron-induced fission of ^{235}U and ^{238}U.

In the preceding example, we assumed that the fission fragments have equal mass, which would appear to be most natural. In general, however, the fission fragments have quite asymmetrical mass distributions (this reduces the effective size of the Coulomb barrier). In fact, masses of the daughter nuclei tend to cluster around nucleon numbers of $A \sim 95$ and $A \sim 140$. Thus far there is no fundamental understanding of this particular clustering. Just after fission, the daughter nuclei are usually left in excited states and they decay to ground states through neutron "evaporation" or emission. Thus, often one or more neutrons are produced along with the larger fission products. Also, in general, the original fission products do not lie on the stability line in the N–Z plane, but, being neutron rich, lie above that line. Subsequent to fragmentation, however, the daughter nuclei decay to the stability line through β^- emission.

CHAIN REACTION

It is clear from the preceding discussion that each nuclear fission produces a large amount of energy. This in itself would not be very interesting, because what is needed for useful applications is a steady supply of energy. What makes fission attractive as a possible source of commercial power is the fact that neutrons are often produced along with the daughter nuclei. For example, in ^{235}U, an average of 2.5 neutrons are produced per nuclear fission. Since such neutrons can induce additional fission, they can, in principle, sustain a continuous process. In such a case, we could have a useful output of energy.

Consider the ratio of neutrons defined by

$$k = \frac{\text{Number of neutrons produced in the } (n+1)\text{th stage of fission}}{\text{Number of neutrons produced in the } n\text{th stage of fission}} \tag{5.11}$$

If the ratio of the number of neutrons produced in the successive stages of fission is less than unity, that is, if $k < 1$, the process is called *subcritical.* It is clear that, in such a case, the fissioning of some sample of material cannot continue indefinitely, and eventually the reaction stops. This condition is therefore not very useful as far as power generation is concerned. If $k = 1$, namely, the number of neutrons inducing fission remains constant at every stage, then the process is called *critical.* In this case, a continued reaction rate is possible. This is the most desirable condition for providing a constant supply of power in a nuclear reactor. When $k > 1$, then more and more neutrons are produced at every stage of fission, causing a runaway chain reaction. This scenario is called *supercritical,* and, in this case, the energy output grows rapidly leading to an uncontrollable explosion. Needless to say, this kind of condition finds application in the design of nuclear weapons.

In a controlled environment, such as a nuclear reactor, the chain reaction can be controlled and put to practical use for power generation.

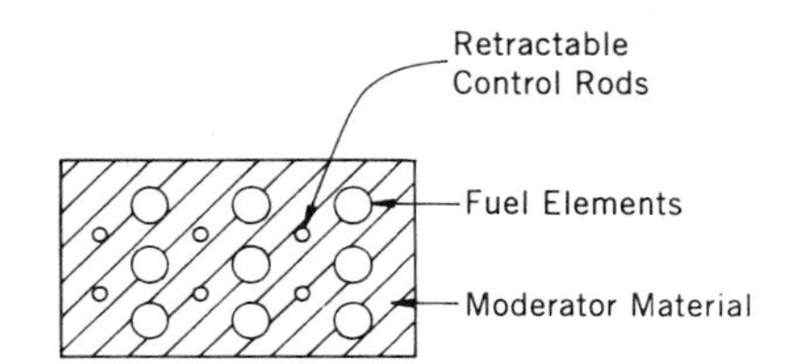

Figure 5.4 Sketch of the elements of the core of a reactor.

Very briefly, a nuclear reactor consists of several components, the most important of which is the core (see Fig. 5.4). The core contains the fissile material, or the fuel elements, the control rods, and the moderator. One can use natural uranium as the fuel in a reactor. Because natural uranium is a mixture of ^{235}U and ^{238}U, however, and ^{235}U has a shorter lifetime (about 7×10^8 yr) than ^{238}U (about 5×10^9 yr), the mixture contains only a small fraction of ^{235}U (the ratio of ^{235}U to ^{238}U in natural uranium is about 1:138). As a result, most of the thermal neutrons incident on such a sample will be captured radiatively by ^{238}U nuclei and will not induce fission. It is for this reason that enriched uranium, which is essentially ^{235}U, is used as fuel in nuclear reactors.

The control rods in a reactor are often made of cadmium, which has a high absorption cross section for neutrons. Therefore, by retracting or inserting the control rods, one can regulate the number of neutrons that are available to induce fission. This mechanism is the key element in maintaining a constant k-value, and therefore a constant power output. The fuel elements are usually surrounded by a moderator, whose main function is to slow down any fast neutrons that may be produced in the course of fission, so that they will have a larger probability of being absorbed and thereby induce more fission (higher energy neutrons have smaller absorption cross sections). It is advantageous to have a moderator material that is inexpensive, and that has a negligible cross section for absorption of neutrons. Heavy water (D_2O), for example, is preferred as a moderator over normal water (H_2O) because the cross section for neutron capture by protons in normal water (to make deuterons) is much larger than that for capture by deuterons in heavy water (to make tritium nuclei).

In a power plant (see Fig. 5.5), the reactor core is immersed in a coolant (often water), which removes the heat energy produced in the core and keeps the core at low enough temperature to prevent a meltdown. (The heat is generated from energy deposited by the fission remnants as they ionize all the material in the core.) The entire setup is surrounded by heavy shielding needed to minimize any leakage of radiation. When the nuclear reactor starts operation, the value of k is set slightly higher than unity, and is maintained at that value until the desired power output is achieved, after which the k-value is lowered to unity. As we have just noted, the

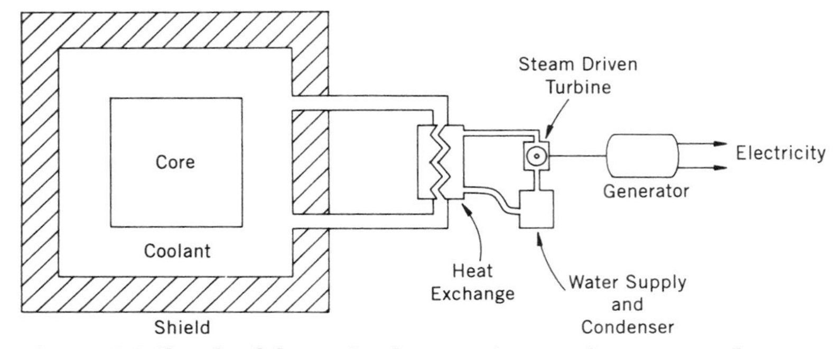

Figure 5.5 Sketch of the main elements in a nuclear power plant.

coolant removes the heat energy produced in the fission process, which can then be used to boil water and produce steam. The steam, in turn, can run turbines that generate electricity. This is, of course, only a very basic outline of the design and functioning of a nuclear power reactor. In practice, the design and construction is far more complicated, especially because of the many safety features needed to avoid accidents.

Finally, it is worth calculating the efficiency with which one can expect to produce energy in a nuclear reactor. As we have seen, a single fission of the ^{235}U nucleus yields an energy of $\sim$200 MeV or 3.2×10^{-11} joules. We also know that one gram of any element contains A_0/A atoms, where A_0 is Avogadro's number. Thus, one gram of ^{235}U has about $6 \times 10^{23}/235 \simeq 3 \times 10^{21}$ atoms. It follows, therefore, that the complete fission of one gram of ^{235}U will yield a total energy of:

$$\begin{aligned} &\simeq 3 \times 10^{21} \times 3.2 \times 10^{-11} \text{ joules} \\ &\simeq 10^{11} \text{ joules} \\ &\simeq 1 \text{ MWD (megawatt-day)} \end{aligned} \tag{5.12}$$

That is, the fission of one gram of ^{235}U can produce one megawatt of power for an entire day. To compare this yield to the energy expected from one gram of coal, we recall that burning one ton of coal yields a thermal energy of 0.36 MWD. Thus, ignoring relative efficiency of converting the energy to electric power, a gram of completely fissioned ^{235}U yields about $\sim 3 \times 10^6$ more energy than a gram of coal.

NUCLEAR FUSION

The data for the binding energy per nucleon have interesting structure in that they show a maximum for medium nuclei with $A \sim 60$. As we have seen, the slow decrease in this value for heavier nuclei is what makes fission possible as a source of energy. For lighter nuclei, the binding energy per nucleon falls much more sharply, indicating that, with the exception of magic nuclei, lighter nuclei are less tightly bound than medium-size nuclei. We can therefore imagine using a process, just the opposite of fission, that can serve as another source for generating energy. Namely, if we combine (fuse) two light nuclei to produce a relatively heavy and tightly bound nucleus, this combination will produce a release of energy due to the difference in binding energies of the initial and the final states. This process is known as nuclear fusion. In fusion, the energy released per nucleon is comparable to that of fission. However, since lighter nuclei contain fewer nucleons, the total energy released per fusion is smaller. On the other hand, since the abundance of light, stable, nuclei in nature is much greater than that of very heavy nuclei, fusion provides an attractive alternative for generating power. In fact, fusion is the mechanism responsible for energy generation in the interior of the sun and the stars.

In principle, for fusion to occur, we just have to bring two light nuclei close enough to each other so that they overlap and fuse, and thereby release energy. However, for this to take place, the Coulomb barrier between the two nuclei must be overcome. The value of the repulsive Coulomb energy is a maximum when the two nuclei are just touching, and has the form

$$V_{\text{Coulomb}} = \frac{ZZ'e^2}{R + R'} \tag{5.13}$$

where Z and Z' are the atomic numbers of the two nuclei, and R and R' are their respective radii. Recalling Eq. 2.13, we can rewrite this as

$$\begin{aligned} V_{\text{Coulomb}} &= \frac{e^2}{\hbar c} \frac{\hbar c ZZ'}{1.2(A^{1/3} + (A')^{1/3})\,\text{fm}} \\ &= \frac{1}{137} \frac{197\,\text{MeV-fm}}{1.2\text{ fm}} \frac{ZZ'}{A^{1/3} + (A')^{1/3}} \\ &\simeq \frac{ZZ'}{A^{1/3} + (A')^{1/3}}\,\text{MeV} \simeq \frac{1}{8} A^{5/3}\,\text{MeV} \end{aligned} \tag{5.14}$$

where A and A' are the nucleon numbers of the two light nuclei, and for the final expression we have set $A \sim A' \sim 2Z \sim 2Z'$. Thus, the Coulomb barrier between two nuclei of $A \simeq 8$ is about 4 MeV. Consequently, for fusion to take place, we must provide kinetic energies of the order of a few MeV to overcome the Coulomb barrier (the exact value clearly depends on the specific values of nuclear mass and charge).

It would therefore appear that a natural way to achieve fusion is by colliding two energetic beams of light nuclei. In such a process, however, most of the nuclei get scattered elastically and, as a result, this process turns out to be an inefficient way of inducing fusion. An alternate method is to heat up the relevant nuclei to high enough temperatures to provide them with sufficient kinetic energy to overcome the Coulomb barrier. To estimate these temperatures, let us assume that each of the nuclei needs a kinetic energy of about 2 MeV (i.e., the Coulomb barrier is roughly 4 MeV). Recalling that room temperature (300 K) corresponds to $\frac{1}{40}$ eV, we obtain that 2 MeV corresponds to

$$\frac{2 \times 10^6}{(1/40)} \times 300 \text{ K} \simeq 10^{10} \text{ K} \tag{5.15}$$

Although this is beyond the typical scale of $\sim 10^7$ K temperatures found inside the sun and the stars, the Maxwellian tail of the spectrum is broad enough to provide the required excursions in energy, and explains why fusion takes place in stellar interiors. There is a variety of fusion reactions that can take place inside stars; we describe below only two of the "burning cycles."

Our sun is quite rich in hydrogen. With a mass of about 10^{30} kg, it consists primarily of about 10^{56} hydrogen atoms. Consequently, we would expect that the main source of energy in the sun would be due to the burning of hydrogen. This happens through the proton-proton cycle as follows:

$$\begin{aligned} {}^1\text{H} + {}^1\text{H} &\longrightarrow {}^2\text{H} + e^+ + \nu_e + 0.42\,\text{MeV} \\ {}^1\text{H} + {}^2\text{H} &\longrightarrow {}^3\text{He} + \gamma + 5.49\,\text{MeV} \\ {}^3\text{He} + {}^3\text{He} &\longrightarrow {}^4\text{He} + 2\,({}^1\text{H}) + 12.86\,\text{MeV} \end{aligned} \tag{5.16}$$

The large amount of kinetic energy released in the last step is due to the fact that the ^{4}He nucleus is doubly magic, and is bound extremely tightly. The kinetic energies are shared among the end products of the reactions, and can be deposited in the stellar material. Thus, we see that in a proton-proton cycle, one effectively burns four hydrogen atoms to obtain:

$$6\,({}^1\text{H}) \longrightarrow {}^4\text{He} + 2\,({}^1\text{H}) + 2\,e^+ + 2\,\nu_e + 2\,\gamma + 24.68\,\text{MeV}$$

or

$$4\,({}^1\text{H}) \longrightarrow {}^4\text{He} + 2\,e^+ + 2\,\nu_e + 2\,\gamma + 24.68\,\text{MeV} \tag{5.17}$$

Note that the atoms inside the sun are in a highly ionized plasma state, and consequently the positrons (e^+) emitted in this cycle can annihilate with the prevalent electrons and contribute to an increase in the total release of energy. Similarly, the produced photons can interact with stellar matter, and deposit their energy. From the fact that the age of the universe is about 10^{10} yr, and from the power output of the sun, we can estimate that the sun will continue to burn for about another 10^9 yr before it runs out of fuel for fusion.

Another fusion cycle that plays a fundamental role within stars is the carbon or CNO cycle. The helium produced in the proton-proton cycle can produce carbon nuclei through the reaction

$$3\,(^4\text{He}) \longrightarrow\ ^{12}\text{C} + 7.27\,\text{MeV} \tag{5.18}$$

Subsequently, we can have the cycle

$$\begin{aligned} ^{12}\text{C} +\ ^{1}\text{H} &\longrightarrow\ ^{13}\text{N} + \gamma + 1.95\,\text{MeV} \\ ^{13}\text{N} &\longrightarrow\ ^{13}\text{C} + e^{+} + \nu_e + 1.20\,\text{MeV} \\ ^{13}\text{C} +\ ^{1}\text{H} &\longrightarrow\ ^{14}\text{N} + \gamma + 7.55\,\text{MeV} \\ ^{14}\text{N} +\ ^{1}\text{H} &\longrightarrow\ ^{15}\text{O} + \gamma + 7.34\,\text{MeV} \\ ^{15}\text{O} &\longrightarrow\ ^{15}\text{N} + e^{+} + \nu_e + 1.68\,\text{MeV} \\ ^{15}\text{N} +\ ^{1}\text{H} &\longrightarrow\ ^{12}\text{C} +\ ^{4}\text{He} + 4.96\,\text{MeV} \end{aligned} \tag{5.19}$$

Thus, in the carbon cycle, we effectively get

$$^{12}\text{C} + 4\,(^1\text{H}) \longrightarrow\ ^{12}\text{C} +\ ^{4}\text{He} + 2\,e^{+} + 2\,\nu_e + 3\,\gamma + 24.68\,\text{MeV}$$

or

$$4\,(^1\text{H}) \longrightarrow\ ^{4}\text{He} + 2\,e^{+} + 2\,\nu_e + 3\gamma + 24.68\,\text{MeV} \tag{5.20}$$

The different burning cycles play an important role in determining the evolution of stars.

Finally, we mention that there is currently a considerable effort worldwide in trying to achieve controlled thermonuclear fusion in the laboratory. In fact, the following processes have already been observed to occur:

$$\begin{aligned} ^{2}\text{H} +\ ^{3}\text{H} &\longrightarrow\ ^{4}\text{He} + n + 17.6\,\text{MeV} \\ ^{2}\text{H} +\ ^{2}\text{H} &\longrightarrow\ ^{3}\text{He} + n + 3.2\,\text{MeV} \\ ^{2}\text{H} +\ ^{2}\text{H} &\longrightarrow\ ^{3}\text{H} +\ ^{1}\text{H} + 4.0\,\text{MeV} \end{aligned} \tag{5.21}$$

The main obstacle to producing fusion on a large scale lies in difficulties with containing the fuel material long enough at the high temperatures needed to penetrate the Coulomb barrier. At present there are two popular methods used for achieving this goal. One is magnetic confinement, wherein a hot plasma of ^{2}H and ^{3}H circulates and fuses within a region of confining electromagnetic fields. The other is inertial confinement, wherein electromagnetic energy (laser light or beams of heavily ionizing ions) is injected into a small region that contains the fuel material. However, much work remains to be done before fusion can be put to practical use.

RADIOACTIVE DECAY

As we have seen, unstable nuclei can often transmute into other nuclei through the emission of α, β, and γ particles. Any such spontaneous transition from one state to another is known as a *radioactive decay,* and in this chapter we describe some of the general properties of such processes.

As we mentioned before, radioactive decay is a statistical process. Namely, if we have a large number of radioactive nuclei, we cannot say specifically which nucleus will decay at a given time. But there is a unique constant probability of decay associated with each nucleus. Thus, if N denotes the number of radioactive nuclei of any specified type, at a given time, and λ is the constant probability for decay per unit time (that is, the decay constant), then the change in the number of our observed nuclei during a time interval dt is given by

$$N(t + dt) - N(t) = dN = -N(t)\lambda\, dt \tag{5.22}$$

The negative sign in the preceding equation represents, as usual, the fact that the number of nuclei decreases as a result of decay. If we assume N_0 to be the initial number of nuclei at $t = 0$, then the number of nuclei $N(t)$ at a later time can be obtained from Eq. 5.22 as follows:

$$\frac{dN}{N} = -\lambda\, dt$$

or

$$\int_{N_0}^{N} \frac{dN}{N} = -\lambda \int_0^t dt$$

or

$$\ln \frac{N(t)}{N_0} = -\lambda t$$

or

$$N(t) = N_0 e^{-\lambda t} \tag{5.23}$$

In other words, for a radioactively decaying system, the number of nuclei that survive decreases exponentially, and vanishes only at infinite times. This is the characteristic law for all such statistical decay processes.

There are several time scales that one can associate with such a system. Let us denote by $t_{1/2}$ the time interval during which half of the nuclei in the sample decay. It then follows that

$$N(t_{1/2}) = \frac{N_0}{2} = N_0\, e^{-\lambda t_{1/2}}$$

or

$$\lambda\, t_{1/2} = \ln\, 2$$

and

$$t_{1/2} = \frac{\ln\, 2}{\lambda} = \frac{0.693}{\lambda} \tag{5.24}$$

If the decay constant can be calculated, then the half-life $t_{1/2}$ can be obtained and compared directly with measured values. Another useful time scale for describing decays is the mean or average life of a radioactive material. This can be calculated using Eq. 5.23

$$\begin{aligned}\langle t\rangle = \tau &= \frac{\int_0^\infty t\, N(t)\, dt}{\int_0^\infty N(t)\, dt}\\ &= \frac{N_0 \int_0^\infty t\, e^{-\lambda t}\, dt}{N_0 \int_0^\infty e^{-\lambda t}\, dt}\\ &= \frac{\lambda^{-2}}{\lambda^{-1}} = \frac{1}{\lambda}\end{aligned} \tag{5.25}$$

where the definite integrals can be found in tables. Consequently, as we mentioned in the previous chapter in connection with barrier penetration, the mean life of the sample is the inverse of the decay constant. Furthermore, it is related to the half-life through the multiplicative constant $\ln 2 = 0.693$.

We noted before that Eq. 5.23 implied that it would take an infinite amount of time for the total sample to disintegrate. Nevertheless, after several half-lives, the number of nuclei decaying is often too small to be detectable. The number of disintegrations per unit time, or the activity, of a material is defined as follows:

$$\mathscr{A}(t) = \left|\frac{dN}{dt}\right| = \lambda N(t) = \lambda N_0 e^{-\lambda t} \tag{5.26}$$

The activity is clearly a function of time and, in fact, also falls off exponentially with time. For example, for ^{226}Ra, whose half-life is 1620 years, we have $t_{1/2} = 1620\,\text{yr} = 1.62\times10^3\,\text{yr} \simeq 1.62\times10^3\times3.1\times10^7\,\text{sec} \simeq 5\times10^{10}\,\text{sec}$.

Consequently,

$$\lambda = \frac{0.693}{t_{1/2}} \simeq \frac{0.693}{5 \times 10^{10}\ \text{sec}} \simeq 1.4 \times 10^{-11}/\text{sec} \tag{5.27}$$

If the radioactive sample at $t = 0$ consists of one gram of ^{226}Ra, then the original number of radioactive nuclei in the sample is given by

$$N_0 \simeq \frac{6 \times 10^{23}}{226} \simeq 2.7 \times 10^{21} \tag{5.28}$$

and the activity of the sample at $t = 0$ is consequently,

$$\begin{aligned} \mathcal{A}(t = 0) &= \lambda N_0 \\ &\simeq 1.4 \times 10^{-11} \times 2.7 \times 10^{21}/\text{sec} \\ &\simeq 3.7 \times 10^{10}\ \text{disintegrations/sec} \end{aligned} \tag{5.29}$$

This initial activity falls off exponentially with time, with the decay constant given in Eq. 5.27.

The natural activity of ^{226}Ra has been used to define a unit of radioactivity. Thus, any sample with 3.7×10^{10} disintegrations per second is said to have a radioactivity of 1 curie:

$$1\ \text{curie} = 1\ \text{Ci} = 3.7 \times 10^{10}\ \text{disintegrations/sec} \tag{5.30}$$

In laboratory samples, the radioactivity is usually much smaller, and one typically uses 1 millicurie = 1 mCi = 10^{-3} Ci, or 1 microcurie = 1 μCi = 10^{-6} Ci as units of measurement. A more rational unit of activity is known as the rutherford. By definition

$$1\ \text{rutherford} = 10^{6}\ \text{disintegrations/sec} \tag{5.31}$$

An activity of a microrutherford in a material therefore corresponds to 1 disintegration per second, and is also referred to as one Bq (or becquerel).

Example 1

Let us suppose that we have a small sample of a radioactive substance that has a mean life τ of 10^3 sec. At some time $t = 0$ we observe 10^6 disintegrations per second. At a later time t, we expect from Eq. 5.26 that the activity would be

$$\mathcal{A}(t) = \mathcal{A}(0)e^{-\lambda t}$$

Thus if we want the number of disintegrations expected in any 10-sec interval centered on t, this would be

$$\Delta N(t) = \int_{t-5}^{t+5} \mathcal{A}(t)\, dt = -\frac{1}{\lambda}\, \mathcal{A}(0) e^{-\lambda t}\Big|_{t-5}^{t+5}$$

$$= \tau \mathcal{A}(0) e^{-\lambda t}\Big|_{t+5}^{t-5} = \tau \mathcal{A}(0) \left(e^{-\lambda(t-5)} - e^{-\lambda(t+5)}\right)$$

Let us suppose that we wish to know $\Delta N(t)$ at $t = 1000$ sec. Then, for a 10-sec interval centered on $t = 1000$, we would predict

$$\Delta N(1000) = \tau \mathcal{A}(0)\left[e^{-995/1000} - e^{-1005/1000}\right]$$

$$= \tau \mathcal{A}(0) e^{-1}\left(e^{5/1000} - e^{-5/1000}\right)$$

$$\simeq \tau \mathcal{A}(0)\, \frac{1}{e}\left(\left(1 + \frac{5}{1000} + \ldots\right) - \left(1 - \frac{5}{1000} + \ldots\right)\right)$$

$$\simeq \tau \mathcal{A}(0)\, \frac{10}{1000e} = \frac{10^3 \times 10^6 \times 10}{10^3 \times 2.7} \simeq 4 \times 10^6 \text{ counts}$$

In fact, the general expression for an arbitrary Δt would be

$$\Delta N(t) = \tau \mathcal{A}(0)\, e^{-t/\tau}\left(e^{\Delta t/2\tau} - e^{-\Delta t/2\tau}\right)$$

which for $\Delta t \ll \tau$ reduces to

$$\Delta N(t) \simeq \tau \mathcal{A}(0) \frac{\Delta t}{\tau}\, e^{-t/\tau} = \mathcal{A}(0) \Delta t\, e^{-t/\tau}$$

Clearly, the expected number of disintegrations for our chosen time interval will drop with time. There is, of course, no a priori way of determining specifically which of our nuclei will disintegrate. We just know that there will be an expected average number of disintegrations. In statistical processes where the probability of any occurrence (p) is small, but there is a large sample of events (N) that can contribute to the process, Poisson statistics can be used to describe the system. For Poisson statistics, when the expected mean is $\Delta N = pN$, then the error or standard deviation on the mean can be shown to be just $\sqrt{pN} = \sqrt{\Delta N}$. (Note that for our chosen interval of $\Delta t = 10$ sec, the probability $p = \lambda \Delta t = 10^{-2} \ll 1$, and therefore Poisson statistics are appropriate.)

Going back to our specific example, where ΔN is 4×10^6, we must now interpret the predicted result as follows. We can state that, in any given experiment performed to count ΔN, we will rarely observe the exact expected mean number of counts ΔN. What we will see is that, in about 60 percent of such experiments (assuming a Gaussian approximation for the error), the observed counting rate will be between $\Delta N - \sqrt{\Delta N}$ and $\Delta N + \sqrt{\Delta N}$. Thus, if we expect $\Delta N = 4 \times 10^6$ counts, then $\sqrt{\Delta N}/\Delta N$

is only 5×10^{-4}, so the relative fluctuations about the mean will be negligible. However, if we wish to look at the counting rate at a somewhat later time of $t = 10^4$ sec, in this case, $\Delta N(t = 10^4)$ will be far smaller:

$$\Delta N(10^4) \simeq 10^6 \times 10 \times e^{-10} \simeq 450$$

and $\sqrt{\Delta N}$ will be ~ 21, and consequently the deviations from the expected value will be more noticeable.

RADIOACTIVE EQUILIBRIUM

Upon decay, a radioactive parent nucleus produces what is called a *daughter nucleus*. The daughter nucleus can either be stable or radioactive. If it is radioactive, then it decays into a granddaughter nucleus, and so on. Thus, with the radioactivity of each parent nucleus there is often associated a series of decays. Each decay in the series has a characteristic decay constant and, therefore, a different half-life. In general, however, the mean life of the parent nucleus is much longer than that of any other member in the decay chain.

Consider a radioactive sample of material where the parent nucleus has a very long lifetime, and therefore the number of parent nuclei barely changes during any small time interval. Let us suppose that the daughter, granddaughter, etc., decay comparatively fast. After a certain lapse in time, a situation may develop where the number of nuclei of any member of the decay chain stops changing. In such a case, one says that radioactive equilibrium has set in. To see when such an equilibrium can occur, let us denote by $N_1, N_2, N_3, \ldots$ the number of nuclei of species $1, 2, 3, \ldots$ in the series, at some specified time, and by $\lambda_1, \lambda_2, \lambda_3, \ldots$, respectively, the decay constants for these members of the decay chain. The equations governing the time-evolution of the populations $N_1, N_2, N_3, \ldots$ can be deduced from the contributions to the change in any species, as follows. The daughter nuclei are produced at a rate of $\lambda_1 N_1$ due to the decay of the parent nuclei (see Eq. 5.22), and they in turn decay at a rate of $\lambda_2 N_2$. The difference between the two gives the net rate of change of the daughter nuclei. For any nucleus in the chain, there will be a similar increase in population from the feed-down and a decrease from decay, except for the parent nucleus, for which there is no feed-down possible. Thus, for the change in the number of parent, daughter, granddaughter nuclei, etc., in a time interval Δt, we can write

$$\begin{aligned}
\Delta N_1 &= -\lambda_1 N_1 \Delta t \\
\Delta N_2 &= \lambda_1 N_1 \Delta t - \lambda_2 N_2 \Delta t \\
\Delta N_3 &= \lambda_2 N_2 \Delta t - \lambda_3 N_3 \Delta t \\
\vdots & \qquad \vdots \qquad\qquad \vdots
\end{aligned} \tag{5.32}$$

Dividing Eq. 5.32 by Δt, and taking the limit of infinitesimal time intervals, we can rewrite relations (5.32) in terms of our species numbers $N_1, N_2, N_3, \ldots$, as follows:

$$\begin{aligned} \frac{dN_1}{dt} &= -\lambda_1 N_1 \\ \frac{dN_2}{dt} &= \lambda_1 N_1 - \lambda_2 N_2 \\ \frac{dN_3}{dt} &= \lambda_2 N_2 - \lambda_3 N_3 \\ \vdots \quad & \quad \vdots \quad \vdots \end{aligned} \tag{5.33}$$

We say that a "secular" equilbrium is reached when

$$\frac{dN_1}{dt} = \frac{dN_2}{dt} = \frac{dN_3}{dt} = \cdots = 0 \tag{5.34}$$

Note that by assumption, τ_1 is very large, and the change in N_1 is therefore very small ($dN_1/dt \approx 0$). Clearly, Eq. 5.34 holds only when

$$\lambda_1 N_1 = \lambda_2 N_2 = \lambda_3 N_3 = \cdots \tag{5.35}$$

or, equivalently, when

$$\frac{N_1}{\tau_1} = \frac{N_2}{\tau_2} = \frac{N_3}{\tau_3} = \cdots \tag{5.36}$$

Consequently, under these conditions, the daughter, the granddaughter, etc., will all be in equilibrium with each other, as well as with the parent nucleus (i.e., their numbers will effectively not change).

NATURAL RADIOACTIVITY AND RADIOACTIVE DATING

There are about 60 known naturally radioactive nuclei. This is a much smaller number than the order of $\sim$1000 radioactive isotopes that have been produced artificially in laboratories. If, at the time of the formation of our planet, all isotopes were almost equally abundant, then their absence in nature can be used to estimate the age of the Solar System. In fact, our Solar System is believed to be about 10 billion years old (10^{10} yr). It is therefore not surprising that during this time most of the radioactive nuclei with shorter lifetimes have completely decayed away.

The naturally occurring radioactive nuclei have atomic numbers mostly between $Z = 81$ and $Z = 92$, and are characterized by substantial neutron excess. Nevertheless, the presence of a large number of protons in these nuclei leads to strong Coulomb repulsion and instability. Such nuclei can decay by successive emission of one or more α-particles (two protons and two neutrons). The resulting daughter nuclei will therefore have an even larger *neutron*-to-proton ratio and will tend to decay through the emission of β^--particles. The granddaughters may still be unstable and decay again through the emission of more α-particles. This chain of α- and β-decays will continue until the nucleus reaches the N–Z stability band (Fig. 2.3). Because the nucleon number of an α-particle is 4, whereas that of β^--particles is zero, the alternate α- and β-decays will define a radioactive nuclear series with atomic mass numbers differing by four units.

The heavy α-emitters can be organized into four series:

$$\begin{aligned} A &= 4n && \text{Thorium series} \\ A &= 4n + 1 && \text{Neptunium series} \\ A &= 4n + 2 && \text{Uranium–Radium series} \\ A &= 4n + 3 && \text{Uranium–Actinium series} \end{aligned} \tag{5.37}$$

where n is an integer. Each of the series is labeled using the historical name of its parent nucleus, which is the longest lived nuclide in the decay chain. (The parent of the "actinium" series is, in fact, ^{235}U.) From the experimental values,

$$\begin{aligned} \tau(\text{thorium } ^{232}\text{Th}^{90}) &\simeq 1.39 \times 10^{10} \text{ yr} \\ \tau(\text{neptunium } ^{237}\text{Np}^{93}) &\simeq 2.2 \times 10^{6} \text{ yr} \\ \tau(\text{uranium } ^{238}\text{U}^{92}) &\simeq 4.5 \times 10^{9} \text{ yr} \\ \tau(\text{"actinium" } ^{235}\text{U}^{92}) &\simeq 7.15 \times 10^{8} \text{ yr} \end{aligned} \tag{5.38}$$

and from the fact that the age of the universe is about 10^{10} yr, we should not expect to find any of the radioactive isotopes of the neptunium series on earth. In fact, we have natural evidence only for the parents of the other three series. It is also curious that isotopes of lead define the stable ends for each of these three series, namely, $^{208}\text{Pb}^{82}$, $^{206}\text{Pb}^{82}$, and $^{207}\text{Pb}^{82}$, corresponding to the thorium, uranium, and the actinium series, respectively. In addition to the heavier nuclei, there also exist a few medium-size nuclei in nature, such as $^{40}\text{K}^{19}$ ($t_{1/2} \sim 1.3 \times 10^9$ yr) and $^{115}\text{In}^{49}$ ($t_{1/2} \sim 5 \times 10^{14}$ yr), which are radioactive.

One of the important applications of radioactivity is in determining the age of organic material that may be thousands of years old. The method is based on the following simple observation. Our atmosphere contains many gases, including ^{14}N and ^{12}C. Furthermore, the atmosphere is constantly being bombarded with high-energy cosmic rays, consisting of protons, photons, and other particles. These high-energy cosmic rays interact with nuclei in the atmosphere and produce lower energy particles. Any slow neutrons produced in these collisions can be absorbed by ^{14}N to produce an isotope of carbon that is radioactive, namely,

$$^{14}N^7 + n \longrightarrow {}^{14}C^6 + p \tag{5.39}$$

Subsequently, ^{14}C decays with a half-life of 5730 years through β^- emission:

$$^{14}C^6 \longrightarrow {}^{14}N^7 + e^- + \overline{\nu}_e \tag{5.40}$$

At any particular time, our atmosphere contains a substantial amount of ^{12}C and a small amount of ^{14}C, both of which can form carbon dioxide (CO_2) molecules. Living organisms, such as plants, consume CO_2 from the atmosphere, and consequently contain both of these carbon isotopes. The intake of CO_2 stops with the death of the organism. Subsequently, ^{14}C, being radioactive, continues decaying, whereas ^{12}C remains stable. As a result, the relative concentration of the two isotopes in any fossil changes with time. Therefore, by measuring directly the relative amounts of ^{14}C and ^{12}C in a fossil, and comparing this result with that in a corresponding living organism, we can estimate the age of the fossil. Alternately, we can compare the activity of ^{14}C in a fossil with that in a living organism and thereby deduce the fossil's age. This technique is known as *radioactive dating,* or *^{14}C dating,* and finds great use in archaeological and anthropological studies. The idea of carbon dating was suggested initially by Walter Libby.

Example 2

As an example, consider a piece of wood, weighing 50 gm, which has an activity of 320 disintegrations/minute for ^{14}C. The corresponding activity in a living plant is 12 disintegrations/minute/gm, and we wish to determine the age of the wood. (The half-life of ^{14}C is $t_{1/2} = 5730$ yr, and $\lambda = 0.693/t_{1/2}$.) We are given that the initial and current activities are

$$\mathcal{A}(t = 0) = 12/\text{min/gm}$$

$$\mathcal{A}(t) = \frac{320}{50}/\text{min/gm}$$

From the definition of activity, we can relate the activities at our two times as follows:

$$\mathcal{A}(t) = \left|\frac{dN}{dt}\right| = \lambda N(t) = \lambda N_0 e^{-\lambda t} = \mathcal{A}(t = 0)e^{-\lambda t}$$

Therefore,

$$\lambda t = \ln \frac{\mathscr{A}(t = 0)}{\mathscr{A}(t)}$$

or

$$t = \frac{1}{\lambda} \ln\left(\frac{12 \times 50}{320}\right) \simeq \frac{5730\text{yr}}{0.693} \times 0.626$$
$$\simeq 5170\ \text{yr}$$

In other words, the piece of wood is about 5170 years old. Recently, carbon dating techniques have greatly improved through the use of nuclear mass spectrometers, which can measure directly very small differences in the concentrations of ^{14}C and ^{12}C in any material. Using sample sizes of about 1 mgm (as opposed to 1 gram in the older counting method), measurements with sensitivity of $\sim 10^{-14}$ in the $^{14}\text{C}/^{12}\text{C}$ ratio have been achieved. In our example, we have ignored variations in the concentrations of ^{14}C stemming from any time dependence in the flux of cosmic rays, or from other, more recent sources, such as atmospheric nuclear testing. Such effects can be observed and must, clearly, be taken into account in radioactive carbon dating.

Problems

V.1 To study neutron absorption cross sections at very low energies, one must often slow down (moderate) energetic (~1-MeV) neutrons that are produced in reactors. Show that paraffin would be a better moderator than aluminum, by specifically calculating the maximum energy that a 1-MeV neutron can transfer in a collision with a proton (within paraffin) as opposed to that with an Aℓ nucleus.

V.2 Calculate the energy released when 1 gm of ^{235}U fissions into ^{148}La and ^{87}Br. Compare this amount to the energy released in fusing deuterium and tritium nuclei in 1 gm of tritiated water with 1 gm of deuterated water (i.e., T_2O and D_2O).

V.3 The counting rate for a radioactive source is measured during one minute intervals every hour, and the resulting counts are: 107, 84, 65, 50, 36, 48, 33, 25, ... Plot the counting rate versus time, and from the graph roughly estimate the mean life and the half-life. Recalling that the expected error on N counts is $\sqrt{N}$, do the data points seem reasonable? (*Hint:* Use "semilog" paper to plot $\log N$ vs. t.)

V.4 A relic from an Egyptian tomb contains 1 gm of carbon with a measured activity of 4×10^{-12} Ci. If live trees contain a ratio of 1.3×10^{-12}

for the number of $^{14}C/^{12}C$ nuclei, how old is the relic? Assume the half-life of ^{14}C is 5730 yr.

V.5 If the lifetime of the proton is 10^{33} yr, how many proton decays would you expect per year in a mass of 10^3 metric tons of water? What would be the approximate number expected in the year 2050?

V.6 Calculate the surface energies and Coulomb energies of the following nuclei:

$$^{228}\text{Th}\,,\ ^{234}\text{U}\,,\ ^{236}\text{U}\,,\ ^{240}\text{Pu}\,,\ ^{243}\text{Pu}$$

Based on your calculations, which nuclei would you expect to fission most easily?

V.7 If the efficiency for conversion of heat to electricity is only 5 percent, calculate the rate of consumption of ^{235}U fuel in a nuclear reactor operating at a power level of 500 MW of electricity.

V.8 In the fission of ^{235}U, the mass ratio of the two produced fission fragments is 1.5. What is the ratio of the velocities of these fragments?

V.9 How much energy is liberated when 1 gm of hydrogen atoms is converted into helium atoms through fusion? Compare this with the energy liberated in the fission of 1 gm of ^{235}U.

V.10 The half-life of radioactive cobalt-60 is 5.26 yr.

(a) Calculate the mean life and the disintegration constant.

(b) What is the activity of 1 gm of ^{60}Co? Express this in curies and rutherfords.

(c) What is the mass of a 10-Ci sample of cobalt-60?

V.11 Suppose that atoms of type 1 decay to type 2, which, in turn, decay to stable atoms of type 3. The decay constants of 1 and 2 are λ_1 and λ_2, respectively. Assume that at $t = 0$, $N_1 = N_0$, and $N_2 = N_3 = 0$. What are the values for $N_1(t)$, $N_2(t)$, and $N_3(t)$ at any later time t?

V.12 The activity of a certain material decreases by a factor of 8 in a time interval of 30 days. What is its half-life, mean life, and disintegration constant?

V.13 For a prolate spheroid (ellipsoid) with eccentricity x, the semimajor axis a and semiminor axis b in Fig. 5.2 are related through $b = \sqrt{1 - x^2}\, a$. If the volume and surface area of the nuclear ellipsoid are given, respectively, as $\frac{4}{3}\pi a b^2$ and $2\pi b\left(b + \frac{a \sin^{-1} x}{x}\right)$, defining $\epsilon = \frac{1}{3}x^2$, show that Eq. 5.5 holds for small values of x. (*Hint:* Assume that the volume does not change under distortion; expand functions of x and keep all terms up to order x^5.) Using this result, roughly, how would you argue that Eq. 5.6 has the right dependence?

V.14 Secular equilibrium can also be defined through the requirement that

$$\frac{d}{dt}\left(\frac{N_2}{N_1}\right) = \frac{d}{dt}\left(\frac{N_3}{N_2}\right) = \frac{d}{dt}\left(\frac{N_4}{N_3}\right) \cdots = 0$$

Assuming $\Lambda_1 \ll \Lambda_2, \Lambda_3, \Lambda_4 \ldots$, show explicitly that you retrieve the first three relations in Eq. 5.35. What happens for the final state of the decay chain? Is this sensible?

Suggested Readings

Bevington, P. R. 1969. *Data Reduction and Analysis for the Physical Sciences.* New York: McGraw-Hill.

Evans, R. D. 1955. *The Atomic Nucleus.* New York: McGraw-Hill.

Frauenfelder, H. and E. M. Henley. 1991. *Subatomic Physics.* Englewood Cliffs, N.J.: Prentice-Hall.

Krane, K. S. 1987. *Introductory Nuclear Physics.* New York: Wiley.

Williams, W. S. C. 1991. *Nuclear and Particle Physics.* London/New York: Oxford Univ. Press.

Chapter **VI**

ENERGY DEPOSITION IN MEDIA

INTRODUCTORY REMARKS

Physics is an experimental science and experiments provide the foundation for our understanding of nature and of physical laws. As we have argued repeatedly, nowhere has the need for experiments been greater than in the development of nuclear and particle physics. In these subatomic domains, scattering of particles from each other provides the primary source of information. The experiments are often quite challenging in their own right, and the experimental techniques can be as fascinating as the underlying structure they are meant to study. In the following three chapters we discuss some of the principles and devices that form the basis of experimentation in nuclear and particle physics. Most modern experiments are based on the application of a variety of exceedingly sophisticated electronic and computer tools. These tools provide the means for automatically preselecting interactions of greatest interest and the handling of enormous volumes of scientific data. We do not cover these important areas of experimentation, but rather restrict ourselves to the more general ideas encountered in the acceleration of probe particles to high energies and in the detection of particles produced in subatomic collisions. We begin with the principles underlying the detection of different kinds of particles, and defer the description of detectors and accelerators to the two following chapters.

ENERGY LOSS

In order to be detected, an object must leave some trace of its presence. That is, it must deposit energy in its wake. Ideally, detectors should help us observe particles without affecting them in any measurable way, but, as we see later, this is not always possible. Independent of the sizes or shapes of particle detectors, their operation is usually based on the electromagnetic interactions of particles with matter. Energetic charged particles, for example, can ionize atoms, and thereby release electrons that can subsequently be accelerated to produce small detectable currents. Most electrically neutral particles can also interact with matter and transfer some or all of their energies to the charged nuclei or to the atomic electrons of the medium, which in turn can yield detectable electric signals. Particles such as neutrinos, which have no electromagnetic interactions, and therefore have very low probabilities for colliding in matter (that is, have small cross sections), are therefore especially difficult to detect. We now discuss some of the more straightforward ways in which particles can deposit their energies in matter.

CHARGED PARTICLES

When a charged particle moves through any medium, it interacts primarily with the atomic electrons within that medium. If the particle has sufficient kinetic energy, it can deposit that energy by ionizing the atoms in its path or by exciting them to higher states; the excited atoms can subsequently drop down to their ground levels through photon emission. When the particle is massive, its interactions with the atomic electrons (Rutherford-like scattering) will not greatly affect its trajectory (see the discussion in Chapter I). The particle can also suffer more catastrophic nuclear collisions, but these have smaller cross sections, and are therefore relatively rare. Consequently, most of the energy that a particle deposits in the medium can be attributed to its collisions with atomic electrons.

A convenient variable that describes the ionization properties of any medium is the stopping power $S(T)$, which is defined as the amount of kinetic energy lost by any incident object per unit path length traversed in the medium (this is often termed ionization-energy loss, or simply "energy loss"). That is,

$$S(T) = -\frac{dT}{dx} = n_{\text{ion}}\overline{I} \tag{6.1}$$

where T is the kinetic energy of the particle, n_{ion} is the number of electron–ion pairs formed per unit path length, and $\overline{I}$ denotes the average energy needed to ionize an atom in the medium. (For large atomic numbers, $\overline{I}$ can be approximated as 10Z in eV units.) The negative sign in Eq. 6.1 simply reflects the fact that the particle's energy decreases as it moves along (that is, the change in kinetic energy between x and $x + dx, dT = T(x + dx) - T(x)$, is negative). Note that for any given medium, the stopping power is, in general, a function of the energy of the incident particle, and it must, of course, also depend on the particle's electric charge. We see later that the dependence on energy becomes very weak for relativistic particles.

Because the stopping power involves only electromagnetic interactions, it can be calculated quite reliably. Hans Bethe and Felix Bloch derived the following expression for $S(T)$ for relativistic particles:

$$S(T) = \frac{4\pi Q^2 e^2 nZ}{m\beta^2 c^2}\left[\ln\left(\frac{2m\beta^2 c^2}{\bar{I}} \cdot \gamma^2\right) - \beta^2\right] \qquad (6.2)$$

where m is the rest mass of the electron, $\beta = v/c$ is the particle's velocity relative to the speed of light in vacuum, γ is the particle's Lorentz factor $(1 - \beta^2)^{-1/2}$, $Q = ze$ is its charge, Z is the atomic number of the medium, and n is the number of atoms per unit volume (equal to $\rho A_0/A$, as given in Eq. 1.40).

In natural α-decay of nuclei, the emitted α-particles have typical kinetic energies of the order of several MeV, and consequently the relativistic corrections in Eq. 6.2 can be ignored, and in this case, we can use the simpler form for $S(T)$:

$$S(T) = \frac{4\pi Q^2 e^2 nZ}{m\beta^2 c^2}\ln\left[\frac{2m\beta^2 c^2}{\bar{I}}\right] \qquad (6.3)$$

However, for energetic particles produced in accelerator experiments, or in the case of electrons from nuclear β-emission, the relativistic corrections are usually substantial, and Eq. 6.2 must be used. (In fact, for electrons, there are additional small correction terms.) The expressions for $S(T)$ have been verified experimentally for different kinds of media and various types of particles, over a wide range of energies.

In light of the arguments presented in Chapter I, it may seem puzzling that energy loss due to scattering from atomic electrons dominates that for scattering from nuclei. The reason for this is that large angular deviations in scattering arise from large changes in vector momenta, which need not be accompanied by significant energy loss. For example, for the case of the elastic scattering of α-particles in the nuclear Coulomb field, there is a significant change in the momentum of the α-particle, but very little transfer of energy to the massive nucleus. On the other hand, scattering from the weakly bound atomic electrons (and ionization) represents an inelastic process that requires energy transfer. To be more specific, a momentum transfer of 0.1 MeV/c to an electron target would require an energy transfer of about 10 keV, while the same momentum transfer to a gold nucleus would correspond to less than 0.1 eV of energy transfer. Consequently, the dependence of energy loss in Eq. 6.2 on the inverse of the mass of the target, supports our previous contention that, ignoring strong nuclear collisions, small-angle scattering from atomic electrons is the dominant mechanism of energy deposition for massive charged particles traversing matter.

Because of the β^{-2} dependence in Eq. 6.2, at low particle velocities, the ionization loss is quite sensitive to particle energy. In fact, this dependence on v^{-2} suggests that particles of different rest mass (M), but same momentum (p), can be distinguished because of their different rates of energy loss. Although $S(T)$ has no explicit dependence on particle mass, for any fixed momentum, the effect of mass comes in through

$$S(T) \propto \frac{1}{v^2} = \frac{M^2\gamma^2}{p^2}$$

Consequently, at low velocities ($\gamma \approx 1$), particles of same momentum but different mass will display significantly different energy loss.

Independent of particle mass, the stopping power decreases with increasing particle velocity, and $S(T)$ displays a rather shallow minimum when $\gamma\beta \simeq 3$ (that is, the minimum occurs at higher momenta for more massive particles). This minimum in Eq. 6.2 is due to the convolution of the decrease in $S(T)$ caused by the β^{-2} dependence (β saturates at $\beta \simeq 1$ at high energies), and the rise caused by the $\ln(\gamma^2)$ term that is due to relativistic effects. When the stopping power is displayed as a function of $\gamma\beta$ or p/Mc, $S(T)$ is almost independent of M, and we can therefore say that $S(T)$ "scales as" $\gamma\beta$ or p/Mc (see Fig. 6.1).

The relativistic $\ln(\gamma^2)$ rise in $S(T)$ for $\gamma\beta > 3$ ($v > 0.96c$) eventually plateaus (saturates) because of the presence of long-range interatomic

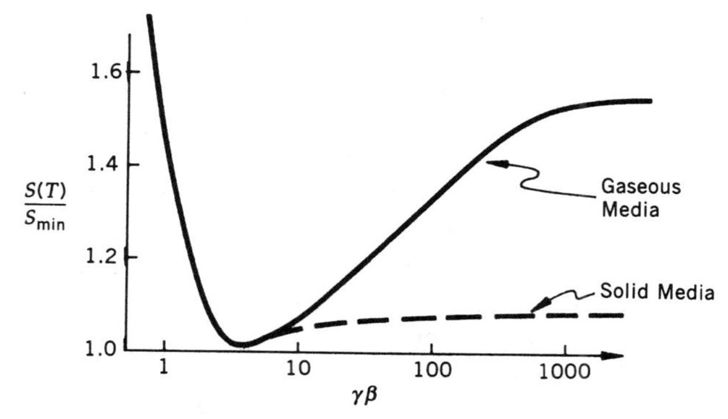

Figure 6.1 Stopping power relative to the minimum value for particles as a function of p/Mc or $\gamma\beta$.

screening effects (ignored in the Bethe-Bloch calculation). The total increase in ionization is rarely found to be greater than 50 percent beyond the value measured for a "minimum-ionizing" particle, namely a particle that has $v \simeq 0.96c$. The relativistic rise is best observed in gaseous media, and is only several percent in magnitude for dense materials. The effect can be used to distinguish different particle types through their small differences in energy loss in gaseous detectors for energies corresponding to $\gamma\beta > 3$.

At very high energies, after the saturation of the relativistic rise, the ionization loss becomes an energy-independent constant, and it is therefore not possible to distinguish particle types purely on the basis of their ionization loss. Except in gaseous media, the stopping power at high energies can be approximated quite adequately by the value when $\gamma\beta \simeq 3$ (see next section). We should also point out that, at very low energies, the stopping power in Eq. 6.2 becomes unphysical (negative), reflecting the fact that ionization loss becomes very small when the velocity of the particle becomes small. In this regime, the details of the atomic structure of the medium become important, and the incident particle can even capture electrons from the medium to form atomic systems of its own.

Once we know the stopping power, we can calculate the expected range, R, of any particle in the medium, that is, the distance it will travel before it runs out of kinetic energy and comes to a halt:

$$R = \int_0^R dx = \int_T^0 \frac{dx}{dT} dT = \int_0^T \frac{dT}{S(T)} \tag{6.4}$$

At low energies, two particles of same kinetic energy, but different mass, can have substantially different ranges. For example, an electron with a kinetic energy of 5 MeV has a range that is several hundred times that of an α-particle of same kinetic energy. At high energies, where the range becomes essentially proportional to energy, the difference in path length for particles of same kinetic energy becomes less pronounced.

UNITS OF ENERGY LOSS AND RANGE

The units of $S(T)$ in Eq. 6.2 are erg/cm in the cgs system. The more common way to specify energy loss is in MeV/cm, or in terms of an equivalent thickness of gm/cm^2, that is, in MeV/(gm/cm^2) of material. Similarly, the range is commonly expressed in cm or in gm/cm^2, where the two units are simply related through the density of the medium. When $\gamma\beta \simeq 3$, the minimum value of $S(T)$ for a particle with $z = 1$ can be evaluated approximately from Eq. 6.2 as follows:

$$S_{\min} \simeq \frac{4\pi e^4 A_0(\rho Z/A)}{mc^2\beta^2} \ln\left(\frac{2mc^2}{\bar{I}}\gamma^2\beta^2\right)$$

$$\simeq \frac{(12)(4.8 \times 10^{-10} \text{ esu})^4(6 \times 10^{23} \text{ atoms/mole})(\rho Z/A)}{(9.1 \times 10^{-28} \text{ gm})(3 \times 10^{10} \text{ cm/sec})^2(9/10)}$$

$$\times \ln\left(\frac{2 \times 0.5 \times 10^6 \text{ eV} \times 9}{10Z \text{ eV}}\right)$$

$$\simeq 5.2 \times 10^{-7}(13.7 - \ln Z)\rho\frac{Z}{A} \text{ erg/cm}$$

The $\ln Z$ term is relatively small (< 4.5), and it varies slowly with Z. Let us therefore use $\langle Z \rangle \simeq 20$ to get an approximate result

$$S_{\min} \simeq 5.6 \times 10^{-6}\rho\frac{Z}{A}\frac{\text{erg}}{\text{cm}} \cdot 6.3 \times 10^{5} \text{ MeV/erg}$$

$$\simeq 3.5\rho\frac{Z}{A} \text{ in units of MeV/cm}$$

or

$$S_{\min} \simeq 3.5\frac{Z}{A} \text{ in units of MeV/ (gm/cm}^2) \tag{6.5}$$

As we mentioned before, Eq. 6.5, which represents the ionization loss for a minimum-ionizing particle of unit charge, can also be used as a high-energy approximation for the ionization loss in most media.

Example 1

The range of a 5-MeV α-particle moving through air is given (in cm) approximately by $R = 0.318T^{3/2}$, where T is in MeV units. If the stopping power of aluminum relative to air is 1600, calculate the range of the α-particle in aluminum in cm and in the equivalent thickness in gm/cm^2.

The range in air is just $0.318 \times 5^{3/2} \simeq 3.56$ cm. Consequently, the range in aluminum foil is $3.56/1600 = 2.225 \times 10^{-3}$ cm. Now, using the density of 2.7 gm/cm^3 for aluminum, this yields an equivalent thickness of material of $(2.225 \times 10^{-3}$ cm$) \times (2.7$ gm/cm$^3) \simeq 6.1 \times 10^{-3}$ gm/cm^2, or 6.1 mgm/cm^2.

Example 2

Using the empirical formula for an electron's range–energy relation at low-energy, namely: $R(\text{gm/cm}^2) = 0.53T \text{ (MeV)} - 0.16$, calculate the energy of an electron that has a range in aluminum of 2.5 gm/cm^2.

The energy in MeV would be

$$T = \frac{1}{0.53}(R + 0.16) = \frac{1}{0.53}(2.5 + 0.16) \simeq 5.0 \text{ MeV}$$

Comparison of this with Example 1 shows that a 5-MeV electron has a range that is about 400 times longer than that of an α-particle of same kinetic energy.

STRAGGLING, MULTIPLE SCATTERING, AND STATISTICAL PROCESSES

In our examples concerning the ranges of particles in matter, we calculated values expected on the basis of phenomenological expressions. On the average, these predictions are quite accurate, but substantial variations are observed from one event to another. The magnitude of the dispersion of individual-particle ranges about the mean depends on the mass of the particle. Thus the ranges of α-particles of same energy have relatively little dispersion (or straggling) compared to that found for electrons stopping in matter.

The fundamental reason for such variations can be attributed to the inherent statistical nature of scattering processes. The energy that is transferred from an incident particle to target particles is not just a fixed and unique quantity, but rather has a range of values that are distributed according to some functional form. Thus, for example, for Rutherford scattering, the distribution function is given by Eq. 1.73. Once a function of this kind is known, we can calculate a mean value and a dispersion about the mean for any chosen variable, such as, for example, the kinetic energy transferred to the target. Any finite dispersion about the mean implies the presence of variations in the process from one interaction to another. (We have already witnessed the presence of similar fluctuations in our discussion of natural radioactivity.) The range of a particle in matter is determined by the sum over a series of independent collisions with atomic electrons in the medium. It should therefore not be surprising that fluctuations in energy transfer in individual collisions can lead to variations in the ranges of particles of the same initial energy.

Another important effect that has statistical origin involves the angular deviations experienced by particles in their Rutherford scattering off atomic electrons in the medium. The consecutive collisions add up in a random fashion and provide some net deflection of any incident particle from its original line of flight. This "multiple-Coulomb scattering" also increases the path length that any particle follows as it traverses a given thickness of material. Because multiple scattering is a random process, the average angular deviation for an ensemble of many particles passing through some thickness L of material must be zero. However, the root mean square (rms), or standard deviation $\theta_{\text{rms}} = \sqrt{<\theta^2>}$ in the space angle due to this "random walk" is finite and equals approximately

$$\theta_{\text{rms}} \simeq \frac{20 \text{ MeV}}{\beta P c} z \sqrt{\frac{L}{X_0}} \tag{6.6}$$

where z is the charge of the incident particle (in units of e) of momentum P (in MeV/c) and velocity βc, and X_0 is the radiation length of the medium (see next section).

Example 3

Calculate the mean kinetic energy transferred ($\langle T \rangle$) to a target at rest in the laboratory, and the dispersion about the mean (ΔT), for a process that can be characterized by the cross section

$$\frac{d\sigma}{dq^2} = e^{-8R^2q^2} \tag{6.7}$$

(This is, in fact, the approximate form of the dependence for scattering of nucleons at small q^2 from a nucleus of radius R in fm, when q^2 is in (GeV/c)^2 units. This R has, of course, nothing to do with the R of the range defined in Eq. 6.4.)

The kinetic energy transferred to the target can be obtained from Eq. 1.70:

$$T = \frac{q^2}{2M}$$

Hence, for the mean and for the second moment of $T(\langle T^2 \rangle)$, we obtain

$$\begin{aligned} \langle T \rangle &= \frac{\int_0^\infty (q^2/2M)e^{-8R^2q^2}\, dq^2}{\int_0^\infty e^{-8R^2q^2}\, dq^2} = \frac{1}{16MR^2} \\ \langle T^2 \rangle &= \frac{\int_0^\infty (q^2/2M)^2 e^{-8R^2q^2}\, dq^2}{\int_0^\infty e^{-8R^2q^2}\, dq^2} = \frac{1}{128M^2R^4} \end{aligned} \tag{6.8}$$

where, to evaluate the preceding integrals, we used the result

$$\int_0^\infty x^n e^{-ax}\, dx = n!/a^{n+1} \tag{6.9}$$

Consequently, the dispersion in T, defined by the square root of the variance, reduces to

$$\begin{aligned} \Delta T &= \left[\langle (T - \langle T \rangle)^2 \rangle\right]^{1/2} \\ &= \left[\langle T^2 \rangle - \langle T \rangle^2\right]^{1/2} = \frac{1}{16MR^2} \end{aligned} \tag{6.10}$$

Thus, for this simple exponential dependence on q^2, the dispersion, or rms variation in $\langle T \rangle$, from scattering to scattering, is just equal to the mean value of T. Because M is almost equal to the atomic weight A in GeV units, and $R \sim 1.2A^{1/3}$, we can write

$$\Delta T = T_{\text{rms}} = \langle T \rangle \simeq (20A^{5/3})^{-1} \text{ GeV} \tag{6.11}$$

From this example we can see once again the very strong dependence of the kinetic energy transfers on target mass. For protons interacting with protons, $\langle T \rangle \sim 0.05$ GeV, and momentum transfers are typically ~ 0.3 GeV/c, but for proton–lead collisions $\langle T \rangle \sim 7$ keV!, and momentum transfers are ~ 0.05 GeV/c. (These results, including our exponential formula for q^2, hold only for elastic reactions when nuclei do not break apart in the course of the collision.)

ENERGY LOSS THROUGH BREMSSTRAHLUNG

Although Eq. 6.2 was derived for the case of Rutherford scattering of massive projectile particles, it also holds surprisingly well for incident electrons. The scattering of electrons in matter is more complicated because electrons have small mass, and consequently relativistic corrections become important for kinetic energies as low as several hundred keV. In addition, electron projectiles can transfer substantial fractions of their energies to the atomic electrons with which they collide, thereby producing what are referred to as δ-rays, or knock-on electrons, which cannot be distinguished from the incident (i.e., scattered) electrons. This indistinguishability requires more delicate quantum-mechanical treatment of the scattering cross section. Despite these complications, Eq. 6.2 still provides an adequate approximation to the ionization loss by electrons for energies in excess of about 1 MeV. (The relativistic rise for electrons is somewhat smaller than it is for massive particles.)

Unlike massive particles, however, electrons usually suffer large accelerations as a result of their interactions with atomic electric fields (and especially with the intense nuclear Coulomb fields). These accelerations can then lead to radiation of electromagnetic waves. Such emission of photons, or bremsstrahlung as it is termed, is an important mechanism for energy loss, especially for ultrarelativistic electrons. (Bremsstrahlung can also become significant for more massive particles, but only beyond 10^{12} eV, or TeV, energy scales.) Thus for the total energy loss by electrons traversing matter we can write schematically,

$$\left(-\frac{dT}{dx}\right)_{\text{tot}} = \left(-\frac{dT}{dx}\right)_{\text{ion}} + \left(-\frac{dT}{dx}\right)_{\text{brem}} \tag{6.12}$$

The ratio of the bremsstrahlung to ionization loss for high-energy electrons can be shown to be approximately equal to

$$\frac{(dT/dx)_{\text{brem}}}{(dT/dx)_{\text{ion}}} \simeq \frac{TZ}{1200mc^2} \tag{6.13}$$

where Z is the atomic number of the medium, m is the rest mass of the electron (projectile), and T is its kinetic energy in MeV. At such high energies, the ionization loss is constant (saturated by the density effect) and given approximately by Eq. 6.5, and radiation dominates the total energy loss in Eq. 6.12. (This is illustrated in Figure 6.2.) According to Eq. 6.13, the radiated energy at high energies is proportional to the energy of the electron, and for this regime it is useful to define the radiation length, X_0, which is the distance that an electron travels before its energy drops to $1/e$ of its original value. From Eqs. 6.5 and 6.13 we obtain

$$\left(\frac{dT}{dx}\right)_{\text{brem}} = -\frac{T}{X_0}, \quad \text{where } X_0 \simeq 170A/Z^2 \text{ (in gm/cm}^2\text{)} \tag{6.14}$$

Dividing X_0 by the density of the medium, or multiplying the right–hand side of Eq. 6.14 by that density, converts the units to energy loss per cm (see Eq. 6.5). At high energies ($\gamma\beta > 3$) it is also useful to write an approximate expression for ionization loss in terms of the radiation length. Defining the critical energy, T_c, as the energy at which energy loss due to collisions (ionization) is the same as that due to bremsstrahlung, we can write

$$\left(\frac{dT}{dx}\right)_{\text{brem}} = \left(\frac{dT}{dx}\right)_{\text{ion}} = -\frac{T_c}{X_0} \tag{6.15}$$

where, from Eqs. 6.5 and 6.14,we get that $T_c \simeq 600/Z$ (in MeV).

Except for smallest Z values, the preceding expressions provide quite satisfactory approximations for calculating ionization-energy loss for any high-energy particle of unit charge, and the radiation loss for high-energy electrons. Substituting X_0 from Eq. 6.14 into Eq. 6.15, and setting A/Z = 2.2 for $Z \simeq 20$, we get that $(dT/dx)_{\text{ion}} \simeq -1.6$ MeV/(gm/cm^2). This high-energy approximation holds to $\sim$ 30 percent accuracy for every medium, except hydrogen.

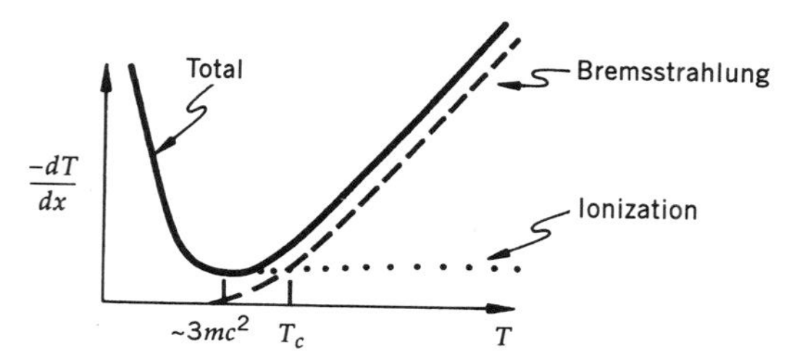

Figure 6.2 Energy loss in matter for electrons as a function of electron energy.

An important consequence of Eq. 6.14 is that, because of bremsstrahlung, high-energy electrons lose their kinetic energy exponentially with the distance traveled in matter. That is, integrating Eq. 6.14 between an initial kinetic energy T_0 and some later value T, provides the relation

$$T = T_0 e^{-x/X_0} \tag{6.16}$$

Thus energetic electrons will radiate most of that energy within several radiation lengths of material. This characteristic behavior is particularly important in the design of electron detectors. More massive ultrarelativistic charged particles, which do not radiate, can lose their energy only through nuclear (strong) collisions or simply through ionization loss.*

Example 4

As an example of the effects of multiple scattering, let us calculate the typical angular deviation of a 5-MeV proton traversing 1 cm of argon gas at atmospheric pressure and 0°C, and compare this to the case of an electron of same kinetic energy.

The radiation length of gaseous argon at the stated conditions is approximately 105 m. The proton is nonrelativistic, and its momentum can therefore be approximated as

$$\sqrt{2MT} \simeq \sqrt{2 \times 1000 \text{ MeV}/c^2 \times 5 \text{ MeV}} \simeq 100 \text{ MeV}/c$$

The proton's velocity can be calculated from

$$\sqrt{\frac{2T}{M}} \simeq \sqrt{\frac{2 \times 5 \text{ MeV}}{1000 \text{ MeV}/c^2}} \simeq 0.1c$$

The electron, on the other hand, is quite relativistic, and its momentum can therefore be taken as

$$\frac{E}{c} = \frac{T + mc^2}{c} \simeq 5.5 \text{ MeV}/c$$

The electron's velocity is essentially equal to c. Thus from Eq. 6.6, for the proton we obtain

*Muons, as we will see, are massive charged particles that do not have strong interactions, and therefore can neither radiate nor deposit their energies through large transfers of momentum to nuclei. Consequently, muons have ranges that are essentially proportional to their incident energies. Shielding radiation-sensitive equipment and personnel from excessive exposure to high-energy muons is therefore an issue of substantial concern at high-energy laboratories.

$$\theta^p_{\text{rms}} \simeq \frac{20}{0.1 \times 100}\sqrt{\frac{0.01}{105}} \simeq 0.02 \text{ rad} = 20 \text{ mrad}$$

and for the electron we get

$$\theta^e_{\text{rms}} \simeq \frac{20}{1 \times 5.5}\sqrt{\frac{0.01}{105}} \simeq 40 \text{ mrad}$$

Consequently, as expected on the basis of their small mass, electrons are scattered much farther away from their initial directions than are the more massive particles. Because low-energy electrons also have far longer ranges than more massive particles of same kinetic energy, they will therefore also exhibit far greater dispersion or straggling in range.

INTERACTIONS OF PHOTONS WITH MATTER

Because photons are electrically neutral, they do not experience the Coulomb force the way charged particles do. We might therefore conclude, incorrectly, that they cannot ionize atoms. In fact, photons are the carriers of electromagnetic force and can interact with matter in a variety of ways that lead to ionization of atoms and to energy deposition in a medium. This is discussed below.

We can describe the attenuation of light (photons, X-rays, or γ-rays) in a medium in terms of an effective absorption coefficient, μ, which reflects the total cross section for photon interactions in the medium. In general, μ will depend on the energy or frequency of the incident light. If $I(x)$ represents the intensity of photons at any point x in the medium, then the change in intensity dI in an infinitesimal thickness of material dx can be written in terms of μ as follow:

$$dI = I(x + dx) - I(x) = -\mu I(x)dx \tag{6.17}$$

where, as usual, the negative sign indicates that the intensity decreases with traversed distance. Integrating the preceding expression from some initial value I_0 at $x = 0$ to the final intensity $I(x)$ at the point x, we obtain

$$\frac{dI}{I} = -\mu\, dx$$

or

$$\int_{I_0}^{I} \frac{dI}{I} = -\mu \int_0^x dx$$

or

$$I(x) = I_0 e^{-\mu x} \tag{6.18}$$

As in the case of other statistical processes, such as radioactive decay, we can define a half-thickness, $x_{1/2}$, as the thickness of material that photons must traverse in order for their intensity to fall to half of the original value. This can be related to μ, as follows. From Eq. 6.18, we can write

$$I(x_{1/2}) = \frac{I_0}{2} = I_0 e^{-\mu x_{1/2}}$$

which implies that

$$\mu x_{1/2} = \ln 2$$

or

$$x_{1/2} = \frac{\ln 2}{\mu} = \frac{0.693}{\mu} \tag{6.19}$$

If $x_{1/2}$ is expressed in cm, then μ must have units of cm^{-1}, and when $x_{1/2}$ is given in terms of $\mathrm{gm/cm}^2$, then μ has units of $\mathrm{cm}^2/\mathrm{gm}$. The value of μ^{-1} is just the mean free path for absorption, or the average distance through which a beam of photons will propagate before their number drops to $1/e$ of the initial value.

We now turn to a brief discussion of the specific processes that contribute to absorption of photons in any medium.

Photoelectric Effect

In this process, a low-energy photon is absorbed by a bound electron, which is subsequently emitted with kinetic energy T_e (see Fig. 6.3). If we call the energy needed to free the atomic electron I_B (this is the negative of the binding energy), and the frequency of the photon ν, then energy

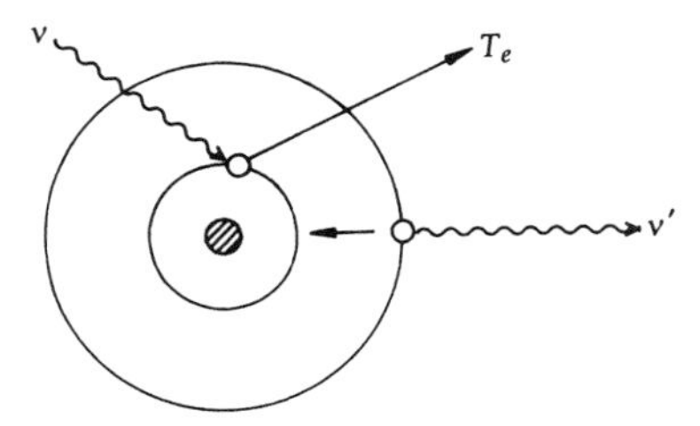

Figure 6.3 Pictorial representation of the photoelectric effect.

conservation requires that the Einstein relation holds, namely,

$$E_\gamma = h\nu = I_B + T_e$$

or

$$T_e = h\nu - I_B \tag{6.20}$$

where I_B sets the scale for the appropriate photon energies that are required for the process to take place. The photoelectric effect has a large cross section in the range of X-ray energies (keV), and ignoring the absolute normalization, is known to scale experimentally as follows:

$$\begin{aligned} \sigma &\sim \frac{Z^5}{(h\nu)^{7/2}} \quad \text{for} \quad E_\gamma < m_e c^2 \\ \sigma &\sim \frac{Z^5}{h\nu} \quad \text{for} \quad E_\gamma > m_e c^2 \end{aligned} \tag{6.21}$$

Thus the process is particularly important in high-Z atoms, and is not very significant above the 1-MeV range of energies. When the emitted electron originates from an inner shell of the atom, one of the outer electrons drops down to fill the lower (more stable) empty level, and the emitted electron is consequently accompanied by an X-ray photon produced in the subsequent atomic transition.

Compton Scattering

Compton scattering can be thought of as equivalent to a photoelectric effect on a free electron. In conventional language, one can think of the process as involving the collision of two classical particles—the photon, with energy $E = h\nu$ and momentum $p = E/c$, and an electron at rest. Alternately, the process can be viewed as follows. The electron absorbs an incident photon, and forms an electronlike system that has an unphysical mass (see Problem VI.8); this virtual (that is, "existing" only for very brief times as determined through the uncertainty relation $\tau \simeq \hbar/\Delta mc^2$, where Δm is the uncertainty in the system's mass) system then deexcites into a physical electron and to a photon of shifted frequency (or energy), as shown in Fig. 6.4.

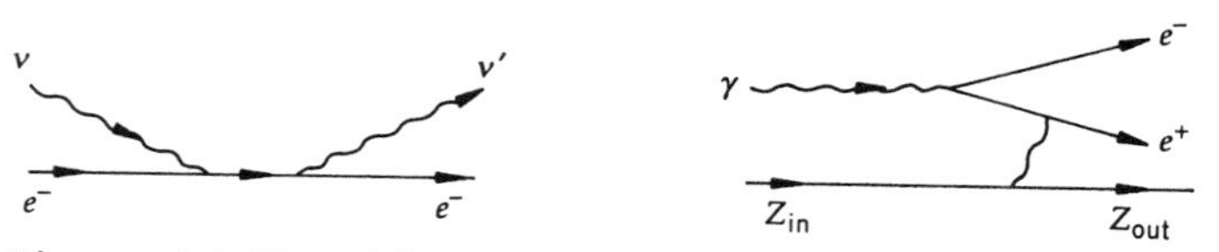

Figure 6.4 Pictorial representations of Compton scattering (left) and of pair production (right).

The kinematics for the scattering assumes that the target electron is free. This means that the results are not expected to hold for incident photons of very low-energy (much below 100 keV), where effects of atomic binding can be important. Treating the photon as a particle of energy $h\nu$ and momentum $h\nu/c$ (zero rest mass), and using fully relativistic momentum–energy expressions for the electron, it is straightforward to show that the kinematic relation between the frequency of the original and the scattered photon (ν'), at a photon scattering angle θ, is given as follows:

$$\nu' = \frac{\nu}{1 + (h\nu/mc^2)(1 - \cos\theta)} \qquad (6.22)$$

where m is the rest mass of the electron. From the preceding expression, we see that, for any finite scattering angle, the energy of the scattered photon is smaller than that of the incident one. The incident photon must therefore transfer some of its energy to the electron, which consequently has a recoil energy that depends on the scattering angle.

Relying on special relativity, the quantization of light (that is the particle properties of photons), and quantum theory, the Compton reaction served as one of the early major confirmations of the veracity of the new ideas of twentieth century physics. Again, ignoring absolute normalization, the cross section for Compton scattering is known to scale as follows:

$$\sigma \sim \frac{Z}{h\nu} \qquad (6.23)$$

where Z is the atomic number of the medium. Compton scattering dominates energy deposition typically in the 0.1- to 10-MeV range of photon energies.

Pair Production

When a photon has sufficient energy, it can be absorbed in matter and produce a pair of oppositely charged particles. Such conversions can only take place when no known conservation laws are violated in the process. In addition to charge and momentum–energy conservation, other quantum numbers may restrict the possible final states. The best known conversion process, referred to as pair production, involves the creation of a positron–electron (e^+e^-) pair and the disappearance of a photon.

A massless photon, however, cannot be converted into a pair of massive particles without violating momentum–energy conservation. This is best seen heuristically as follows. Let us suppose that the photon has a very small rest mass (far smaller than the mass of an electron). Now, in the photon's rest frame, the energy is its rest mass, namely, close to zero, while for the final state, the minimum energy is given by the sum of the rest masses of the two particles, which by assumption is substantial. It follows therefore that a process such as pair production can only be observed in a medium in which, for example, a recoiling nucleus can absorb any momentum (but very little energy!) required to assure momentum–energy conservation. Since the mass of the positron equals that of the electron, the threshold for e^+e^- pair production is essentially $h\nu \simeq 2mc^2 = 2 \times 0.511$ MeV ~ 1.022 MeV (see Problem VI.9).

The pair production cross section scales essentially as Z^2, where Z is the atomic number of the medium. The cross section rises rapidly from threshold, and dominates all energy-loss mechanisms for photon energies $\gtrsim 10$ MeV. At very high energies (> 100 MeV), the e^+e^- pair cross section saturates, and can be characterized by a constant mean free path for conversion (or by a constant absorption coefficient) that is essentially equal to the electron radiation length of the medium,

$$X_{\text{pair}} = (\mu_{\text{pair}})^{-1} \simeq \tfrac{9}{7} X_0 \tag{6.24}$$

A natural question to ask is, What happens to the positrons that are created in the conversion of photons in matter? Because positrons are known to be the antiparticles of electrons, after production, they traverse matter, much as electrons do, and deposit their energies through ionization or through bremsstrahlung. Once a positron loses most of its kinetic energy, however, it captures an electron to form a hydrogenlike atom, referred to as *positronium*, where the proton is replaced by a positron. Unlike hydrogen, positronium atoms are unstable, and decay (annihilate) with lifetimes of about 10^{-10} sec to form two photons:

$$e^+ + e^- \longrightarrow \gamma + \gamma \tag{6.25}$$

The process of annihilation produces photons of equal energy, back-to-back in the laboratory. To conserve momentum–energy, each photon carries away exactly 0.511 MeV. Thus pair annihilation provides a very clean signal for detecting positrons, as well as for calibrating the low-energy response of detectors.

The three processes that we have just discussed provide independent contributions to the absorption of photons in any medium. We can therefore write the total absorption coefficient as the sum of the three separate coefficients,

$$\mu = \mu_{\text{pe}} + \mu_{\text{Comp}} + \mu_{\text{pair}} \tag{6.26}$$

The independent contributions, and their sum, are shown as a function of photon energy in Fig. 6.5.

Finally, referring back to our discussion of Rutherford scattering in Chapter I, we can relate any absorption coefficient to the scattering cross section as follows. We argue that an object scattered out of the beam produces a drop in the beam intensity or an equivalent increase in the counting rate for scattering. According to Eq. 1.39, the fraction of the incident beam that is scattered, or lost, is proportional to the cross section per

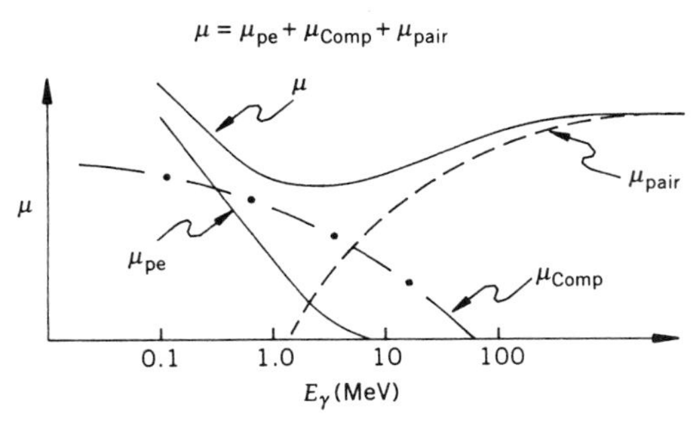

Figure 6.5 Sketch of photon absorption coefficients as a function of energy for relatively high-Z material.

nuclear scatterer (σ) and to the thickness of target material (dx):

$$\frac{dn}{N_0} = \frac{A_0}{A} \rho \sigma \, dx \tag{1.39'}$$

This is just the negative of dI/I, or the fraction of the beam that is attenuated or absorbed, as given in Eq. 6.18. Equating the two terms, we obtain the following relationship between a cross section and an absorption coefficient:

$$\mu = \rho \frac{A_0}{A} \sigma = n\sigma \tag{6.27}$$

where μ has units of cm^{-1}, when n is the number of scattering centers per cm^3, and σ is expressed in cm^2. When μ is given in units of cm^2/gm, then n corresponds to the number of atoms per gram of material.

Example 5

The total absorption coefficient for 5-MeV photons in lead is about 0.04 cm^2/gm. If the density of lead is taken as 11.3 gm/cm^3, what is the half-thickness of lead for these γ-rays? What thickness of lead would be required to reduce the intensity of such photons to 0.06 of the initial value?

The absorption coefficient in cm^{-1} is $\mu = 0.04\ cm^2/gm \times 11.3\ gm/cm^3 = 0.45\ cm^{-1}$. Hence, $x_{1/2} = 0.693/\mu = 0.693/0.45\ cm^{-1} \simeq 1.53$ cm. In other words, this is the thickness of lead that will reduce the photon intensity to half of its original value. To find the thickness needed to reduce the intensity to 0.06 of the initial value, we use Eq. 6.18:

$$\frac{I}{I_o} = e^{-\mu x}$$

or

$$0.06 = e^{-\mu x}$$

or

$$\ln(0.06) = -\mu x$$

or

$$x = -\frac{\ln(0.06)}{\mu} = -\frac{\ln(0.06)}{0.45\ \mathrm{cm}^{-1}} \simeq 6.2\ \mathrm{cm}$$

If we characterize the range of 5-MeV photons in matter by the inverse of the absorption coefficient (that is, by the absorption length), we then obtain a typical penetration length or range in lead of $1/(0.45\ \mathrm{cm}^{-1}) \simeq 2.2$ cm, or equivalently $2.2\ \mathrm{cm} \times 11.3\ \mathrm{gm/cm}^3 \simeq 25\ \mathrm{gm/cm}^2$. Comparing this with the ranges of electrons and α-particles of similar energy, it is clear that γ-rays are far more penetrating at such low energies.

Example 6

What is the cross section that corresponds to an absorption coefficient of $0.45\ \mathrm{cm}^{-1}$ for photons in lead?

The relationship we need is given in Eq. 6.27, namely,

$$\sigma = \frac{\mu}{n} = \frac{A}{A_0}\frac{\mu}{\rho}$$

Using $A_0 = 6.02 \times 10^{23}$, $A = 207.2$ gm, and $\rho = 11.3\ \mathrm{gm/cm}^3$, we obtain the cross section

$$\sigma = \left(\frac{207.2\ \mathrm{gm}}{6.02 \times 10^{23}}\right)\left(\frac{0.45\ \mathrm{cm}^{-1}}{11.3\ \mathrm{gm/cm}^3}\right) \simeq 1.37 \times 10^{-23}\ \mathrm{cm}^2 = 13.7\ \mathrm{b}$$

Example 7

The radiation length of lead at high energy is 5.6 mm. What is the value of the absorption coefficient, and what is the cross section for e^+e^- pair production on a lead target?

From Eq. 6.24 we can calculate the absorption coefficient for pair production as $\mu \simeq (7/9)X_0^{-1} \simeq 1.39\ \mathrm{cm}^{-1}$, and, following Example 6, we can calculate the cross section to be $\sigma \simeq 42.3$ b. This can be compared with the total inelastic nuclear (strong) cross section for nucleon interactions with lead at high energies, which is about 1.6 b. Consequently, we conclude that the mean free path for nuclear collisions in lead is about 15 cm, as opposed to the 0.6-cm radiation length that characterizes electromagnetic interactions. At high energies, far less material is therefore required to stop photons or electrons than particles that interact primarily through the strong force.

INTERACTIONS OF NEUTRONS

As we have already mentioned, neutrons are in most respects very similar to protons. They are the constituents of nuclei, and have essentially the same mass, same nucleon number, and spin as protons. They are, however, electrically neutral, and consequently, just like photons, cannot interact directly through the Coulomb force. (Although neutrons have small magnetic dipole moments, these moments do not produce substantial interactions in media.)

Neutrons do not sense the nuclear Coulomb force, and as a result even slow neutrons can be scattered or captured by the strong nuclear force. When low-energy neutrons interact inelastically, they can leave nuclei in excited states that can subsequently decay to ground levels through the emission of photons or other particles. Such emitted γ-rays or other particles can then be detected through their characteristic interactions with matter. Elastically scattered neutrons can transfer some of their kinetic energy to nuclear centers, which in recoiling can also provide signals (e.g., ionization) that can be used to reveal the presence of neutrons. In the elastic scattering of neutrons from nuclei, just as for the case of ionization loss, it is more difficult to transfer a sizable part of a neutron's kinetic energy to a nucleus if the nuclear mass is large (recall from Eq. 1.70 that for any momentum transfer q, the transfer of kinetic energy goes as $\frac{1}{2}(q/M)q$ with nuclear mass M). As we have already suggested in Problem V.1, this is the reason that hydrogen-rich paraffin is often used as a moderator to slow down energetic neutrons.

When neutrons are produced in collisions, they can be quite penetrating, especially if their energies are in the range of several MeV and there are no hydrogen nuclei available for absorbing their kinetic energies. The neutron shine, or "albedo," at accelerators and reactors is often a major source of background to experiments, and can only be reduced through use of appropriate moderators and materials that have large neutron-absorption cross sections (e.g., boron, which captures low-energy neutrons: $^{10}B + n \longrightarrow {}^{7}Li + \alpha$).

INTERACTION OF HADRONS AT HIGH ENERGIES

All particles that interact through the strong nuclear force are known collectively as *hadrons*. Neutrons, protons, π mesons, and K mesons are the most common hadrons. We discuss the intrinsic properties of such particles in Chapter IX, but describe here the overall character of their interactions.

Protons are, of course, the nuclei of hydrogen atoms, and are therefore easiest to accelerate and to use as particle beams (see Chapter VIII). When proton beams interact with other protons, or with larger target nuclei, they can produce π mesons, K mesons, neutrons, and other hadrons. At low beam energies (below $\sim$ 2 GeV), the interactions between pions and nucleons, kaons and nucleons, and between two nucleons, differ quite markedly. At such low energies, the collision cross sections between any two hadrons also change rapidly (oscillate) with energy. All this is because some hadronic systems resonate at specific energies while others do not. Beyond 5 GeV, the total cross sections for hadron–hadron interactions change (drop) only slightly with increasing energy. They reach minimum values, typically 20–40 mb ($\sim \pi R^2$), at $\sim$ 70–100 GeV, and then appear to increase logarithmically with increasing beam energy.

Hadronic collisions, in the main, involve very small momentum transfers, small production angles, and ranges of the order of $\sim$ 1 fm. Central collisions, involving large momentum transfers are quite rare, but very interesting from the point of view of developing an understanding of the structure of hadrons. Typical momentum transfers in hadronic reactions are of the order of $q^2 \simeq 0.1\ (\text{GeV}/c)^2$. The mean multiplicity, or the average number of particles (usually pions) produced in a typical hadronic collision, grows logarithmically with incident energy: from $\sim$ 3 particles at 5 GeV to $\sim$ 12 at 500 GeV, with great fluctuations around the mean occurring from one event to another. Thus when hadrons interact with matter, they break apart the nuclei, produce mesons and other hadrons, that can, in turn, interact again and deposit energy in the medium. Any energy deposited by the primary or the secondary particles can then be used to estimate the incident energy of the hadron (see our discussion of calorimeters in Chapter VII).

Problems

VI.1 What is the minimum thickness of aluminum in cm that is needed to stop a 3-MeV α-particle? What about the thickness needed to stop a 3-MeV electron? (Use the approximate range–energy relationships provided in Examples 1 and 2.)

VI.2 About how much steel in cm is required to stop a 500-GeV muon if the muon deposits energy only via ionization loss? (Use Eq. 6.5 to calculate your result.) Would you need a comparable amount of material to stop 500-GeV electrons? What about 500-GeV protons?

VI.3 Multiple scattering error often limits the ability to measure the direction of motion of a charged particle. To what accuracy can the incident angle of a 500-GeV muon be measured after the particle traverses one meter of iron?

VI.4 Typically, what fraction of a beam of 100-GeV photons will be transmitted through a 2-cm-thick lead absorber?

VI.5 The capture cross section for thermal neutrons on ^{27}Al is 233 mb. On the average, how far can a beam of such neutrons penetrate a slab of aluminum ($\rho = 2.7$ gm/cm^3) before half of the beam is absorbed. (See relation (1.39′) in this chapter.)

VI.6 Protons and α-particles of 20 MeV pass through 0.001 cm of aluminum foil. How much energy do they deposit within the foil?

VI.7 Compare the stopping power of electrons, protons, and α-particles in copper, for particle velocities of $0.5c$.

VI.8 Calculate the mass (i.e., $\sqrt{s}$) of the virtual electron in Fig. 6.4 for an incident photon of wavelength of 1.25×10^{-10} cm. What is the approximate lifetime of such an object? Repeat your calculation for a wavelength of 1.25×10^{-12} cm.

VI.9 Consider the collision of a photon with a target of mass M that is initially at rest in the laboratory. Show that the minimum laboratory energy that a photon must have to produce an e^+–e^- pair is $E_\gamma = 2m_e c^2(1 + (m_e/M))$. (*Hint:* Equate the expression for s given in Eqs. 1.64 and 1.65.) Thus the threshold for pair production is essentially $2m_e c^2$.

VI.10 What is the mean free path for nuclear collisions of 10-GeV protons in liquid hydrogen if the proton–proton total cross section is 40 mb? (Assume a liquid hydrogen density of 0.07 gm/cm^3.)

VI.11 Prove the kinematic relation given in Eq. 6.22.

Suggested Readings

Fernow, R. C. 1986. *Introduction to Experimental Particle Physics*. London/New York: Cambridge Univ. Press.

Kleinknecht, K. 1986. *Detectors for Particle Radiation*. London/New York: Cambridge Univ. Press.

Knoll, G. F. 1989. *Radiation Detection and Measurement*, 2nd Ed. New York: Wiley.

Leo, W. R. 1987. *Techniques for Nuclear and Particle Physics Experiments*. New York/Berlin: Springer-Verlag.

Chapter VII

PARTICLE DETECTION

INTRODUCTORY REMARKS

The investigation of nuclear and particle collisions or decays relies upon detectors for measuring the products of such interactions. Although subatomic particles are certainly too small to be observed through purely visual means, we can use the mechanisms for energy deposition we have just described in the previous chapter to detect such particles. We now discuss some of the techniques that have been developed for studying the particles in the subatomic domain. We describe only the very simplest prototype detectors. Nevertheless, the principles underlying their performance are similar to those used in the most sophisticated devices.

IONIZATION DETECTORS

Ionization detectors are devices designed to measure the ionization produced when an incident particle traverses some medium. If the number of detected electrons and positive ions is to reflect the energy deposited in the material, then any produced electron–ion pairs must be kept from immediately recombining into atoms. This can be done by applying a sufficiently high electric field across the medium. This field will separate the charges, start their drift toward their respective electrodes, and thereby keep them from recombining.

The basic ionization detector consists of a chamber that is filled with a suitable medium that can be easily ionized. The chamber has a cathode and an anode that are held at some high relative voltage (HV), and the device is characterized by a capacitance (C) that is determined by the geometry of the electrodes. The operating medium should be chemically stable (or inert) so that the moving ionization electrons are not easily captured by the molecules of that medium. The medium should not be very sensitive to radiation damage so that its response to incident particles does not change markedly with use. In addition, the medium should have a low value of ionization potential ($\bar{I}$) in order to maximize the amount of ionization produced per energy deposited by any incident particle.

As we have mentioned, when a charged particle traverses a sensitive region of a detector, it ionizes the medium and produces electron–ion pairs. These start drifting immediately along the electric field lines: electrons to the anode and the positive ions to the cathode. As the charges drift, they induce signals on the electrodes, which provide small currents that flow through some resistor R (see Fig. 7.1). This, in turn, produces a voltage drop that can be sensed with an amplifier A. The amplifier signal can be analyzed to obtain a pulse height that can be related to the amount of produced ionization. The amount of produced ionization depends primarily on the density and atomic structure of the ionizable medium and, of course, on the energy and charge of the incident particle. The amount of ionization that is detected, however, is determined by many technical factors, foremost among which is the nature and size of the applied electric field, or the high voltage (see Fig. 7.2).

When the voltage difference between electrodes is small, the electrons and ions can recombine soon after they are produced, and only a small fraction of the produced electrons and ions reach their respective electrodes. This provides an output signal that corresponds to fewer electron–ion pairs than are actually produced in the medium. The range of operating voltage where this occurs is referred to as the recombination region. As the voltage difference is increased beyond the point where

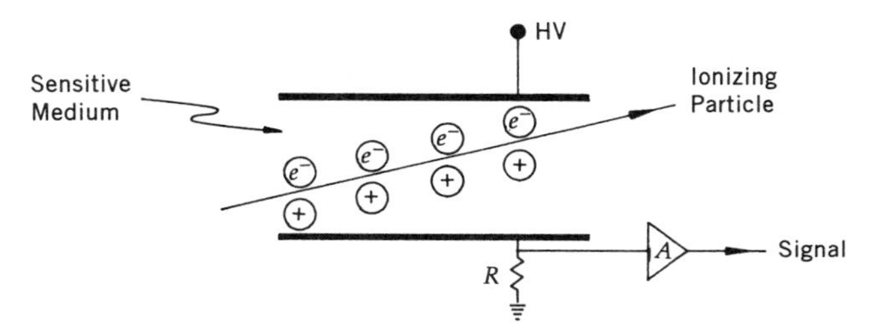

Figure 7.1 Basic elements of an ionization detector.

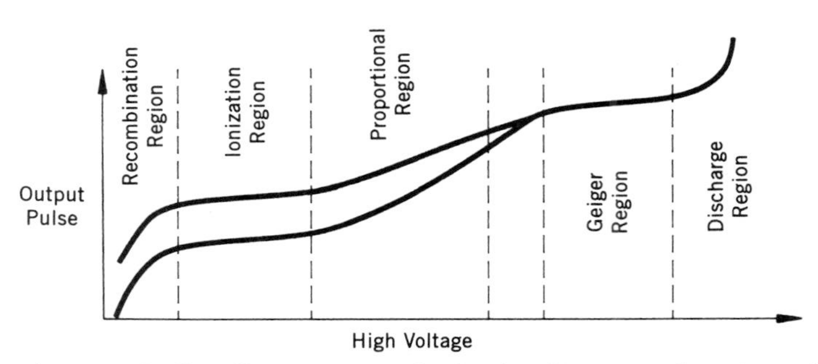

Figure 7.2 Signal response to ionization loss as a function of imposed voltage for heavily ionizing (top curve) and minimum ionizing particles (lower curve). In the Geiger region, the output does not depend on HV, nor on the amount of deposited energy or initially produced ionization.

dissociated electron–ion pairs can recombine, we obtain a signal that reflects the total amount of produced ionization. This operating range is called the *ionization region*. Increasing the voltage further provides the initially freed electrons with sufficient acceleration and energy so that they can ionize additional atoms of the medium. This increased ionization is often referred to as *signal amplification* or *multiplication*. The output signal in this operating region of voltages is larger but proportional to the initially produced ionization, and for obvious reasons, this operating range is referred to as the proportional range. (It should be noted that proportional does not necessarily mean that the signal increases linearly with voltage.) Increasing the voltage even further yields an avalanche of electron–ion pairs. In this mode, referred to as the Geiger-operating range, the energy of the original ionization electrons increases sufficiently rapidly so that they can immediately excite or ionize more atoms, thereby providing more freed electrons or photons from deexcitation of atoms. This, in turn, produces more electron–ion pairs, and eventually leads to a discharge, that is, to a highly amplified output signal whose size is independent of the amount of original ionization. Finally, increasing the voltage beyond the Geiger range leads to a breakdown that generates a continuous discharge of the medium, with the chamber no longer being sensitive to any incident ionization. Most detectors are operated as ionization, proportional, or Geiger counters, depending on the circumstances, and most detectors use gas as the operating medium.

IONIZATION COUNTERS

Ionization chambers, or *ionization counters*, operate at relatively low voltages, and therefore provide no amplification of the original signal. Consequently, the output pulses for single minimum-ionizing particles tend to be quite small and usually require special low-noise amplifiers for attaining efficient operating performance. For heavily ionizing nuclear fragments, however, or for a flux of many particles, the fully integrated signals can be substantial and easily detected. Ionization chambers are not very sensitive to voltage variations, and provide very linear output response for a wide range of input signals. Because there is no inherent amplification of signal, or discharge, in the operating medium, these types of counters do not require much time to recover from large currents, and can therefore be used in environments with high interaction rates. In addition, because there is no amplification, they provide excellent energy resolution, which is limited primarily by electronic noise and by the inherent fluctuations in the production of the initial ionization. Liquid-argon ionization chambers have been used with great success as "sampling" detectors in high-energy calorimetric measurements of energy deposition (to be discussed later). Solid-state devices, pioneered in nuclear physics, are now used commonly as ionization counters in high-energy experiments. Gaseous ionization chambers are useful for monitoring high radiation levels; they were also used in the past to measure, for example, ranges of α-particles from radioactive decays of nuclei.

Let us illustrate how an ionization counter could be employed to determine the range of 5.25-MeV α-particles emitted in ^{210}Po decay. The chamber can be chosen to be a precisely manufactured round-bottom flask with a radius of about 6 cm. The inside wall can be silver coated to serve as one of the two electrodes. A small sample of Po (of about 10 μCi) can be suspended from a grounded insulated wire at the center of the flask. An operating gas can be admitted, under pressure if need be, and the flask sealed. (Alternatively, the chamber can be simply a parallel-plate device that has the α-particle source imbedded on one of the surfaces.) A potential can be applied to the silvered surface, and the current monitored through an amplifier, as sketched in Fig. 7.1. If the ionizing medium is air, then we expect to produce a small but detectable current. The value of $\bar{I}$ for air is about 30 eV; thus, the number of electron–ion pairs that will be produced by one α-particle will be about

$$n = \frac{5.25 \times 10^6 \text{ eV}}{30 \text{ eV}} = 1.75 \times 10^5 \tag{7.1}$$

The activity of the source is

$$\mathcal{A} = 10\ \mu\text{Ci} = (10 \times 10^{-6}) \times (3.7 \times 10^{10}) = 3.7 \times 10^5\ \alpha\text{-particles/sec} \tag{7.2}$$

where we have used the definition of Ci from Eq. 5.30. Hence, the number of charged pairs produced per second is

$$N = n\mathcal{A} = (1.75 \times 10^5) \times (3.7 \times 10^5/\text{sec}) \simeq 6.5 \times 10^{10}/\text{sec} \tag{7.3}$$

Consequently, if both the positive and the negative charges are collected, this will provide a current

$$J = Ne = 6.5 \times 10^{10}/\text{sec} \times 1.6 \times 10^{-19}\ \text{C}$$
$$= 1.04 \times 10^{-8}\ \text{C/sec} = 1.04 \times 10^{-8}\ \text{A} \qquad (7.4)$$

Currents of this size can be readily measured. (For the parallel-plate geometry, the current would be only 5 nA, because only half of the α-particles would be emitted into the sensitive region.)

The range measurement proceeds as follows. The current is monitored as a function of decreasing gas (air) pressure. As long as the density of the air is high enough to stop the α-particles, the observed current, reflecting the total ionization produced by the α-particles, remains constant. When the pressure drops below that critical value, the α-particles do not lose all their kinetic energy within the gas volume, and therefore produce fewer electron–ion pairs in the sensitive region. As a result, the current drops, and keeps decreasing as the pressure is decreased further. For an α-particle range of 6 cm in air, at a temperature of 25°C, the critical pressure P_{crit} is found to be 51 cm of Hg. Thus, at any other temperature and pressure, the range can be calculated assuming the scaling of the simple gas law. In particular, for the standard conditions $T = 288$ K and $P = 76$ cm of Hg, we obtain

$$R = R_{crit}\frac{P_{crit}}{P} \cdot \frac{T}{T_{crit}} = 6.0\ \text{cm}\frac{51\ \text{cm Hg}}{76\ \text{cm Hg}} \cdot \frac{288\ \text{K}}{298\ \text{K}} = 3.9\ \text{cm} \qquad (7.5)$$

Although, in principle, the observed current provides an absolute measure of the total energy deposited in the form of ionization, it is always wise to calibrate ionization chambers with sources or signals of known energy. This is particularly important when counting rates are very high and individual pulses must be counted rapidly in succession. Under such conditions, small concentrations of impurities in the detector medium (often at less than a part per million level) can cause loss of electron signal through attachment. That is, some of the electrons drifting toward the anode can be attracted to contaminant (electronegative) molecules that can form negative ions, which drift far slower than electrons, and therefore do not contribute to the fast output pulse that is produced by the electrons.

PROPORTIONAL COUNTERS

Gaseous proportional counters usually operate in high electric fields of the order of 10^4 V/cm and achieve typical amplification factors of about 10^5. Such fields can be obtained using thin (diameters of 10–50 μm) metallic wires as anode field electrodes in a cylindrical chamber geometry. Because the fields are most intense near the axial anode wires, this is where the multiplication of charge, that is, secondary ionization, also takes place. For a large variety of gases, the output signals, even for minimum-ionizing particles, are quite large. Also, these detectors can operate over a relatively wide range of HV settings. Although proportional chambers can be used for measuring absolute energy deposition (pulse heights), their reliance on the multiplication of ionization in the medium makes them quite sensitive to the dependence of the output signal on the magnitude of the operating voltage.

Georges Charpak and his collaborators have developed a variant of the proportional counter in the multiwire proportional chamber (MWPC) that has found primary application as a position detector in high-energy physics experiments. The idea is illustrated in Figs. 7.3 and 7.4. The principle is to have a plane of anode wires positioned precisely, with typical wire-spacings of about 2 mm. Such planes can be sandwiched between two similar cathode planes (or, alternatively, thin stretched aluminum foil can be used for the cathodes). Typically, a space of $\lesssim 1$ cm is left between cathode and anode planes. These doublets are then enclosed in some superstructure—usually with thin Mylar sheet windows on the outside. Operating gas can then be admitted into the regions between electrodes. Several doublet planes, with different orientations of anode wires, can be sandwiched together. Charged particles passing through the gaseous medium will produce ionization along their paths, which will produce pulses on the anode wires that are closest to the trajectories. The anode wires, each with its own amplifier, function essentially as independent proportional counters, and can therefore be used to localize the position of any charged particle to an accuracy of the order of the spacing between the anode wires.

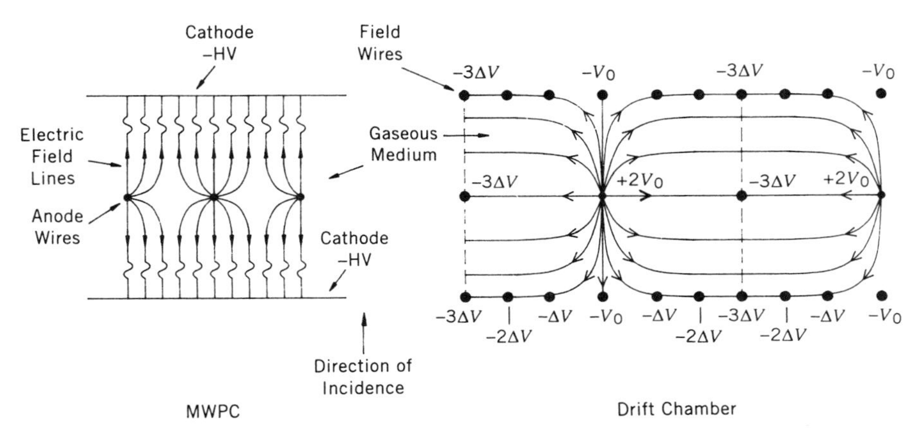

Figure 7.3 Electric field structure in a multiwire proportional chamber and in a multiwire drift chamber.

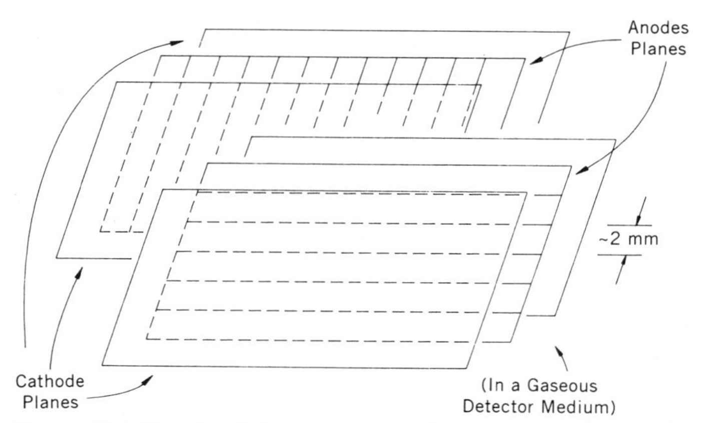

Figure 7.4 Sketch of the geometry of multiwire proportional chamber planes.

A set of MWPC planes placed before and after a region that has an applied magnetic field can be used to obtain the change in the trajectory of a charged track as it passes through that region, which, in turn, provides the momentum of the particle. The principle is shown in Fig. 7.5. Note that particles of different incident momentum will be dispersed, or fanned out, in θ. The effect is very similar to the dispersion of light through a prism. Thus, the position of a particle at the exit of the magnet is determined by its momentum. A system of the kind shown in Fig. 7.5 can therefore be used as a spectrometer to analyze the momentum distribution of particles in a beam.

By modifying the electric field structure in a planar multiwire chamber, the position measuring accuracy can be improved substantially. The

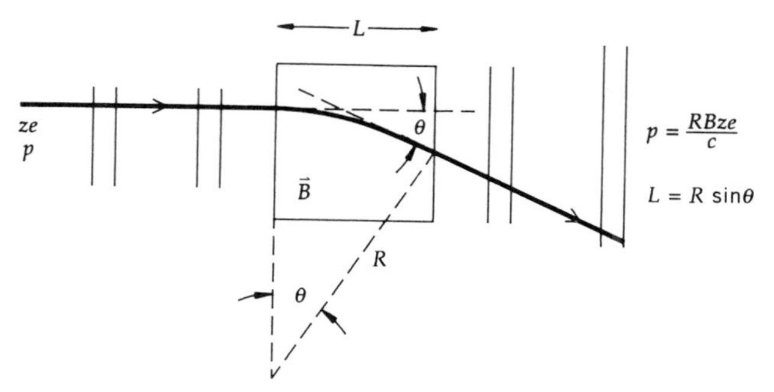

Figure 7.5 Measuring the momentum of a charged particle using MWPCs, assuming the particle's trajectory traverses a region of constant magnetic field $\vec{B}$ (normal to the plane). Reconstructing $\boldsymbol{\theta}$, and knowing L and B, provides p.

modified structure is shown in Fig. 7.3, where we have sketched the electric field lines in both a MWPC and in a drift chamber. The idea in the drift chamber is to provide a relatively constant electric field ($E = -\Delta V/\Delta x$) in each cell in a direction transverse to normal incidence. This can be achieved by grading the potential uniformly along neighboring cathode field wires, and using additional "field-shaping" wires interspersed between the anode wires, as shown. The additional wires provide corrections needed to attain approximately constant electric fields between the anode wires. For most of a cell, the field is such that the ionization electrons drift along the electric field lines at a slow and constant velocity (about 50 μm/ns or 50 mm/μs), until they arrive within a few diameters of the anode wire, where, in a short time, multiplication of signal takes place in the intense electric field. By timing the arrival of the pulse at the wire relative to some external fast signal, we can obtain the distance of drift, and from that a precise position of the incident particle. For example, it is relatively straightforward to achieve precisions of ~ 200 μm in the measurement of position for interanode spacings or drift distances of ~ 1 cm.

GEIGER-MÜLLER COUNTERS

A *Geiger-Müller counter*, or simply a *Geiger counter*, is an ionization detector that operates in the Geiger range of voltages, namely, at a voltage high enough so that any produced ionization causes a gaseous discharge, independent of the initial energy deposited in the medium. To appreciate the advantage of a Geiger counter, let us consider the simple example of an electron with a kinetic energy of 0.5 MeV that loses all its energy within the counter. Suppose that the gaseous medium is helium, with an average ionization energy of 42 eV. The number of electron–ion pairs that will be produced in the gas is

$$N = 0.5 \times 10^6 \text{ eV}/42 \text{ eV} \simeq 12{,}000 \tag{7.6}$$

If the detector operates as an ionization chamber, and has a capacitance of $\approx 10^{-9}$ F $= 1$ nF, then the resulting voltage signal would correspond to

$$V = Q/C = Ne/C = (12 \times 10^3)(1.6 \times 10^{-19} \text{ C})/10^{-9} \text{ F} \simeq 2 \times 10^{-6} \text{ V} \tag{7.7}$$

which is very small, indeed. On the other hand, if the detector operates in the Geiger mode, because of multiplication, the expected number of ion pairs would be $\sim 10^{10}$, independent of electron energy. Consequently, the voltage pulse in this case would be a large and easily detectable ~ 1.6 V.

The technical advantage of a Geiger counter is its simplicity of construction and its insensitivity to small voltage fluctuations. It is very useful for general measurement of nuclear radiation, but it has two important disadvantages. First, there is no information whatsoever on the nature of the ionization that caused the pulse. Second, because of the large avalanche induced by any ionization, a Geiger counter takes a long time to recover between successive pulses (about 1 ms), and so it has a substantial dead-time, which means that it cannot be used when counting rates are high.

SCINTILLATION DETECTORS

The ionization produced by charged particles can excite atoms and molecules in the medium to higher energy levels. When these atoms and molecules deexcite, they emit light that, in principle, can be detected and provide evidence for the traversal of the charged particle. *Scintillators* are kinds of materials that, after the passage of a charged particle, provide detectable photons in the visible part of the light spectrum. There are primarily two types of scintillators in common use in nuclear and particle physics: the organic or plastic scintillators, and the inorganic or crystalline scintillators. Although the physics of light emission is different in the two kinds of scintillators, and somewhat complicated, it is nevertheless well understood, but we will not discuss it here in any detail. Organic scintillators, such as anthracene or naphthalene, tend to emit ultraviolet light in their molecular deexcitation. Unfortunately, light of such frequency is rapidly attenuated, and consequently "wavelength shifter" material has to be admixed with the scintillator to permit detection of photons. That is, the initially produced light interacts with the wave-shifter material, which shifts the light to the visible part of the spectrum. Inorganic crystals, such as NaI or CsI, are usually doped with activators that can be excited by electron–hole pairs produced by charged particles in the crystal lattice; these dopants can then deexcite by photon emission.

Organic scintillators have fast decay times (typically $\sim 10^{-8}$ s), while inorganic crystals are usually far slower ($\sim 10^{-6}$ s), although some also have fast components in their response. Plastic scintillators are therefore more appropriate for use in high-flux environments. It takes substantially more energy to produce a detectable photon in a scintillator than an electron–ion pair through ionization (typically, by a factor of 10), and because inorganic scintillators produce more light than organic scintillators, they are consequently better for applications at low energies.

In the pioneering days of nuclear studies, different phosphors were used routinely in experiments, and viewed by eye. The light produced in scintillators is usually very weak and normally cannot be seen in this manner. For scintillation light to be detected, the scintillator material must be transparent to its own radiation, that is, it cannot have a short attenuation length at the frequencies of interest. In addition, because of the low intensity of the emitted light, the photon signal must somehow be amplified in order to be counted. The signal amplification is most commonly achieved using photomultiplier tubes (PMT) that view the scintillator either directly or through light guides.

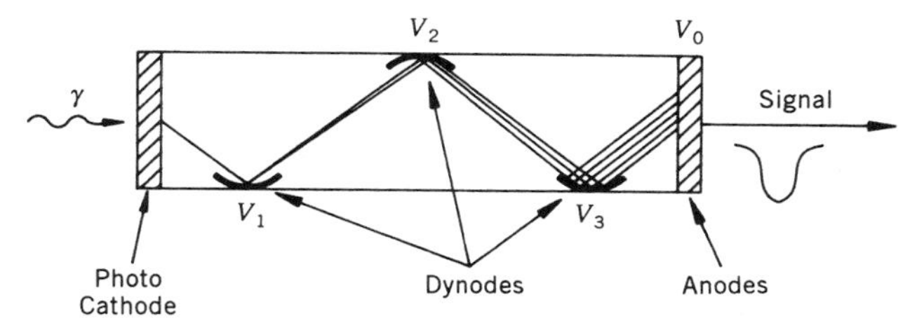

Figure 7.6 Sketch of the main elements of a photomultiplier tube. The most common tubes have ~ 5 cm diameters and are about ~ 20 cm long.

A photomultiplier tube converts a weak photon signal to a detectable electric pulse. The device consists of several components (see Fig. 7.6). First, right after a thin entry window, is a photocathode, which is made of material in which the valence electrons are weakly bound and have a high cross section for converting photons to electrons via the photoelectric effect. As a result, any photon striking the photocathode will have a high probability for releasing an electron. (Cathode diameters are typically 2–12 cm, but far larger PMTs are also available.) Next, within the tube, there is a series ("stages") of dynodes made of material of relatively low work function. These electrodes are operated at ever increasing potential ($\Delta V \sim$ 100–200 V between dynodes), which can be provided, for example, through a regulated DC power supply and a resistor–divider chain. The dynodes accelerate the initial electrons to the next stage, and multiply them through secondary emission at each dynode. There are typically 6–14 dynode stages in PMTs, with a total gain, or electron amplification factor, in the range of $\sim 10^4$–10^7 (usually the multiplication factor is ~ 3–5 per dynode). The voltage is applied to the electrodes through pins embedded at the back end of the glass PMT; these pins are connected directly to the dynodes that are located within the vacuum region of the tube.

The quantum conversion efficiency of the photocathodes is typically ~ 0.25 in the usual range of operating wavelengths (~ 400 nm). The output signal is usually taken from the anode or last dynode of the PMT. Except for statistical fluctuation, this signal is linearly proportional to the amount of light incident on the photocathode. Although there is some spread in the time of arrival of the signal due to different electron transit times (different paths and different velocities), this is usually only several nanoseconds; the output signals are therefore quite narrow, and, in conjunction with fast plastic scintillators, can be used very effectively for triggering any detector system on interesting events and for timing intervals between successive signals.

A scintillator used in conjunction with a photomultiplier is consequently an excellent detector of charged particles, and of any photons or neutrons that interact within the scintillator material. As an example, let us consider the β-decay of ^{60}Co:

$$^{60}\text{Co}^{27} \longrightarrow {}^{60}\text{Ni}^{28} + e^- + \overline{\nu}_e \tag{7.8}$$

The ^{60}Ni nucleus in this decay is, in fact, left in an excited state and decays to the ground level through two successive photon emissions: one of 1.17 MeV to the first excited level, which is followed by a 1.33 MeV photon to the ground level. Let us suppose that the ^{60}Co sample is mounted on the front face of a NaI(Tℓ) crystal (thallium-activated NaI scintillator), and that a PMT is attached to the opposite side of the crystal. Because the anticipated signals are small, the crystal and PMT must be properly wrapped to prevent external light from leaking into the detector. Also, because crystals such as NaI are often hygroscopic, they must be well sealed to prevent deterioration through absorption of moisture. (Plastic scintillators do not have this disadvantage.)

When the photons from the deexcitation of ^{60}Ni enter the scintillator, they can interact through the photoelectric effect, through Compton scattering or through e^+e^- pair production. Any photon that is converted to a photoelectron through the photoelectric effect generally deposits all of its energy within the scintillator in the form of ionization produced by the emitted electron. The intensity of the subsequently produced scintillation light is therefore proportional to the energy of the original photon. On the other hand, photons that undergo Compton scattering usually do not deposit all their energy within the scintillator, unless that scintillator block is exceedingly large. That is, although the scattered electrons often deposit their entire energy, the scattered photons tend to escape from the scintillator. (The radiation length of NaI is about 2.6 cm, while that of plastic scintillator is about 40 cm. It is therefore not surprising that, for detectors several centimeters on a side, a fraction of the energy of the incident photon is not converted into ionization but, instead, leaves the detector.) Pair production is exceedingly unlikely for low-energy photons, but, when it occurs, the produced electron and positron deposit their kinetic energy in the scintillator, and eventually the positron annihilates with an atomic electron, yielding two 0.511-MeV photons.

Consequently, ignoring the low-energy electron in reaction (7.8), the energy deposited in the NaI will, in general, have two kinds of contributions. First, the full energies of any of the photons that convert into photoelectrons, and, second, a continuous spectrum of energies deposited by the Compton-scattered recoil electrons. The scintillation light and the output from the PMT will therefore have signals equivalent to the deposition of 1.17 MeV, 1.33 MeV, and a continuum of energies below these peak values. (If enough 0.511-MeV photons are produced from the annihilation of e^+e^- pairs, they can also provide photoelectrons, and therefore a very useful calibration signal at 0.511 MeV.) The PMT output signals can be passed through a discriminator to eliminate the small signals from random-noise pulses produced through thermal electron emission from the cathode and dynode surfaces. After discrimination, the pulses can be digitized and displayed on a pulse-height analyzer (see Fig. 7.7). Because of fluctuations in ionization loss, differences in the efficiency of light collection, and fluctuations from electron multiplication, the 1.17-MeV and the 1.33-MeV energy

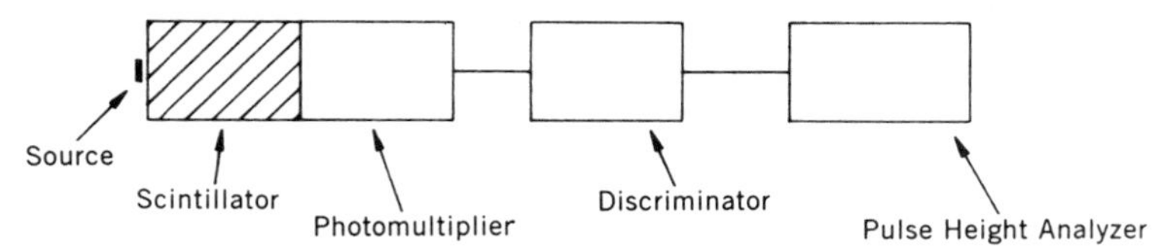

Figure 7.7 Block diagram of apparatus needed for measuring the decay products in the decay of ^{60}Co.

signals will not be sharp, but will have a shape that reflects the experimental resolution of the detector system. The expected counting rate as a function of pulse height is shown in Fig. 7.8. (The resolution for NaI(Tℓ) crystals used in this energy range is about 10 percent.) We should point out that our simplified discussion has ignored the possibility of the simultaneous observation of the summed signals from the two emitted photons. We have left this common ramification to Problem VII.6.

TIME OF FLIGHT

We have already remarked that a scintillation counter viewed with a PMT can provide excellent time resolution. In fact, with care, time resolutions of about 2×10^{-10} sec (0.2 nsec) can be readily achieved. (This corresponds to a spatial resolution of ~ 6 cm for particles traveling at the speed of light.) Thus arrays of scintillation counters can be used to measure the time of flight (TOF) of particles and thereby obtain their velocities. Such TOF measurements have important applications in providing discrimination between particles of similar momentum but different mass that may be produced in a collision. For example, measuring the momentum (p) of a charged particle in a magnetic field (see Fig. 7.5), as well as its time of flight (t) for reaching some scintillation counter located at a distance L from the point of origin of the particle, determines the velocity and therefore the

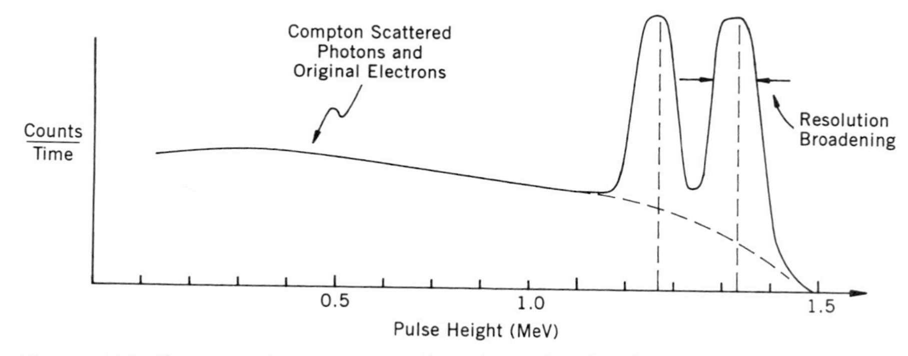

Figure 7.8 The counting rate as a function of pulse height expected for the products in the decay of ^{60}Co.

rest mass of the particle. Let us assume that the momentum measurement is very precise, and investigate the limitation of the TOF technique.

The TOF just corresponds to the distance traveled divided by the speed of the particle: $t = L/v$. Consequently, the difference in the flight time of two particles of mass m_2 and m_1 will be

$$\Delta t = t_2 - t_1 = L\left(\frac{1}{v_2} - \frac{1}{v_1}\right) = \frac{L}{c}\left(\frac{1}{\beta_2} - \frac{1}{\beta_1}\right) \tag{7.9}$$

For our known momentum p, this can be rewritten as

$$\Delta t = \frac{L}{c}\left[\frac{E_2}{pc} - \frac{E_1}{pc}\right] = \frac{L}{pc^2}\left[\left(m_2^2c^4 + p^2c^2\right)^{1/2} - \left(m_1^2c^4 + p^2c^2\right)^{1/2}\right] \tag{7.10}$$

In the nonrelativistic limit, this reduces to the classical expression

$$\Delta t = \frac{L}{p}(m_2 - m_1) = \frac{L}{p}\Delta m \tag{7.11}$$

which, for the most interesting (and most demanding) regime of $m_2 \simeq m_1 = m$ and $v_2 \simeq v_1 = v = \beta c$, becomes

$$\Delta t = \frac{L}{\beta c}\frac{\Delta m}{m} = t\frac{\Delta m}{m} \tag{7.12}$$

Also, using $v = L/t$, we can write

$$\Delta v = v_2 - v_1 = -\frac{L}{t^2}\Delta t = -v\frac{\Delta t}{t} = -\frac{v^2}{L}\Delta t \tag{7.13}$$

Consequently, at low energies ($\beta \sim 0.1$), for a time resolution $\Delta t \sim 2 \times 10^{-10}$ sec, and for a flight path of $L \simeq 10^2$ cm, we obtain a very respectable resolution in velocity:

$$|\Delta v| = \frac{v^2}{L}\Delta t \simeq \frac{(0.1 \times 3 \times 10^{10})^2}{10^2} \times 2 \times 10^{-10} \simeq 2 \times 10^7 \text{ cm/sec} \tag{7.14}$$

And because $-\Delta v/v = \Delta t/t = \Delta m/m$, this means that, using TOF, it is possible to discriminate between low-energy particles of same momentum but different mass to better than ~ 1 percent level of accuracy. Clearly, the relative mass resolution deteriorates linearly with increasing momentum, and improves with increasing path length.

Now, for the relativistic limit, let us rewrite Eq. 7.10 as

$$\Delta t = \frac{L}{c}\left[\left(1 + \frac{m_2^2c^2}{p^2}\right)^{1/2} - \left(1 + \frac{m_1^2c^2}{p^2}\right)^{1/2}\right]$$

$$\simeq \frac{L}{c}\left[1 + \frac{m_2^2c^2}{2p^2} - \left(1 + \frac{m_1^2c^2}{2p^2}\right)\right] \simeq \frac{Lc}{2p^2}\left(m_2^2 - m_1^2\right) \qquad (7.15)$$

which, for $m_1 \simeq m_2$ and $v_1 \simeq v_2$, becomes

$$\Delta t = \frac{Lc}{2}\left(\frac{m_2^2}{p^2} - \frac{m_1^2}{p^2}\right) = \frac{Lc}{2}\left[\frac{m_2^2}{m_2^2\gamma_2^2v_2^2} - \frac{m_1^2}{m_1^2\gamma_1^2v_1^2}\right]$$

$$\simeq \frac{Lc}{2}\left[\frac{1 - v_2^2}{v_2^2} - \frac{1 - v_1^2}{v_1^2}\right] \simeq \frac{Lc}{v^2}\,\frac{v_1 - v_2}{v} = -\frac{Lc}{v^2}\frac{\Delta v}{v} \qquad (7.16)$$

In this case ($v \simeq c$), we see that assuming our conditions of $\Delta t \simeq 2 \times 10^{-10}$ sec and $L \simeq 10^2$ cm, we obtain for the resolution in velocity

$$|\Delta v| \simeq \frac{c^2}{L}\Delta t \sim \frac{(3 \times 10^{10})^2}{10^2} \times 2 \times 10^{-10} \simeq 2 \times 10^9 \text{ cm/sec} \qquad (7.17)$$

Although the resolution is still ~ 10 percent of the particle velocity, unfortunately, the relative mass resolution for $v \simeq c$ is no longer as good. In fact, from Eq. 7.16 we can deduce that

$$\Delta t \simeq \frac{Lc}{2}\frac{(m_2 - m_1)(m_2 + m_1)}{p^2} \simeq \frac{Lcm}{p^2}\Delta m = Lc\frac{m^2}{p^2}\frac{\Delta m}{m} \simeq \frac{L}{c\gamma^2}\frac{\Delta m}{m}$$

and

$$\frac{\Delta m}{m} = \frac{c\gamma^2}{L}\Delta t = \gamma^2\frac{\Delta t}{t} \qquad (7.18)$$

Consequently, for momenta of $\gtrsim 3$ GeV/c, and masses of ~ 1 GeV/c^2, we have $\gamma \gtrsim 3$, and the resolution in mass discrimination is essentially lost. Our example is, of course, only true if the flight path cannot be greatly increased beyond ≈ 100 cm. For fixed-target experiments that involve highly relativistic particles this can be a possible alternative; however, for most large-collider experiments, increasing the flight path would also mean increasing the size of the overall detection system (see next chapter), which could be very costly. It should also be recognized that the flight path cannot be increased without limit if the particles of interest decay with short lifetimes.

As we have implied previously, TOF can also be used to obtain momenta of low-energy neutrons or of any photons that interact within our scintillation counter. In such cases, the collision time has to be obtained from other means, as, for example, from the interaction time as given by a pulse caused by an incident beam particle. The difference in time between a signal in the scintillation counter and some "start" time can provide the TOF for a neutral particle.

CHERENKOV DETECTORS

When a charged particle moves with uniform velocity in vacuum, it does not emit radiation. However, if it travels in a dielectric medium of index of refraction $n > 1$, and with a speed greater than the speed of light in that medium (i.e., $v > c/n$ or $\beta > 1/n$), then it emits what is known as Cherenkov radiation (after Pavel Cherenkov, who first observed the effect in 1934). The direction of the emitted light can be calculated classically using Huygen's wave construction, and can be attributed to the emission of coherent radiation from the excitation of atoms and molecules in the path of the charged particle. The effect is completely analogous to the "shock" front produced by a supersonic aircraft. The emitted light has a spectrum of frequencies, with the most interesting component being in the blue and ultraviolet band of wavelengths. The blue light can be detected with relatively standard photomultiplier tubes, while the ultraviolet light can be converted to electrons using photosensitive molecules that are mixed in with the operating gas in some ionization chamber (e.g., MWPC).

The angle of emission for Cherenkov light is given essentially by

$$\cos\theta_c = \frac{1}{\beta n} \tag{7.19}$$

and the intensity of the produced radiation per unit length of radiator is proportional to $\sin^2\theta_c$. Consequently, if $\beta n > 1$, light will be emitted, and, if $\beta n < 1$, θ_c is complex and no light can be observed. The Cherenkov effect therefore provides a means for distinguishing two particles of same momentum but different mass. For example, protons, kaons, and pions of 1-GeV/c momentum have $\beta = 0.73$, 0.89, and 0.99, respectively. Consequently, to observe Cherenkov light from these particles would require media of different refractive index. In particular, for protons to emit light, we would need a threshold $n > 1.37$, kaons would require $n > 1.12$, and pions $n > 1.01$. Now suppose that we arrange two Cherenkov counters in series, one filled with water ($n = 1.33$) and the other filled with gas under pressure so that it has $n = 1.05$. If we allow a mixture of protons, kaons, and pions to traverse the counters, protons would not provide a signal in either detector, kaons would radiate Cherenkov light only in the water vessel, while pions would register in both counters. This would therefore provide a way of discriminating between particles that have different Cherenkov thresholds. When counters are used in this manner, they are termed threshold counters. (Most of the large experiments searching for proton decay rely on Cherenkov light to identify the end products of the decay, e.g., $p \to e^+\pi^0$.)

We can also see from Eq. 7.19 that we can discriminate between particle types on the basis of the observed angle of emission of the Cherenkov light. That is, for some fixed n value, the cone angle for light emitted by pions will be greater than that from kaons and protons. Cherenkov counters that are sensitive to different emission angles are known as *differential counters*.

Finally, recent developments in this area have centered on the ultraviolet (UV) part of the emission spectrum ($\sim$ 5 eV). As we just mentioned, the UV photons, through photoionization, can produce electrons that can be detected using MWPCs. At high energies, several UV photons can be emitted by a single charged particle. These photons will be distributed in a cone at angle θ_c about the incident charged track. Consequently, in any ionization-sensitive device positioned transverse to the incident line of flight, the electrons that are produced by these UV photons will be distributed in a ring pattern. Detectors that rely on this principle are known as ring-imaging Cherenkov counters (or RICH counters), and they are particularly useful in experiments in which many particles are produced in any given collision.

SEMICONDUCTOR DETECTORS

The formation of an electron–hole pair in a semiconductor such as silicon or germanium requires an energy of only about 3 eV; consequently, when these crystals are used as solid-state ionization chambers, they can provide large signals for very little energy deposition in the medium. Solid-state devices can therefore be particularly advantageous for applications at low energies. These devices were, in fact, developed initially in nuclear physics for high-resolution measurements of energy, and for obtaining ranges and stopping power of nuclear fragments. More recently, silicon strip detectors have gained wide acceptance in both nuclear and particle physics for precise measurement of position of charged particles.

Because the number of free charge carriers produced in semiconductors is so large, and both the electrons and holes have high mobility, very thin wafers of crystal (about 200–300 μm) suffice for attaining good signals even for minimum-ionizing particles. The performance of these detectors is quite linear in that the output signal is proportional to the ionization loss, provided that an imposed electric field within the medium is large enough to prevent recombination of the charge carriers. This can be achieved by using very pure semiconductors of high-resistivity, and operating these detectors as diodes with a reverse bias of about 100 V. The semiconductor wafer is sandwiched between very thin conducting electrodes (thickness of tens of μg/cm^2), which can be deposited in electrically separated stripes (or other patterns) on the surface of the wafer. Detectors 5×5 cm^2 in area are quite common; they often have 20–50 μm stripes, and are used in series (just like planes of MWPCs) to determine charged-particle trajectories to position accuracies of the order of several micrometers in the transverse direction. Such devices can be used to measure small-impact parameters, and thereby determine whether some charged particle originated from a primary collision or was the decay product of a particle that traveled a small distance from the original interaction, and then decayed.

Two silicon detectors positioned in series can be used to determine the kinetic energy and velocity of any low-energy particle or nuclear fragment, and therefore its rest mass. This determination is made by placing a very thin wafer in front of a thicker detector that can stop that particle. The velocity is deduced from the stopping power measured in the thin wafer, and the mass from the range or from the total kinetic energy loss in the thicker crystal (or array of thin wafers).

CALORIMETERS

Momenta of charged particles can be measured in a relatively straightforward fashion using magnetic spectrometers (see, e.g., Fig. 7.5). In certain situations, however, magnetic measurement may not be viable. For example, precise magnetic measurements become difficult and expensive at very high energies because they require either large magnetic fields in extended regions of space, or very long lever arms for measuring small changes in the angular trajectories of particles passing through magnets, or both. Also, at times, specific design considerations may preclude the use of an analyzing magnet in an experiment. In addition, magnets cannot be used for measuring energies of neutral particles (e.g., neutrons or photons). Under circumstances when magnetic measurements are not of greatest value, we can turn to calorimetric detectors that rely purely on the measurement of total energy depostion in a medium. A *calorimeter* is a device that absorbs the full kinetic energy of a particle, and provides a signal that is proportional to that deposited energy. One of the simplest calorimeters we can imagine is the device that we described earlier in this chapter for measuring the ranges of α-particles. Large calorimeters were developed during the early 1960s, especially for application in experiments dealing with high-energy cosmic rays, and they have recently become exceedingly important tools for measuring energies of particles produced at large accelerators.

We know that when high-energy photons traverse matter they do not deposit energy until they convert into electron–positron pairs. The produced electrons and positrons deposit their energies, as usual, by ionizing atoms; however, when they are very energetic, they lose most of their energy through bremsstrahlung. When these bremsstrahlung photons have high energies, they can, in turn, convert to electron–positron pairs, which can radiate more photons, etc. This electromagnetic "shower" develops into a sea of low-energy photons, electrons, and positrons that eventually deposit all their remaining energies in the material.

Similarly, hadrons can deposit energy in matter through a series of successive interactions. Because hadrons are relatively massive, however, and cannot radiate much of their energy through bremsstrahlung, they lose their energy through multiple nuclear collisions. Thus, an incident hadron might produce several π mesons in its first nuclear collision; these pions will subsequently collide with downstream nuclei, producing more particles, etc., until the incident high energy is converted to many charged particles of low energy that eventually deposit that energy by ionizing atoms within the medium. Because in most materials (especially of $Z > 10$) the mean free path for nuclear interactions is substantially greater than that for electromagnetic interactions (see Example 7 in Chapter VI), it takes far more thickness of material to generate a hadronic shower than an electromagnetic one. Consequently, calorimeters designed for measuring deposition of electromagnetic energy are physically far thinner than those designed for absorbing hadrons.

We should also mention that hadronic showers can have large fluctuations in their energy deposition. This is because such collisions often involve production of unstable particles that have neutrinos as decay products. Since neutrinos have exceedingly small interaction probabilities, they escape detection and thereby reduce the energy that a hadron deposits in the detector material. This occurs on a statistical basis, and consequently compromises the energy measurement for hadrons. Another important source of fluctuation in the energy deposited by hadrons is due to the production of neutral pions. These particles decay immediately into two photons (lifetime of about 10^{-16} sec), and the produced photons initiate their own showers. Since the energy deposited by photons is electromagnetic, it is therefore deposited relatively locally, and, as a result, the detected energy can be very sensitive to the detailed structure of the detector (see below). Because of such complications, the calorimetric energy resolution that can be obtained for hadronic showers is expected to be worse than that for electromagnetically interacting particles.

We have discussed several mechanisms for energy deposition in matter, and have already described some of the most common ones that are often applied in the detection of particles. Similar principles can be used in calorimetry. For example, relying on the production of ionization in a medium, calorimeters can function as ionization chambers (e.g., liquid-argon calorimeters); or, through the production of scintillation light, they can be scintillation-sensitive detectors (e.g., NaI); or they can rely entirely on the production of Cherenkov radiation (lead glass). In principle, a particle's energy could also be determined from the heating of a detector or from the acoustic energy deposited, and such techniques have, in fact, been explored in the past. Calorimeters can be constructed from homogeneous media (e.g., crystalline scintillators or lead glass), or they can be sampling detectors. Sampling calorimeters contain mainly uninstrumented absorber material that is interspersed with active sampling devices to sense the energy of a developing shower (see Fig. 7.9). Homogenous detectors, which are usually sensitive to the full energy deposition, tend to have best resolution, but they are invariably very expensive. Sampling detectors, because they are subject to additional sampling fluctuations, usually have worse resolution, but are often easier and less expensive to construct. It is relatively straightforward to make large sampling calorimetric systems (with thousands of individual channels) with energy resolution in $\Delta E / E$ of about $0.20 / \sqrt{E}$ for detection of electromagnetic components, and about $1.0 / \sqrt{E}$ for the hadronic components (where E is in GeV units), but anything beyond that is challenging. (The improvement in relative resolution with increase in energy can be attributed to the expected decrease in sampling fluctuations.)

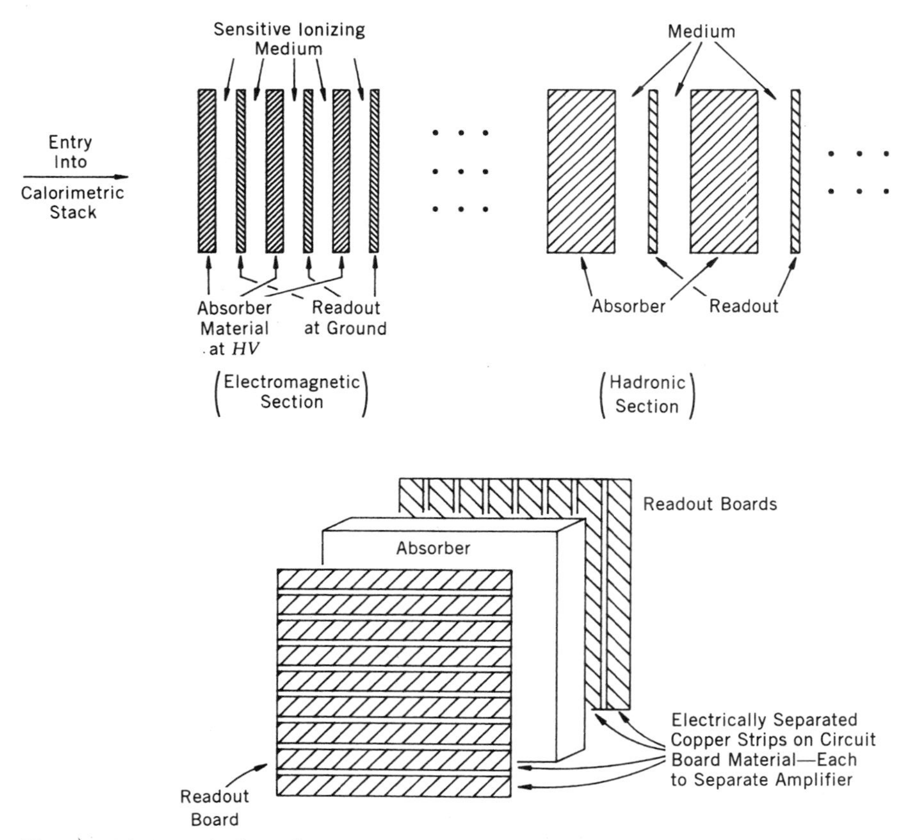

Figure 7.9 A typical stack structure in a sampling calorimeter.

LAYERED DETECTION

Collisions at high energies involve the production of many and different kinds of particles. In any one event we might have electrons, muons, neutrinos, and a large number of pions. Some particles may be stable, others short-lived. To unravel the event structure, and to find out about the underlying physics of the collision, one is often required to measure all these species, and usually with substantial precision. Modern spectrometer systems, especially at colliders, which we discuss in the following chapter, are designed in a layered manner, with every layer having a unique function (see Fig. 7.10). For example, in the region closest to the intersection point, we could place several very thin silicon microstrip detectors to provide precise spatial information for trajectories of charged particles. (The reason for having "thin" material is to minimize any multiple scattering error and to minimize the conversion of photons in the silicon.) Such microstrip detectors also provide sensitivity to the characteristic decays of short-lived particles.

For a collider configuration, for example, the next layer of instrumented detection could have several layers of drift chambers surrounding the silicon system. Such chambers are often located in an axial or solenoidal magnetic field, which can be provided by a thin superconducting coil that surrounds the chambers. (Again, the reason for "thin" is to minimize energy loss and multiple Coulomb scattering within the coil, which worsens the resolution.) The signals from the silicon detectors and from the drift

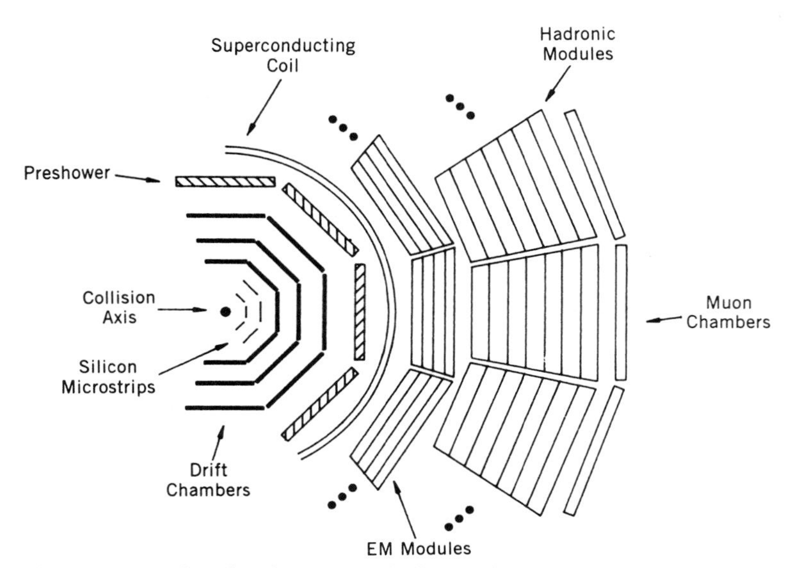

Figure 7.10 Sketch of a "typical" layered detector for high-energy physics experiments at colliders.

chambers provide momentum information for any traversing charged particles that emanate from the interaction point.

The next step in detection might involve segmented "preshower" counters, which usually consist of ~ 3 radiation lengths of absorber followed by scintillation counters. (The scintillation counters can also provide information on TOF.) The inner tracking chambers usually have far less material, and consequently do not produce much photon conversion or bremsstrahlung of any original electrons. The preshower counters, on the other hand, can provide first evidence for the presence of electromagnetically interacting particles through the presence of large pulse heights from photons that convert into e^+e^- pairs or electrons that produce showers in the material in front of the counters. (The magnetic-field coil can function as part of the radiator/absorber material when it precedes the preshower counters.)

With the trajectories of charged particles determined, and some indication available for the presence of photons and electrons, the next step usually involves electromagnetic (EM) calorimetry. Typically, EM calorimeters are made of about 25 radiation lengths of high-Z material, which usually corresponds to about one mean free path for hadronic interactions. Consequently, electrons and photons deposit essentially all their energies within the EM modules, and hadrons only start to interact there. After the EM calorimeters, follow the far thicker hadronic calorimeters in which hadrons deposit most of their energy. (The EM sections are about 30 cm thick, while the hadronic modules are usually about 150 cm thick, depending somewhat on the type of absorber and readout used.)

The particles that penetrate through the calorimetry are primarily neutrinos and high-energy muons (with energy greater than that given by the range within the intervening material). Muons that penetrate the calorimeters can be momentum analyzed again outside of the calorimeters, and their trajectories can be traced back and matched for consistency with the tracking information available from within the calorimeters. This leaves only neutrinos not accounted for. Their presence can be inferred from a lack of overall momentum balance in the event (especially balance in the transverse direction). In order to be sensitive to the presence of missing energy that may be carried off by neutrinos, the detectors have to be designed to surround the intersection region over as much of the 4π solid angle as is possible, and to provide a minimum of loss in coverage from the presence of structural elements, and the like. Needless to say, this poses very great technical challenges to the experimenter.

Problems

VII.1 A radioactive source emits α-particles of kinetic energy of 4 MeV. What must be the value of an applied magnetic field so that the radius of curvature of the orbit of the α-particle is 10 cm? (Does your answer depend on the kind of medium into which the α-particle is emitted?) Do the same calculation for an electron of same kinetic energy.

VII.2 The mass of a K^+ is 494 MeV/c^2 and that of a π^+ is 140 MeV/c^2. If the rms time resolution of each of two scintillation counters that are 2 m apart is 0.2 nsec, calculate to better than 10 percent accuracy the momentum at which the system will just be able to resolve a π^+ from a K^+ (by one standard deviation). (*Hint:* see Eq. 7.10.)

VII.3 What are the Cherenkov angles for electrons and pions of momentum of 1000 MeV/c for a radiator with $n = 1.4$? What will be the ratio of the number of radiated photons for incident electrons and pions?

VII.4 About 10^6 electron–ion pairs are produced by a charged particle traversing a counter. If the typical ionization potential of the medium is $\bar{I} = 30$ eV, in principle, how well can you measure the deposited energy using a Geiger counter, an ionization counter with a gain of unity, and a proportional counter with a gain of 10^6 that has gain variations of 5 percent?

VII.5 If you wish to measure the momentum of a 10-GeV/c singly-charged particle to 1 percent accuracy, in a 2-T field, using a 1-m-long magnet, how well do you have to know the exit angle (see Fig. 7.5)? If you use MWPCs to measure that angle, about how far do you have to separate two planes that have 2-mm interwire anode spacings to achieve your goal? Now suppose that you use, instead, silicon microstrip detectors of 25-μm spacing. What would the separation distance of two such planes have to be to achieve the same goal?

VII.6 Sketch the pulse height spectrum that you might expect in the case of ^{60}Co decay (Eq. 7.8) if both photons were emitted simultaneously, namely within the time resolution of the detector.

Suggested Readings

Fernow, R. C. 1986. *Introduction to Experimental Particle Physics*. London/New York: Cambridge Univ. Press.

Kleinknecht, K. 1986. *Detectors for Particle Radiation*. London/New York: Cambridge Univ. Press.

Knoll, G. F. 1989. *Radiation Detection and Measurement*, 2nd Ed., New York: Wiley.

Leo, W. R. 1987. *Techniques for Nuclear and Particle Physics Experiments*. New York/Berlin: Springer-Verlag.

Chapter VIII

ACCELERATORS

INTRODUCTORY REMARKS

Accelerators are some of the most remarkable tools of modern science. They are precision instruments constructed on a gargantuan scale. They have to track and accelerate particles that traverse millions of kilometers in just matters of seconds, and maneuver and constrain particle motion to accuracies of the order of 1 μm. They can provide sufficient numbers of energetic particles to vaporize macroscopic targets with single pulses of beam. Because of their immensity and their challenging complexity, and because of their symbolic reflection of the intellectual aspirations and creativity of mankind, modern accelerators have been likened by Robert R. Wilson to the great Gothic cathedrals of medieval Europe. Their impact on nuclear and particle physics, the fields for which they were initially developed, has, of course, been pivotal. They have served as the microscopes for probing nuclear and particle structure, and, in fact, were it not for the development of accelerators, the fields of nuclear and particle physics would still be in their infancies.

After the pioneering experiments of Rutherford and his colleagues revealed the presence of a nucleus within the atom, it became clear that higher energy scattering experiments could provide an invaluable probe of the nucleus. For example, it was understood that with sufficient energy to penetrate the Coulomb barrier, projectiles could break nuclei apart to reveal their constituents. It was also recognized that the more energy a particle had, the more deeply it could probe within the nucleus. This is simply a consequence of the fact that large momentum transfers correspond to small distances, and vice versa. To study the short-distance behavior of nuclei and of elementary particles therefore requires the availability of high-energy beams that can be used to impart large momentum transfers either to target particles or to other beams of particles.

Although high energy particles are available in the cosmic rays, their fluxes are quite low, and their energies clearly cannot be controlled. In fact, the excitement brought about by discoveries of new phenomena in experiments with cosmic rays only added impetus to the development of techniques for accelerating charged particles.

The increase achieved in accelerator energies over the past 60 years has been astounding. The first accelerators that were constructed (around 1930) provided beams of particles with energies of hundreds of keV, while the latest modern accelerators being designed for operation around the turn of the twenty-first century will have beam energies of about 10^8 greater than that. Because of the advent of colliding-beam techniques, the effective increase in beam energy (that is, considering the energy available in the center of mass) has gone up by an even more spectacular factor of about 10^{13}! Such changes correspond to differences of about 10^8 in the sensitivity to distance scales that can be studied with the next-generation accelerators, which expect to provide sensitivity in the range of 10^{-18} cm. Nowadays, besides being used in nuclear and particle physics, accelerators are in demand in a variety of applications, ranging from experiments in condensed matter physics, the electronics industry, biomedical and geophysical areas, to food processing and sewage treatment. Accelerator science is therefore no longer just an appendage of nuclear and particle physics, but is a separate intellectual discipline in its own right.

There is a variety of ways available for accelerating charged particles, and the methods used for any specific application depend upon the kinds of probes that are required, their energies, the desired beam intensities, and, of course, on any economic constraints. We now sketch briefly some of the key historical developments in particle acceleration during the past 60 years.

ELECTROSTATIC ACCELERATORS

Cockcroft-Walton Machines

The Cockcroft-Walton machine is the simplest type of accelerator. It is based on passing ions through sets of aligned electrodes that are operated at successively higher fixed potentials. Usually, voltage-doubling circuits are used for generating the high electric fields. The machine consists of an ion source (often hydrogen gas) located at one end, and a target at the other, with the electrodes arranged in between. First, electrons can either be added to or stripped from the atoms of interest in order to produce ions for acceleration; the resulting ions are then passed through the series of accelerating regions. The kinetic energy gained by an ion of charge q, passing through the voltage difference V, is given simply by $T = qV$. John Cockcroft and Ernest Walton were the first to successfully apply these principles to particle acceleration, and used their device to disintegrate lithium nuclei using protons of about 400 keV. Cockcroft-Walton machines are limited to about 1-MeV energies because of voltage breakdown and discharge that takes place much beyond voltages of 1 MV. Currently, Cockcroft-Walton accelerators are available commercially, and are often used as the first step high-current injector (of the order of 1 mA) in the multistage process of accelerating particles to high energies.

Van de Graaff Accelerator

The energy gained by a particle (ion) accelerated in a DC voltage machine is directly proportional to the applied voltage; consequently, clever construction of the high-voltage source is of crucial importance. That is precisely what the Van de Graaff generator, named after Robert Van de Graaff, does. The principle used in this case is quite simple. From the fact that the charge on any conductor resides on its outermost surface, it follows that if a conductor-carrying charge touches another conductor that envelops it, then, irrespective of its potential, it will transfer all its charge to the outer conductor. This can be used to advantage to increase the charge on any conductor, and consequently to create a higher voltage.

In the Van de Graaff accelerator, charge is carried on a conveyor belt into a large metallic dome, where it is picked off, as shown in Fig. 8.1. The conveyer belt is made of insulating material, and goes over motor-driven rollers (R). A "sprayer" (S), connected to a discharge-voltage terminal, sprays positive ions to the conveyer belt (electrons go to P). (Basically, the high voltage ionizes the gas, and the ions are collected on the conveyer belt.) The points where charges are sprayed or injected onto the belt are known as *corona points*. The conveyer belt takes the positive charges up to the dome, which is maintained at a positive voltage. The energy needed to do this work is provided by the motors. At the upper end of the conveyor there is a collector C, which collects the positive charges that are transferred to the dome. Typically, this technique can produce accelerating potentials of up to $\sim$ 12 MV. (A tandem generator is a modification of the Van de Graaff generator, where negative ions enter the accelerating tube from one side, are first accelerated to the positive HV terminal, where they are stripped and made positive, and then accelerated as positive ions down to ground potential on the other side of the terminal. This effectively doubles the acceleration energy to $\lesssim$ 25 MeV.)

The Van de Graaff has an evacuated tube through which ions from the ion source are accelerated to strike the target. This accelerating tube is constructed with equipotential metallic rings embedded within the insulated tube. The entire device operates within a pressurized chamber, containing

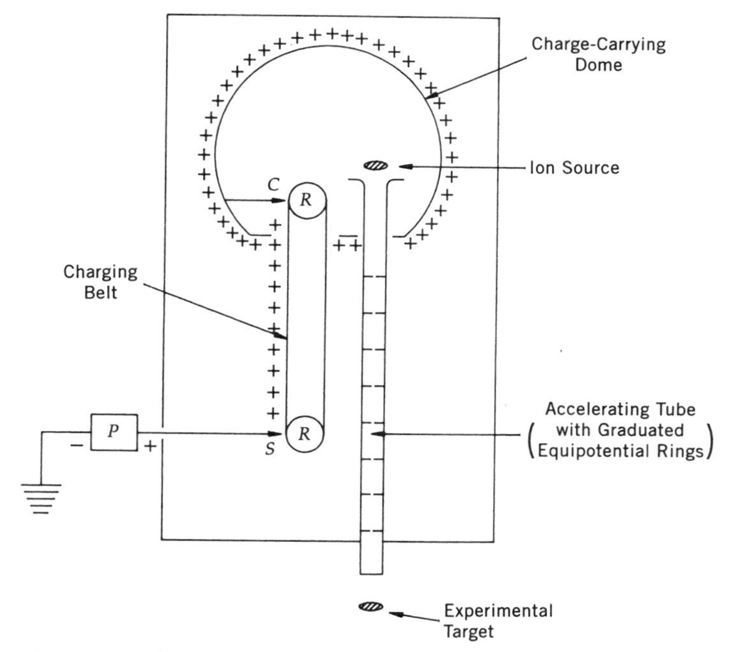

Figure 8.1 The principle of a Van de Graaff accelerator.

some inert gas that does not breakdown easily (often SF_6). Normally, the pressure of the gas inside the Van de Graaff is about 15 atm, and the limit for the highest energy in such a machine comes from the voltage at which there is electrical breakdown and discharge in the gas.

RESONANCE ACCELERATORS

Cyclotron

Fixed-voltage machines have an inherent limitation to the energy they can produce because of voltage breakdown and discharge. An alternate method, which uses the resonance principle, is more important for accelerating particles to higher energy.

The cyclotron (or cyclic accelerator), first built by Ernest Lawrence, is the simplest of the machines that use this principle (see Fig. 8.2). The accelerator is constructed out of two hollow evacuated D-shaped metal chambers (referred to as Ds). These are connected to an alternating high-voltage source. The entire system is placed inside a strong magnetic field perpendicular to the Ds. The principle of operation of the cyclotron is as follows.

Although the hollow Ds are connected to the high-voltage source, because of the shielding effect of the metal chamber walls, there is no electric field within the Ds. Consequently, a strong alternating electric field exists only in the gap between the Ds. A source for producing ions is placed

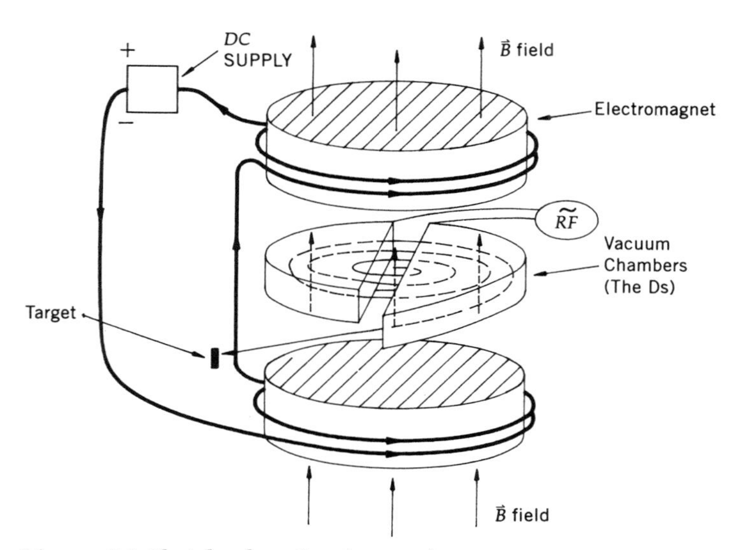

Figure 8.2 Sketch of motion in a cyclotron.

in the gap between the Ds, and depending on the sign of the voltage at that moment, any ion in the gap is attracted toward one of the Ds. The trajectory of the ions, however, is circular because of the bending effect of the magnetic field. Once an ion is inside the D, it stops sensing the electric force, but continues in its circular motion because of the presence of the static magnetic field. After a half circle, however, when the ion is about to emerge from the D, the direction of the voltage can be changed and the ion can be accelerated again before it enters the other D. Similarly, when it is about to exit from the second D, the applied voltage can again be reversed and the particle accelerated further. If the frequency of the alternating voltage source is just right, then the charged particle can be accelerated continuously, and move in ever increasing radial orbits, until it is extracted to strike a target (by, for example, suddenly turning off the B field).

For nonrelativistic motion, the frequency appropriate for the alternating voltage can be easily calculated from the fact that, for a circular orbit, the magnetic force provides the centripetal acceleration. That is,

$$m\frac{v^2}{r} = q\frac{vB}{c}$$

or

$$\frac{v}{r} = \frac{qB}{mc} \tag{8.1}$$

We know that, for circular motion at constant speed, the angular frequency ω is related to the radius and circular velocity of the orbit:

$$\omega = \frac{v}{r} \tag{8.2}$$

We can therefore express the frequency of the motion as

$$\nu = \frac{\omega}{2\pi} = \frac{qB}{2\pi mc} = \frac{1}{2\pi}\left(\frac{q}{m}\right)\frac{B}{c} \tag{8.3}$$

Clearly, keeping the acceleration in phase with the particle motion requires that the frequency of the electric field be the same as ν. This frequency is referred to as the *cyclotron resonance frequency*, and is the origin of the label "resonance accelerator" for this kind of machine. Equation 8.3 therefore provides a means for determining the frequency of the accelerating fields as a function of other parameters. Note that the maximum energy that a charged particle has when it is extracted at a radius $r = R$ is given by

$$\begin{aligned} T_{\text{max}} &= \frac{1}{2}mv_{\text{max}}^2 = \frac{1}{2}m\omega^2R^2 \\ &= \frac{1}{2}m\left(\frac{qB}{mc}\right)^2 R^2 = \frac{1}{2}\frac{(qBR)^2}{mc^2} \end{aligned} \tag{8.4}$$

Equation 8.4 relates the magnitude of the magnetic field and the size of the magnet that is needed to accelerate a particle to any given energy. In a typical cyclotron, $B \lesssim 2$ T, the alternating voltage applied to the Ds is ~ 200 kV, at a frequency of ~ 10–20 MHz. The maximum proton energy that can be attained in such cyclotrons is about 20 MeV (for Ds of $R \sim 30$ cm), as is shown in the following example.

As we increase the energy of charged particles, they become relativistic, and the frequency relation in Eq. 8.3 starts failing. Consequently, a fixed-frequency cyclotron cannot accelerate ions to relativistic energies. For electrons, relativistic effects set in at lower energies, and consequently, such simple cyclotrons are not useful for accelerating electrons. Synchronous accelerators are needed for attaining relativistic energies.

Example 1

For a cyclotron operating at an extraction radius $R = 0.4$ m, and a magnetic field of $B = 1.5\text{ T} = 1.5 \times 10^4$ G, the frequency of the alternating source to accelerate protons, and the maximum energy gained by these protons, can be calculated from Eqs. 8.3 and 8.4 as follows:

$$\begin{aligned} \nu &= \frac{qB}{2\pi m_p c} = \frac{1}{2\pi}\frac{4.8\times 10^{-10}\text{ esu} \times c \times 1.5\times 10^4\text{ G}}{m_p c^2} \\ &\simeq \frac{4.8\times 10^{-10}\text{ esu} \times 1.5\times 10^4\text{ G} \times 3\times 10^{10}\text{ cm/sec}}{6.28\times 10^3\text{ MeV}(1.6\times 10^{-6}\text{ erg/MeV})} \\ &\simeq 22.8\times 10^6\text{ sec}^{-1} = 22.8\text{ MHz} \end{aligned}$$

$$T_{\max} = \frac{1}{2}\frac{(qBR)^2}{m_p c^2} = \frac{1}{2}\frac{(4.8 \times 10^{-10} \text{ esu} \times 1.5 \times 10^4 \text{ G} \times 40 \text{ cm})^2}{(1000 \text{ MeV})(1.6 \times 10^{-6} \text{ erg/MeV})}$$
$$\simeq \frac{(3 \times 10^{-4})^2}{3.2 \times 10^{-3}} \simeq 2.8 \times 10^{-5} \text{ erg} \simeq 17 \text{ MeV}$$

Note that we have used appropriate cgs units in evaluating the preceding expressions, and consequently the results can also be assumed to be in cgs units. This implies that 1 esu-Gauss is equivalent to 1 erg/cm, which is consistent with Problem II.4.

Linac or Linear Accelerator

Linear accelerators, as the name implies, accelerate particles along linear trajectories rather than in circular orbits. These accelerators are also based on the resonance principle, and operate as follows. A series of metal tubes, called *drift tubes*, are located in a vacuum vessel and connected successively to alternate terminals of a radio frequency oscillator, as shown in Fig. 8.3. Let us suppose that at some time the fields are as shown in the figure. Positive ions from the source will be accelerated by the electric field toward the first drift tube. If the alternator can change its direction before the ions pass through that tube, then they will be accelerated again on their way between the exit of the first and entry into the second tube, and so on. However, as the particles accelerate, their velocities increase, and consequently if the drift tubes are all of the same length the phase between a particle's position and the potential on the next tube may not keep in step (that is, the next gap may not accelerate). To avoid this, the drift tubes are made longer along the path so that one radio-frequency (RF) alternator can correctly accelerate any particle all the way along the path.

Because electrons become relativistic at relatively low energies, electron linear accelerators act on a slight variation of the principle just described. The electron source is usually a hot wire filament that effectively boils off electrons. These electrons are accelerated through a positive potential grid and rapidly become relativistic. Bunches of these electrons are then passed through accelerating tubes that are fed with microwave power delivered by klystron amplifiers. Electrons radiate easily as they get accelerated (this is referred to as *synchrotron radiation*), and therefore much power is needed to

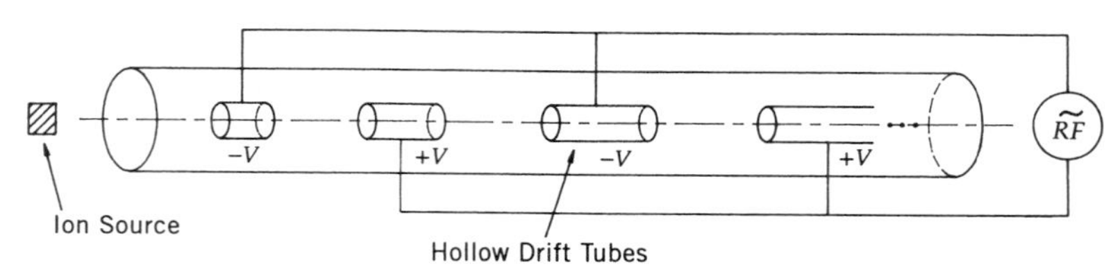

Figure 8.3 A linear ion accelerator.

increase their energy. This power is supplied by the microwave fields that travel in step with electrons in specially shaped iris-loaded waveguides. The longest linear accelerator is the two-mile Stanford Linear Accelerator Center (SLAC) Linac, and it accelerates electrons to energies of 50 GeV.

SYNCHRONOUS ACCELERATORS

As we have discussed before, if we want to accelerate particles to very high energies, we must take effects of relativity into account. For relativistic energies, the equation of motion for a charged particle of mass m and charge q in a magnetic field $\vec{B}$ becomes

$$\frac{d\vec{p}}{dt} = q\frac{\vec{v} \times \vec{B}}{c} \tag{8.5}$$

or

$$m\gamma\frac{d\vec{v}}{dt} = m\gamma\vec{v} \times \vec{\omega} = q\frac{\vec{v} \times \vec{B}}{c} \tag{8.6}$$

where, in the last step, we equated the centripetal force with the Lorentz force. Now, with $|\vec{v}| \approx$ constant $= c$, the resonance relation follows from Eq. 8.6 (remember that both the magnetic field and the axis of circular motion are perpendicular to the direction of motion):

$$\omega = \frac{qB}{m\gamma c}$$

or

$$\nu = \frac{\omega}{2\pi} = \frac{1}{2\pi}\left(\frac{q}{m}\right)\left(1 - v^2/c^2\right)^{1/2}\frac{B}{c} \tag{8.7}$$

Consequently, for this relation to hold during an acceleration cycle, either the alternating frequency has to decrease or the magnetic field has to increase, or both must happen, as $v \to c$. Machines where the magnetic field is held constant but the frequency is varied are called *synchrocyclotrons*. On the other hand, machines where the magnetic field is changed, irrespective of whether the frequency is changed, are known as *synchrotrons*. In electron synchrotrons, the frequency is held constant and the magnetic field is varied, whereas in proton synchrotrons, both the frequency and the magnetic field are varied.

With relativistic effects taken into account, Eq. 8.7 can be used to obtain the parameters for accelerating particles to any desired energy. Let us first rewrite Eq. 8.7 in terms of the momentum of the accelerated particle and the radius of the final orbit. For $v \simeq c$, we can also express the frequency of the motion as follows:

$$\nu = \frac{1}{2\pi}\frac{v}{R} \simeq \frac{c}{2\pi R} \tag{8.8}$$

Writing $p = m\gamma v \simeq m\gamma c$, we can now obtain from Eq. 8.7 our usual relativistic relationship between p, R, and B (see also Fig. 7.5):

$$\frac{c}{2\pi R} = \frac{1}{2\pi}\left(\frac{q}{m}\right)\frac{1}{\gamma}\frac{B}{c}$$

or

$$R = \frac{pc}{qB} \tag{8.9}$$

It is convenient to write Eq. 8.9 in the mixed units of accelerator science, namely,

$$R \simeq \frac{p}{0.3B} \tag{8.9'}$$

where p is in GeV/c, B is in teslas, and R in meters, and where we have assumed that q corresponds to the magnitude of the charge of a single electron.

Now we can see that, independent of the nature of the accelerating fields, any given momentum will be limited (often financially!) by the product of the radius of the final orbit and the largest magnetic field that can be supplied. At present, realistic bending magnets (dipoles) are limited to fields of strength $\lesssim 2$ T for conventional electromagnets and to $\lesssim 9$ T for superconducting dipole magnets. In order to accelerate protons to momenta of about 30 GeV/c, for example, the radius of the orbit using conventional magnets has to be at least

$$R \simeq \frac{p}{0.3B} \simeq \frac{30}{(0.3)(2)} = 50 \text{ m} \tag{8.10}$$

If the accelerator were a synchrocyclotron with a cyclotron type of electromagnet, ignoring the difficulties of construction, the cost of even the steel would be prohibitive: for the volume of the required magnet we can use $\pi R^2 t$, where t is the thickness of the poles (~ 1 meter each); using the density of ~ 8 gm/cm^3 for steel, this would yield a mass of $\sim 2 \times 10^8$ lb, which, at current prices, would mean that the cost of the steel alone would exceed \$100M! A synchrocyclotron option is therefore impractical for energies much beyond several hundred MeV.

Synchrotrons in the GeV (or higher) range of energies have magnets positioned in a ringlike fashion (see Fig. 8.4). Particles with energies of hundreds of MeV are injected (often from a linac) into a narrow vacuum chamber that passes through all the magnet apertures. At the start, the fields are set to low values that correspond to the momenta of the injected particles. The beam particles are constrained by the magnetic field to move in essentially circular orbits within the confines of the vacuum tube. To accelerate the particles to higher energy, RF power stations are placed in some convenient locations within the ring of magnets. Every time the particles pass through the RF cavities, they gain typically MeVs of energy from the electric fields. For this to work, the phases of the accelerating fields for different cavities must keep in step with each other as well as with the motion of the bunched beam. The magnetic field must also increase steadily to assure that, as the momenta of the particles increase, the radius of their orbit remains essentially fixed, in accord with Eq. 8.9.

Most of the space along the ring is taken up by the bending magnets. Space is also needed, however, for general servicing of the vacuum system, the powering and cooling of the magnets, for the RF stations, for the injection and extraction systems, and for other kinds of beam elements; a synchrotron, therefore, has many "straight" sections along the ring, where there is no bending. The accelerated particles consequently travel along circular arcs within the magnets, and in straight lines between them. In order to reach the design energies, the beam particles must traverse the RF cavities, and therefore the entire ring, millions of times. A fundamental question is therefore whether it is, in fact, possible to maintain the beam within the small vacuum chamber for so many traversals. This is the issue we discuss in the next section.

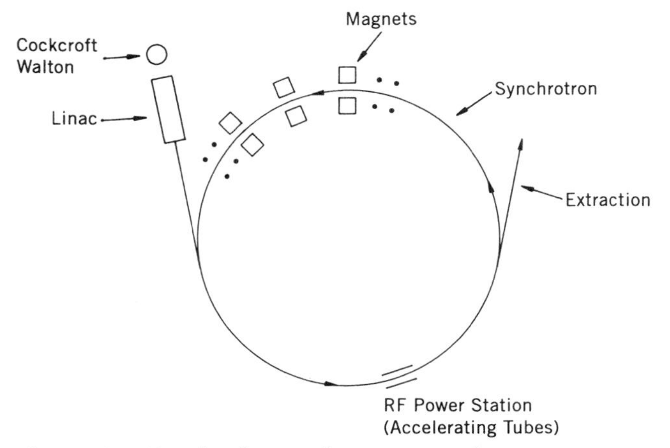

Figure 8.4 Sketch of a synchrotron complex.

PHASE STABILITY

Before proceeding further with our description of different accelerators, we first introduce the principle of phase stability. Let us consider the operation of a synchrocyclotron, namely a cyclotron with a fixed magnetic field and a variable accelerating electric RF field between the two Ds. There is always some finite spread in the time of arrival of individual beam particles into the region of the RF field between the Ds. Referring to Fig. 8.5, let us label as synchronous, or "in time," the particle that arrives at the time (or phase) τ of the accelerating cycle. This particle will experience an electric field E_0, which will accelerate that particle while it is in the gap between the Ds. A particle that arrives earlier in the gap will therefore sense a somewhat larger field $E_>$. Thus the earlier particle will experience a larger accelerating force (qE) while in the gap. This will increase its next orbit radius within the cyclotron D, and shift the particles's next reentry into the gap between Ds to a later time (toward that of the synchronous particle). On the other hand, a particle that arrives later than the synchronous one will experience a smaller acceleration due to the smaller $E_<$. This will reduce the orbit radius in the D, and again shift the arrival time at the next acceleration stage toward that of the synchronous particle. On the following cycle, the synchronous particle will once again sense the same field E_0, while the other particles, if they still arrive later, will be accelerated less, or, if they arrive early will be accelerated more. A similar situation will obtain for all the following cycles. Finally, particles that arrive at some random times relative to τ will experience varying accelerations or even decelerations. Consequently, the self-correcting effect of the field cycle will lead to the grouping of particles into bunches in time centered on the synchronous particle. This is the "RF structure" that one hears about in accelerators.

A similar corrective effect occurs for vertical particle motion in the cyclotron's magnetic field. At large radii (that is near the edges) all dipole, or bending, magnets have substantial fringe fields. This is shown in Fig. 8.6.

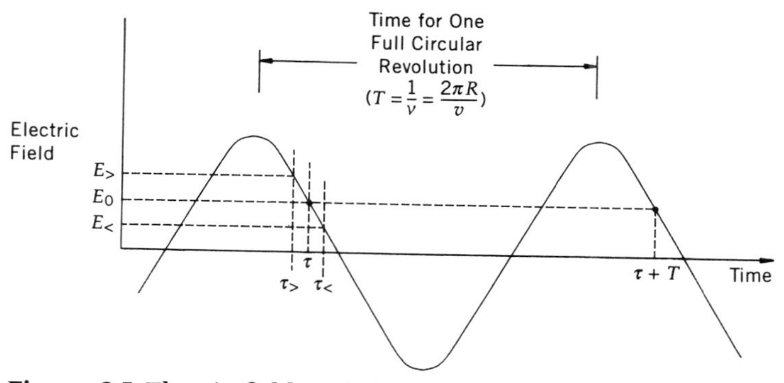

Figure 8.5 Electric field variation with time in the gap between the Ds of a synchrocyclotron.

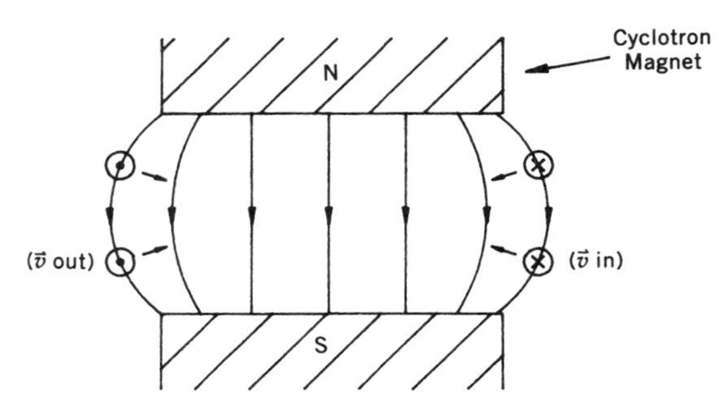

Figure 8.6 Magnetic field in a synchrocyclotron and its effect on a charge circulating near the edge of the magnet.

If our beam of particles is supposed to travel in the bending plane (horizontal circles in the figure), then because of natural angular divergence in the beam, some particles will have trajectories that move them out of the median plane. These particles will, however, experience a vertical restoring Lorentz force ($\vec{v} \times \vec{B}$) that will tend to counter their divergence. Namely, it will deflect the particles back toward the median plane. The larger the divergence, the greater will be the vertical restoring or focusing force. There is no vertical correction for a particle moving in the median plane, but, for any finite divergence, there will be a vertical component to the force.

For motion in a synchrotron, the fields in the apertures of the individual dipole magnets also provide corrections to the transverse motion. Just as in the case of synchrocyclotrons, fringe fields along the edges of the magnets (parallel to the motion) focus the particles vertically toward the median plane. Horizontally, because of the dispersive nature of bending magnets, particles of higher momentum traverse the vacuum pipe at largest radii, and particles of smaller momentum move at smaller radii. Because $v \simeq c$, higher momentum particles take longer to circle the ring than smaller momentum particles, which, once again, provides the possibility for corrective acceleration at the locations of RF cavities, and for the kind of phase stability described in Fig. 8.5.

All these restoring forces induce very small oscillations in the motion about the mean trajectory for both the transverse direction (known as betatron oscillations) and for the longitudinal (energy or time) dimension (known as synchrotron motion), and make it possible to maintain particles in their orbits for long times. The entire concept, especially the corrective nature of the accelerating RF field, is referred to as the principle of phase stability; it was discovered independently by Edwin McMillan and Vladimir Veksler, and serves as the basis for stable operation of most modern high-energy accelerators.

In a proton synchrotron complex, the particles are usually first accelerated in a Cockcroft-Walton (to about 1 MeV), and then in a linac (to several hundred MeV), before they are injected into the synchroton. As we described in the previous section, most synchrotrons have large radii, with a large number of magnets that are positioned in a ring along a circular acceleration path. The magnetic field is increased at a constant rate, usually from $\sim$ several hundred gauss to the maximum value, which depends on whether the magnets are warm-temperature electromagnets or superconducting, as well as on the circumference of the accelerator and the fraction of the ring that is filled with magnets. The radio frequency of the accelerating fields is usually modulated between 0.3 MHz and 50 MHz, depending on the energy of the injected particles, the number of accelerating RF stations in the ring, the extraction (final) energy, etc. (A sketch of a typical accelerator complex was given in Fig. 8.4.)

Because protons do not radiate very much, most proton synchrotron rings are essentially filled with the bending magnets required to maintain particles of the maximum energy in their orbit. The size of the ring is therefore determined primarily by the machine energy. Electrons, on the other hand, emit substantial amounts of synchrotron radiation (because of their small mass and centripetal acceleration in the magnets), which is inversely proportional to the radius of the orbit. (There is $\sim$ 1.5 GeV lost per turn for 30-GeV electrons in an orbit of $R = 50$ meters! This amount of energy is difficult to supply through standard accelerating systems that provide $\lesssim$ 10 MeV/m. The energy loss scales as γ^4, and is therefore insignificant for protons.) Electron synchrotrons, therefore, usually have larger radii than proton synchrotrons of same energy; this is needed both to reduce the amount of radiation and to provide adequate RF power for acceleration. The largest proton synchrotron ring is currently the one at the Fermi National Accelerator Laboratory (Fermilab). It has a circumference of about four miles and contains two rings of magnets: One uses conventional magnets and accelerates protons from $\sim$ 8 GeV to $\sim$ 150 GeV, and the other ring has superconducting magnets, and accelerates the protons from 150 GeV to $\sim$ 900 GeV. (The 8-GeV protons are taken from a "booster" synchrotron. That is, the beam from the Fermilab Linac is injected into the booster, and from the booster, after acceleration, the 8-GeV protons are injected into the "main" ring, and eventually the 150-GeV protons are injected into the Tevatron ring.)

STRONG FOCUSING

In high-energy accelerators, the focusing provided by the fringe fields of dipole magnets is very weak, and greater magnetic gradients are required for keeping large fluxes of particles within their orbits long enough to accelerate them to their full energies. Stronger focusing can be attained through the use of quadrupole rather than dipole magnets. These magnets serve essentially as lenses in optics, as is illustrated in Fig. 8.7. Imagine a positively charged particle entering the field region along the axis of the magnet ($x = y = 0$). We see that, in this case, the superposition of the magnetic field lines is such that no deflection will be produced. Now

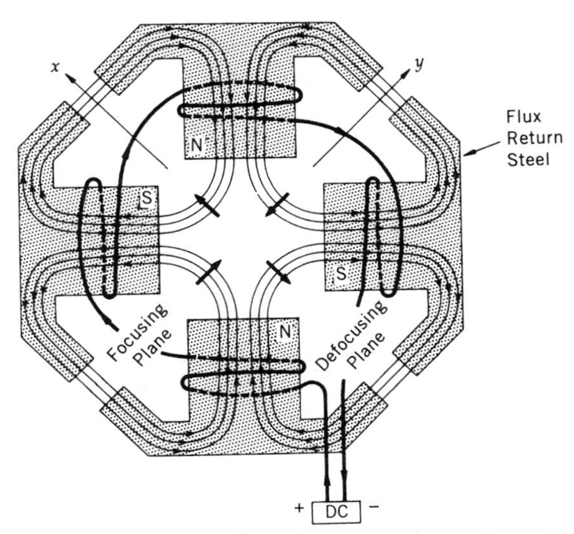

Figure 8.7 Focusing/defocusing properties of a quadrupole magnet for positively charged particles entering the plane of the paper. The sense of the windings on the electromagnet is indicated by the arrows showing the direction of positive current flow. The path of the magnetic field lines through the flux-return iron is also indicated.

suppose that the particle enters along $x = 0$, but $y \neq 0$; here, for both positive and negative y-values, as the particle traverses the region of magnetic field, it will be deflected toward the center of the magnet aperture (to smaller $|y|$). The larger the $|y|$ of the particle, the stronger the magnetic field, and the greater the deflection. Consequently, positively charged particles entering these regions of the quadrupole field will be focused. For particles traversing the magnet along $y = 0$, but $x \neq 0$, the effect is opposite, namely, for both positive and negative x-values, the particles will be deflected away from the center of the magnet, or be defocused. Because of the way the field in quadrupole magnets changes with position (that is, the fixed gradient, or constant change of field with position—equal and opposite for the two orthogonal planes), the magnet will focus particles in one plane and defocus them in the other plane. It can be shown that by placing such magnets in alternating sequence along the beam, particles can, in fact, be focused in both planes. This effect is known as the *principle of alternate gradient* or *strong focusing*.

The principle of strong focusing, proposed independently by Ernest Courant, Stanley Livingston and Harland Snyder, and by Nicholas Christofilos in the early 1950s, was first applied in the construction of the 30-GeV proton accelerator, the Alternating Gradient Synchrotron (AGS), at Brookhaven National Laboratory in the late 1950s. The AGS design was based on the shaping of pole pieces (and therefore the fields) in dipole magnets to provide large alternating gradients. All high-energy synchrotrons now use quadrupole magnets along with dipole magnets to guide particles as they circle the ring, and keep gaining energy at the RF power stations. The use of such separate functions for dipoles (to bend and maintain particles in their orbits of fixed radius) and quadrupoles (to correct positions of particles within an orbit) was first developed for the accelerator at Fermilab.

Just as dipoles have weak focusing properties (moments corresponding to quadrupole terms in their field structure), so do quadrupoles have higher order components in their fields. Correction coils (especially for sextupole effects, but also for even higher octupole terms) are needed in high-energy synchrotrons to assure that beams remain stable for the full acceleration cycle, and do not "blow up," and leave their vacuum vessel. After the beams reach their final energy, they can be extracted to target stations, or can be made to collide with other beams of particles.

We do not discuss the variety of injection and extraction techniques, nor the manner in which beams are brought to external targets in experimental areas. The techniques employed for such purposes are based essentially on the same kinds of electromagnetic tools and principles we have already sketched, namely, the use of dipoles, quadrupoles, and RF cavities. This is another important subbranch of accelerator science that has wide applications in nuclear and particle physics, as well as in other disciplines.

COLLIDING BEAMS

As we know, the ultimate figure of merit in any high-energy scattering experiment is not just the laboratory energy of a colliding particle, but rather the energy that is available for producing more particles, namely, the energy in the center of mass of the collision. We have already discussed some of these issues previously, but let us now review again several of the salient points. Let us assume that a particle of rest mass m and total energy E collides with a stationary particle of equal mass. The energy that is available in the center of mass of the collision is given by our expression for $\sqrt{s}$ (Eq. 1.64):

$$E_{\mathrm{CM}}^{\mathrm{TOT}} = \sqrt{s} = \sqrt{2m^2c^4 + 2mc^2E} \tag{8.11}$$

which for very high energies becomes

$$E_{\mathrm{CM}}^{\mathrm{TOT}} \simeq \sqrt{2mc^2E} \tag{8.12}$$

This is the part of the incident energy that is available for converting energy into new particles; the rest of the incident energy cannot be used because it is required to maintain the motion of the center of mass, that is, to preserve conservation of momentum in the collision. We see therefore that in a collision of an accelerated particle with a fixed target, the energy in the center of mass increases only as the square root of the accelerator energy. Consequently, to make massive objects such as the W and Z bosons with masses $\simeq 90\ \text{GeV}/c^2$, requires enormous laboratory energies (just the threshold for single Z production in p-p collisions is about 4 TeV).* Except for special purposes, interacting beam particles with fixed targets would therefore appear to be a rather inefficient way to utilize the full machine energy. On the other hand, if it were possible to accelerate two separate beams of particles, and make them collide head-on, then the center of mass of the collision would be stationary in the laboratory, and the entire energy of the beams could be available for producing new particles. This is the idea behind the development of colliding-beam accelerators.

There are different kinds of colliding-beam machines. The beams can have particles of the same type, for example, heavy ions on heavy ions, p on p or e^- on e^-, or of opposite type (particles and their antiparticles), for example, $\bar{p}$ on p or e^- on e^+, or of different type, for example, e^- on p. Both beams can have same energy, or different energy. (Asymmetric energies are often used for technical reasons that are either associated with the detection of short-lived particles, or when it is not possible to make beams of same momentum—as in the case of e^-–p collisions, where electron energies are restricted to relatively low values because of the great synchrotron-radiation loss and the consequent financial implications of trying to reach the very top energies available for protons. Naturally, in such asymmetric collisions, the center of mass is not stationary in the laboratory frame.) Also, colliders can have a single ring or two independent rings of magnets. Clearly, independent systems must be used for all but antiparticle on particle colliders, since in the latter case the orbit of a particle moving along one direction in the ring can be maintained while the antiparticle is moving in the same vacuum pipe but in the opposite direction. Whether there is a single ring or two rings of magnets, the two beams can be accelerated at the same time, and maneuvered into colliding orbits at selected intersection regions that contain detector systems. The beams pass through each other and interact at the collision regions until their intensity is substantially reduced (this often takes many hours—the

*It could be argued that using more massive targets, such as lead nuclei or, better yet, blocks of lead, would be advantageous. Unfortunately, this would not help increase the energy in the center of mass, except for interactions that are characterized by distances of the order of the size of lead nuclei (~ 6 fm) or lead blocks (cm). That is, to produce Z and W objects, we have to have large momentum transfers and therefore collisions at distances of the order of the Compton wavelength for these particles (about 10^{-2} fm), and having larger targets has little relevance since the collisions of interest are between the beam and target protons (or, in fact, between constituents within these protons), and not between a proton and an extended target.

reduction is due to beam–beam collisions as well as to interactions of the beams with any remnant molecules within the vacuum pipe), at which time they are removed safely from the collider, and the acceleration cycle restarted (the acceleration and filling part of the cycle is far shorter than the collision part).

The colliders that appear to be appropriate for accelerating protons are the kind of ring synchrotrons that we discussed in the previous pages. For electron–positron colliding accelerators, two options seem possible. One is, again, the synchrotron variety, and the other is a relatively new concept that has been developed at SLAC, which involves linear colliders. Here, two accelerators are constructed to aim beams at each other. One accelerates electrons, and the other positrons. The full beams are then made to collide head on, in a single-shot fashion. Such linear colliders require large beam currents and small transverse beam dimensions (μm in size) in order to produce enough collisions to make them competitive with ring machines, which have the advantage that they operate with stored beams that allow multiple passes of particles past each other in the intersection regions.

The use of colliding beams, especially when it involves antiparticles, requires substantially more beam "gymnastics." Antiparticles must first be produced, extracted, stored, accumulated, and then accelerated in quantities to provide a sufficient number of interactions with the larger flux of opposing particles. This means that all particles must be stored for far longer times than in normal accelerators. The usual synchrotron focusing mechanisms are not sufficient, and must often be enhanced through "cooling" techniques that reduce the transverse momenta of particles in orbit, and prevent the beams from blowing up. The principle of stochastic cooling, developed by Simon Van der Meer, is used to assure that beam fluxes remain adequate. The idea is to sense the transverse position of a particle at some point in the ring, and send that information along a chord so that it arrives before the particle does (this is possible because the particle must traverse the longer path along the arc of the ring). With the help of RF fields, this information is then used to correct the transverse position of the particle, and prevent its escape from orbit.

The largest collider currently being planned for proton collisions on protons is the Large Hadron Collider (LHC), which is to be located at the CERN Laboratory, outside of Geneva, Switzerland. The recently cancelled U.S. Superconducting Supercollider (SSC) was to have had two rings of magnets placed in an essentially circular pattern, each of which were to be about 90 km in circumference. Each ring was to contain about 4000 dipoles with superconducting coils (operating at about 7 T). In addition, there were to be about 1000 quadrupoles per ring, and other correction coils. The top energy was designed to be 20 TeV per beam, as opposed to the LHC design of about 8 TeV. Figures 8.8 and 8.9 provide an idea of the scale of the complexity of the SSC machine design.

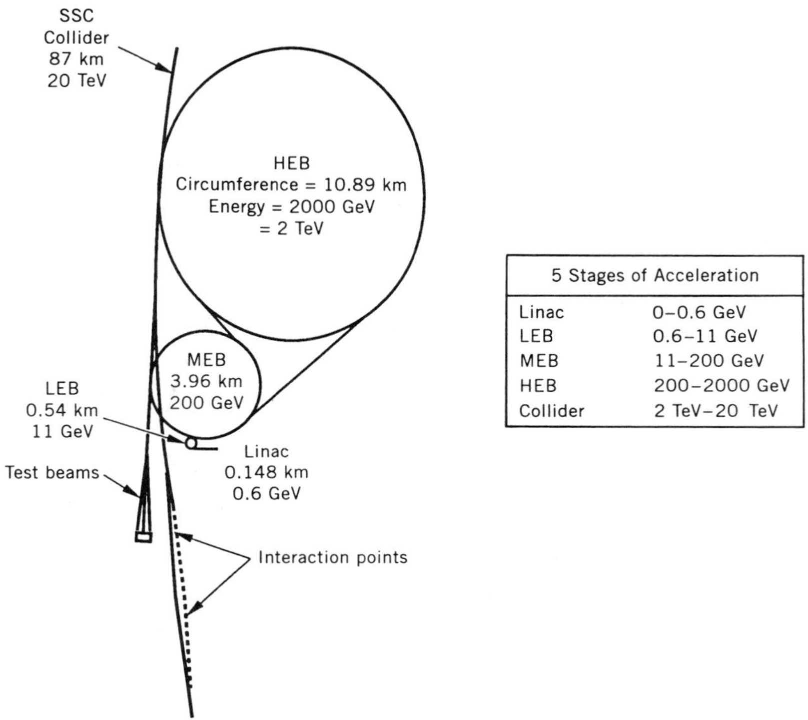

5 Stages of Acceleration	
Linac	0–0.6 GeV
LEB	0.6–11 GeV
MEB	11–200 GeV
HEB	200–2000 GeV
Collider	2 TeV–20 TeV

Figure 8.8 The injector complex that was to be used for the Superconducting Supercollider (SSC). LEB, MEB, and HEB refer, respectively, to the low-energy, medium-energy, and high-energy boosters or synchrotrons.

Problems

VIII.1 Protons are accelerated in a cyclotron by an electric field with oscillating frequency of 8 MHz. If the diameter of the magnet is 1 m, calculate the value of magnetic field and the maximum energy that the protons can reach.

VIII.2 To achieve an energy of 20 TeV, each of the SSC colliding main rings was to contain about 4000 dipole magnets that were 16 meters long, and that operated at ~ 7 T. This means that over half of the ~ 60-mile-long SSC tunnel was to be taken up by dipoles. If you were to build a single synchrotron for use in fixed-target collisions of equivalent energy in the center of mass ($\sqrt{s} = 40$ TeV), and used a similar magnet design, how long would your tunnel have to be?

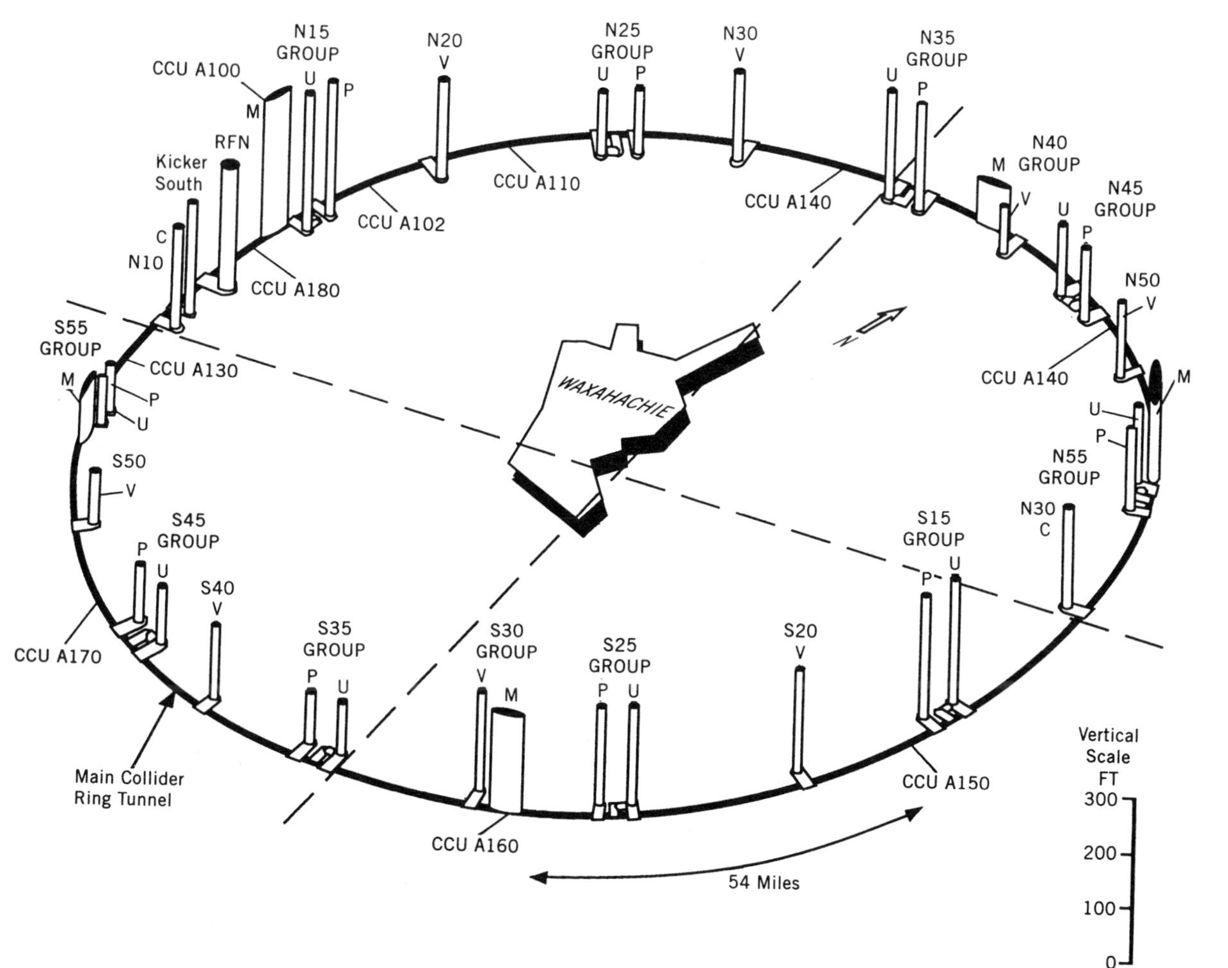

Figure 8.9 Various access shafts and utility stations, and their locations below ground at the recently defunct SSC.

VIII.3 If the capacitance of a Van de Graaff accelerator terminal is 250 $\mu\mu$F (pF), and if it operates at a voltage of 4 MV, what is the total charge on the terminal? If the charging belt can carry a current of 0.2 mA, how long does it take to charge up the accelerator to 4 MV?

VIII.4 Starting with cgs units, show that Relation (8.9′) follows from Relation (8.9).

VIII.5 Suggest a mechanism whereby an accelerated beam could be extracted from a circular accelerator and directed onto an external target.

Suggested Readings

Edwards, D. A. and M. J. Syphers. 1993. *Introduction to the Physics of High Energy Accelerators*. New York: Wiley.

Livingston, M. S. and J. Blewett. 1962. *Particle Accelerators*. New York: McGraw-Hill.

Livingston, M. S. 1969. *Particle Accelerators: A Brief History*. Cambridge, Mass.: Harvard Univ. Press.

Wilson, R. R. 1980. Sci. Am. Jan. **242**: 42.

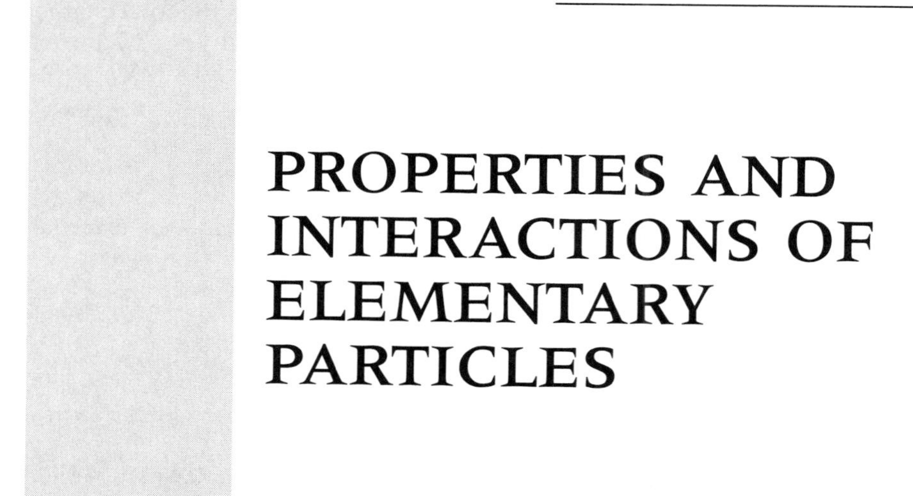

Chapter IX

PROPERTIES AND INTERACTIONS OF ELEMENTARY PARTICLES

INTRODUCTORY REMARKS

After the discovery of the neutron in 1932, it was thought that the electron, the proton, and the neutron were the fundamental constituents of all matter. Subsequent experiments, involving cosmic rays as well as accelerator beams, revealed that there was a host of other particles that could be regarded as equally fundamental. In connection with nuclear β-decay, we already mentioned the families of leptons and some of their properties. In addition, we also know of the existence of hadrons such as π mesons, K mesons, ρ mesons, hyperons, and their many other excited states. All these can collectively be referred to as elementary particles. Usually, an elementary particle is thought to be an object without any substructure, namely a point particle. Experimentally, however, we can probe structure only up to a given length scale that is limited by the energy of the probe. Consequently, our definition of what is elementary or fundamental is always tentative, and must rely on experimental verifications at ever higher energies. Thus, for example, if we want to examine the structure of matter for length scales of $\Delta r \lesssim 0.1$ fm, then the transverse momentum transfers (Δp_T) supplied by the probe must be at least of the order

$$\Delta p_T \sim \frac{\hbar}{\Delta r} = \frac{\hbar c}{(\Delta r)c} \simeq \frac{197 \text{ MeV} - \text{fm}}{(0.1 \text{ fm})c} \sim 2000 \text{ MeV/c} \qquad (9.1)$$

In other words, to be sensitive to small length scales, the energy of the particles to be used as probes must be very high. Because of this need, the study of elementary particles has also come to be known as high-energy physics.

Whenever a higher-energy accelerator becomes available, we can probe deeper into the structure of matter and, possibly, find that what was once considered elementary is not really so. This has, in fact, been the story of the proton, the neutron, the π mesons, the K mesons, and so forth. Our current understanding of which particles should be considered as elementary is very different from that of several decades ago. Nevertheless, we will start with the traditional (historical) perspective, and turn to the more modern view of elementary particles in a later chapter. We will begin at the one GeV/c^2 mass scale, and discuss the properties of elementary particles and their interactions from a purely phenomenological point of view.

FORCES

We are quite familiar with the classical electromagnetic and gravitational forces. We know that every particle, whether with or without rest mass, is subject to gravitational attraction. (The observed bending of light in a gravitational field shows conclusively that particles do not need to have rest mass but rather energy to experience the pull of gravity.) On the other hand, only particles that carry electric charge sense the direct Coulomb field. Both the Coulomb and the gravitational forces are long ranged. The photon is the carrier of the Coulomb interaction, and from the fact that the electromagnetic force has infinite range, we can conclude that the photon must be massless. The carrier of the gravitational interaction is the conjectured graviton, which is also believed to be massless. From our discussion of nuclear phenomena, we learned that there are two more forces that have importance in the subatomic domain. There is the strong force, which, as we have seen, is responsible for the binding of nucleons inside a nucleus, and the weak force, which appears in decay processes such as the β decay of nuclei. These forces have no classical analogues and, unlike the electromagnetic and the gravitational interactions, are exceedingly short ranged. Thus, it seems that we can point to four fundamental forces in nature:

1. Gravitation
2. Electromagnetism
3. Weak Force
4. Strong Force

Because, in principle, all the forces can act at the same time, one could ask how is it possible to determine which force contributes in any particular process? The answer is that the forces can be distinguished through the strengths of their interaction. We can estimate the relative magnitudes of these four forces in a heuristic way by considering their effective potentials. Although such potentials are fundamentally nonrelativistic in concept, they provide a useful guide for a rough comparison. Consider two protons separated by a distance r. The magnitudes of the Coulomb and of the gravitational potential energies for the two particles are

$$\begin{aligned} V_{\mathrm{em}}(r) &= \frac{e^2}{r} \\ V_{\mathrm{grav}}(r) &= \frac{G_N m^2}{r} \end{aligned} \tag{9.2}$$

where G_N is Newton's constant $(6.7 \times 10^{-39} \hbar c (\mathrm{GeV}/c^2)^{-2})$, and m is the mass of the proton. It is more instructive to write the potential energies in the Fourier-transformed momentum space (see Eq. 1.77) where, except for an overall normalization, they take on the form

$$\begin{aligned} V_{\mathrm{em}}(q) &= \frac{e^2}{q^2} \\ V_{\mathrm{grav}}(q) &= \frac{G_N m^2}{q^2} \end{aligned} \tag{9.3}$$

Here q refers to the magnitude of the momentum transfer that characterizes the interaction. The absolute values of the potential energies for the interactions appear to decrease quadratically with momentum transfers.

We note that while the individual potentials have momentum dependence, the ratio of V_{em} and V_{grav} is, in fact, independent of the momentum scale, and we can evaluate this to be

$$\begin{aligned} \frac{V_{\mathrm{em}}}{V_{\mathrm{grav}}} &= \frac{e^2}{G_N m^2} = \left(\frac{e^2}{\hbar c}\right) \frac{1}{(mc^2)^2} \frac{\hbar c \cdot c^4}{G_N} \\ &\simeq \left(\frac{1}{137}\right) \frac{1}{(1\ \mathrm{GeV})^2} \frac{10^{39}\ \mathrm{GeV}^2}{6.7} \simeq 10^{36} \end{aligned} \tag{9.4}$$

where we have substituted 1 GeV/c^2 for the mass of the proton and the value of $\alpha = \frac{1}{137}$, the electromagnetic fine-structure constant, for $e^2/\hbar c$. Equation 9.4 shows that for elementary particles the gravitational force is inherently much weaker than the electromagnetic force.

Next, let us recall that since both the strong and the weak forces are short ranged, they can be described phenomenologically by Yukawa potentials and have the forms

$$
\begin{aligned}
V_{\text{strong}} &= \frac{g_s^2}{r} e^{-(m_\pi c^2 r)/\hbar c} \\
V_{\text{wk}} &= \frac{g_{\text{wk}}^2}{r} e^{-(m_W c^2 r)/\hbar c}
\end{aligned} \tag{9.5}
$$

Here g_s and g_{wk} represent the coupling constants (charges) for the strong and the weak interactions, and m_π and m_W represent the masses of the force-mediating (or exchanged) particles in the two cases. Once again, we can transform the potentials just given to momentum space, and, except for an overall normalization constant, obtain

$$
\begin{aligned}
V_{\text{strong}} &= \frac{g_s^2}{q^2 + m_\pi^2 c^2} \\
V_{\text{wk}} &= \frac{g_{\text{wk}}^2}{q^2 + m_W^2 c^2}
\end{aligned} \tag{9.6}
$$

The values of the coupling constants can be estimated from experiments, and they are $g_s^2/\hbar c \simeq 15$ and $g_{\text{wk}}^2/\hbar c \simeq 0.004$. We know from our discussion in Chapter II, that we can think of the π meson ($m_\pi \simeq 140$ MeV/c^2) as the mediator of the strong force. Also, from weak-interaction processes at low energies (e.g., β-decay), we can phenomenologically estimate $m_W \simeq 80$ GeV/c^2. Consequently, we can compare the magnitude of the Coulomb potential energy to that for the strong and the weak interactions. We note, however, that now there is an explicit dependence on the momentum scale in the ratio. Since we are considering the interaction of two protons, it is natural to choose the momentum scale to correspond to that of the proton mass. Thus, choosing $q^2c^2 = m^2c^4 = (1\text{ GeV})^2$, we obtain

$$
\begin{aligned}
\frac{V_{\text{strong}}}{V_{\text{em}}} &= \frac{g_s^2}{\hbar c}\frac{\hbar c}{e^2}\frac{q^2}{q^2 + m_\pi^2 c^2} = \frac{g_s^2}{\hbar c}\frac{\hbar c}{e^2}\frac{m^2c^4}{m^2c^4 + m_\pi^2 c^4} \\
&\simeq 15 \times 137 \times 1 \simeq 2 \times 10^3 \\
\frac{V_{\text{em}}}{V_{\text{wk}}} &= \frac{e^2}{\hbar c}\frac{\hbar c}{g_{\text{wk}}^2}\frac{m^2c^4 + m_W^2 c^4}{m^2c^4} \\
&\simeq \frac{1}{137}\frac{1}{0.004}(80)^2 \simeq 1.2 \times 10^4
\end{aligned} \tag{9.7}
$$

This shows once again that the strong force is stronger than the electromagnetic force, which in turn is stronger than the weak force, and that gravitation is the weakest of all the forces. For larger momentum scales of order $\sim m_W$, the weak and electromagnetic energies and strengths become more comparable, and suggest the possibility for a unification of the two forces at very high energies. We note, however, that because our phenomenological estimates are only qualitative, the ratios of the effective potentials as given in Eq. 9.7 should not be taken too literally.

The difference in the forces also manifests itself in the interaction time characterizing a particular process. Thus, for example, the typical time scale for a strong reaction is about 10^{-24} sec, which is roughly the time it takes a light signal to traverse a proton's dimension, namely 1 fm. On the other hand, typical electromagnetic reactions for elementary particles occur in time intervals of the order of 10^{-20}–10^{-16} sec, whereas the typical time scales for weak decays are about 10^{-13}–10^{-6} sec.* In the GeV range of energies, the properties of the four fundamental forces are therefore quite different, and they can be used to classify the character of the elementary particles.

ELEMENTARY PARTICLES

Before it was fully appreciated that quarks were the fundamental constituents of nuclear matter, all the known elementary particles were grouped into four classical categories that depended on the nature of their interactions. This grouping is sketched in Table 9.1.

All particles, including photons and neutrinos, participate in gravitational interactions. The photon can also interact directly with any particle that has electric charge. All charged leptons participate both in the weak and electromagnetic interactions, and neutral leptons, of course, have no direct electromagnetic coupling. (Recall that this is what made it so difficult to observe the neutrino in β-decay.) Leptons do not sense the strong force. All hadrons (mesons and baryons) respond to the strong force and appear to participate in all the interactions. We will subsequently discuss the differences between mesons and baryons; their common characteristic, however, is that they appear to have substructure, and a size of the order of one femtometer.

All the particles in nature can be classified as either bosons or fermions. The basic difference between them lies in the statistics that they obey. Bosons obey Bose-Einstein statistics, whereas fermions satisfy Fermi-Dirac statistics. This difference is reflected in the structure of their wave functions. For example, the quantum mechanical wave function for a system of identical bosons is symmetric under the exchange of any pair of particles

---

*Again, we wish to stress that these are only typical time scales. Specific transition rates have varying contributions from spin effects and density of final states factors ("phase space") that can have large bearing on lifetimes. As we mentioned previously, the lifetime of the neutron, for example, is ~ 900 sec, which is far from the norm for weak interactions.

Table 9.1 Sample of different types of elementary particles

Particle	*Symbol*[a]	*Range of Masses*
Photon	γ	0
Leptons	$e^-, \mu^-, \tau^-, \nu_e, \nu_\mu, \nu_\tau$	$\lesssim 10$ eV/c^2–1.4 GeV/c^2
Mesons	$\pi^+, \pi^-, \pi^0, K^+, K^-, K^0, \rho^+, \rho^-, \rho^0, \ldots$	135 MeV/c^2–several GeV/c^2
Baryons	$p, n, \Lambda^0, \Sigma^+, \Sigma^-, \Sigma^0, \Delta^{++}, \Delta^0, N^{*0}, Y_1^{*+}, \Omega^-, \ldots$	940 MeV/c^2–several GeV/c^2

[a] The superscripts denote the electric charges carried by various particles.

(or their coordinates). That is,

$$\Psi_B(x_1, x_2, x_3, \ldots, x_n) = \Psi_B(x_2, x_1, x_3, \ldots, x_n) \tag{9.8}$$

On the other hand, the quantum mechanical wave function for a system of identical fermions is antisymmetric under the exchange of any pair of particle coordinates, namely

$$\Psi_F(x_1, x_2, x_3, \ldots, x_n) = -\Psi_F(x_2, x_1, x_3, \ldots, x_n) \tag{9.9}$$

The Pauli exclusion principle is therefore automatically built into the antisymmetric fermionic wave function, thereby forbidding a pair of identical fermions to occupy the same quantum state. This follows because, for $x_1 = x_2$, the wave function in Eq. 9.9 would equal its negative value, and would therefore vanish. It can be shown that all bosons have integer values of spin angular momentum, while fermions have half-integral spin values.

In a later section we describe how the spin of an elementary particle can be determined, but here let us just state that the photon and all the mesons are known to be bosons, whereas the leptons and all the baryons are fermions. Also, as we have indicated earlier, every known particle has a corresponding antiparticle. The antiparticle has the same mass as the particle, but otherwise opposite quantum numbers. Thus, the positron (e^+) is the antiparticle of the electron, and carries a negative lepton number and a positive charge. The antiproton ($\overline{p}$) has one unit of negative charge and one unit of negative baryon number, in contrast to the proton, which is positively charged and has a positive baryon or nucleon number. Certain particles cannot be distinguished from their own antiparticles. For example, the π^0, which has no electric charge, is also its own antiparticle. It is clear that for a particle to be its own antiparticle, it must, at least, be electrically neutral. However, not all electrically neutral particles are their own antiparticles. As we have seen earlier, the neutron has no electric charge, yet the antineutron is distinctly different because of its negative baryon number and the opposite sign of its magnetic moment from that of the neutron. Similarly, the neutrino and the K^0 meson, although charge neutral, have distinct antiparticles. Except where it is redundant, antiparticles are denoted by the same symbol as the particles, but with opposite charge or a bar over that symbol. Several examples are

$$\begin{aligned} \overline{e^-} &= e^+ \\ \overline{\pi^0} &= \pi^0 \\ \overline{\Sigma^-} &= \overline{\Sigma}^+ \\ \overline{K^+} &= K^- \end{aligned} \tag{9.10}$$

QUANTUM NUMBERS

As we emphasized in our treatment of nuclear phenomena, much of our physical intuition does not help in trying to understand the subatomic domain, and we have to rely on experimental observation as guide. The properties of the elementary particles and their interactions are even more mystifying, and most of our insight comes from scattering one off another or observing transitions among them. Since, as we have remarked, there are many elementary particles, there are also many processes that one can study. It is clear, therefore, that to derive any meaningful conclusions from such observations, we must organize the results in some coherent manner. Here our classical experience does help. Classically, we know that a process or a reaction can take place if it is kinematically allowed, and if it does not violate any recognized conservation law. Thus, for example, we are quite certain that a reaction that violates charge conservation will never take place. This certainty is based upon years of past studies and the development of a reliable theory for electromagnetic interactions. We believe that similar conservation principles hold in the subatomic domain, except that here we do not know all the relevant laws because we do not have a complete theoretical understanding of all the forces. Consequently, to formulate general principles, we must deduce from experiment the type of quantum numbers that are conserved and the conservation laws that are appropriate for each of the interactions of the elementary particles. One of the clearest results observed in reactions of elementary particles is that the number of fermions is always conserved (that is, if we count a fermionic antiparticle as a fermion, but with a negative fermion number), whereas the number of photons and mesons is not. This suggests that the conservation of fermion number is a fundamental feature of all interactions, and will be elaborated upon below.

Baryon Number

From differences in the magnitudes of observed transition rates, or from the absence (upper limits) of kinematically allowed processes, we can often infer the presence of possible conservation laws. As an example, consider the decay

$$p \to e^+ + \pi^0 \tag{9.11}$$

Since the proton is far more massive than the sum of the pion and the positron masses, and since the decay just given satisfies the conservation of electric charge, one might expect this process to take place. Experimentally, however, one does not observe proton decay. (In fact, the upper limit on the probability for Reaction (9.11) is a negligible $\lesssim 10^{-40}$/sec.) This suggests, therefore, that there is some conservation principle that forbids this decay. In fact, we can account for this lack of decay quite simply if we assert that baryons carry an additive, conserved, quantum number known as baryon or nucleon number that is equal to 1 for all baryons (of course, antibaryons will have baryon number $B = -1$), whereas, photons, leptons, and mesons have zero baryon number. It is now clear that, if baryon number is conserved in all processes, then since the proton is the lightest baryon, it should not decay.

Lepton Number

We can similarly postulate a quantum number for the leptons. All leptons carry lepton number $L = 1$, whereas the photon and hadrons carry no lepton number. The introduction of the lepton quantum number is necessitated by many experimental observations. One simple example is the process

$$e^- + e^- \rightarrow \pi^- + \pi^- \tag{9.12}$$

At high energies, this reaction is kinematically allowed, and it certainly satisfies charge conservation; nevertheless, it is not observed. Lepton number conservation would, naturally, prevent this process from taking place.* In fact, one finds that reactions such as

$$\begin{aligned} \mu^- &\rightarrow e^- + \gamma \\ \mu^- &\rightarrow e^- + e^+ + e^- \end{aligned} \tag{9.13}$$

although kinematically allowed, have also never been observed. It is from such experimental findings that we arrive at the conclusion that there must be different kinds of lepton numbers within the family of leptons (see Table 9.2). Thus, the electron and its neutrino (recall our earlier discussion about the fact that each charged lepton has its own distinct neutrino)

Table 9.2 Lepton number assignments

	Electron Number L_e	*Muon Number* L_μ	τ*-lepton Number* L_τ	$L = L_e + L_\mu + L_\tau$
e^-	1	0	0	1
ν_e	1	0	0	1
μ^-	0	1	0	1
ν_μ	0	1	0	1
τ^-	0	0	1	1
ν_τ	0	0	1	1

*We note that conservation of lepton number would also explain the absence of proton decay in the $e^+\pi^0$ channel of Eq. 9.11.

have an electron lepton number $L_e = 1$, whereas the other leptons have $L_e = 0$. The muon and its neutrino have muon lepton number $L_\mu = 1$, whereas the other leptons have $L_\mu = 0$, and similarly for the τ lepton and its neutrino. The lepton number of any particle can therefore be expressed as the sum of the electron number, the muon number, and the τ-lepton number. Leptons can therefore be split up naturally into three families, namely, (e^-, ν_e), (μ^-, ν_μ), (τ^-, ν_τ), with each family number conserved in high-energy processes. This would explain, for example, why the muon decays as follows:

$$\mu^- \to e^- + \overline{\nu}_e + \nu_\mu \tag{9.14}$$

It is worth noting that, although the proton decay in Eq. 9.11 violates both baryon number and lepton number, the combination $B - L$ is conserved in the process. This has, in fact, great implications for model building.

Strangeness

In early studies of cosmic-ray showers, it was found that some particles, which have since been identified with the K mesons, the Σ and the Λ^0 baryons, were produced strongly (that is, with large cross sections of the order of millibarns), but had lifetimes characterizing the weak interactions, namely $\sim 10^{-10}$ sec. These particles were always produced in pairs, namely a K in association with a Σ or Λ^0. All this was certainly puzzling, and led naturally to a suspicion that a new quantum number might be associated with such particles. When specific reactions were studied, such as

$$\pi^- + p \to K^0 + \Lambda^0$$

with the Λ^0 and K^0 subsequently decaying:

$$\begin{aligned} \Lambda^0 &\to \pi^- + p \\ K^0 &\to \pi^+ + \pi^- \end{aligned} \tag{9.15}$$

the Λ^0 was always produced in association with a K^0 and never just with a π^0. The Λ^0 was also observed to be produced in association with a K^+, but not a K^-:

$$\begin{aligned} \pi^- + p &\to K^+ + \pi^- + \Lambda^0 \\ \pi^- + p &\nrightarrow K^- + \pi^+ + \Lambda^0 \\ \pi^- + p &\nrightarrow \pi^- + \pi^+ + \Lambda^0 \end{aligned} \tag{9.16}$$

Similarly, for the reaction

$$\pi^+ + p \to \Sigma^+ + K^+$$

with the Σ^+ and K^+ subsequently decaying:

$$\begin{aligned} \Sigma^+ &\to n + \pi^+ \\ K^+ &\to \pi^+ + \pi^0 \end{aligned} \tag{9.17}$$

it is observed that the Σ^+ is always produced in association with a K^+, and never with just a π^+. Again, Σ^+ baryons can also be produced in association with K^0 mesons, but with an additional π^+ required to conserve electric charge. Similarly, Σ^- baryons can be produced in association with K^+ mesons in $\pi^- p$ collisions, but $\Sigma^+ K^-$ final states are not observed:

$$\begin{aligned} &\pi^+ + p \to \Sigma^+ + \pi^+ + K^0 \\ &\pi^- + p \to \Sigma^- + K^+ \\ &\pi^- + p \not\to \Sigma^+ + K^- \\ &\pi^- + p \not\to \Sigma^- + \pi^+ \end{aligned} \tag{9.18}$$

The production cross sections for reactions such as those given in Eqs. 9.15 and 9.17 for pion momenta of about 1 GeV/c were measured to be about 1 mb, whereas the total cross sections for $\pi^\pm p$ scattering were known to be about 30 mb. Thus, it was clear that these production processes were strong. The subsequent decays of these particles were also studied, and revealed that the Λ^0 traveling at a speed of about $0.1c$, decayed after a flight path of about 0.3 cm. Consequently, the lifetime of this baryon was deduced to be about

$$\tau_{\Lambda^0} \simeq \frac{0.3 \text{ cm}}{3 \times 10^9 \text{ cm/sec}} = 10^{-10} \text{ sec}$$

And similar lifetimes were observed for these other "strange" particles, leading to the conclusion that the decays involved weak interactions (see Fig. 9.1).

The puzzle of associated production was clarified by Murray Gell-Mann and Abraham Pais, who proposed that these particles carried a new additive quantum number, which they called *strangeness*, which had to be conserved in strong production processes, but was violated in the weak decays. All the ordinary mesons and baryons (as well as the photon) are assumed to be nonstrange ($S = 0$). Now, in any associated-production reaction, with the initial state having no strangeness, the strangeness of the final-state particles must also add up to zero. From the analysis of such reactions we can deduce that the strangeness of the K^+ and K^0 must be opposite to that of the Σ^+, Σ^0, Σ^-, and Λ^0. In fact, if we arbitrarily choose

$$S(K^0) = 1 \tag{9.19}$$

it follows that

$$S(K^+) = S(K^0) = 1 \tag{9.20}$$

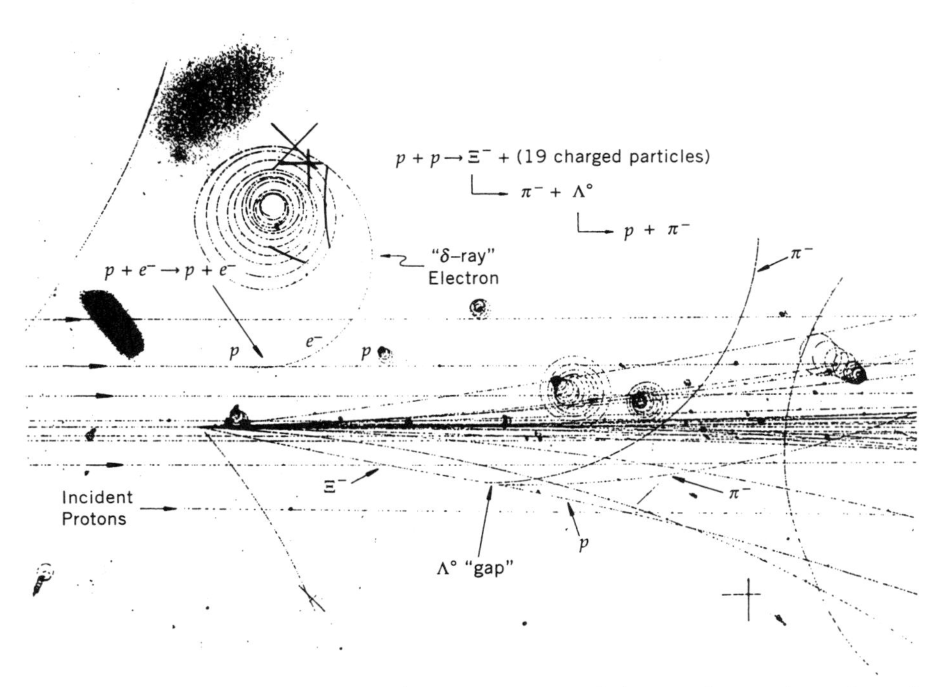

Figure 9.1 A photograph of particle interactions and decays observed in a liquid hydrogen bubble chamber at Fermilab. The tracks correspond to trajectories of charged particles that ionized the hydrogen atoms and thereby produced bubbles (local boiling) along their paths in the superheated liquid. A magnetic field in the chamber causes the curvature of the particle trajectories. The chamber is sensitive for about 1 msec, which is long compared to the lifetimes of some of the particles that can be observed to decay. The incident particles are 400 GeV protons. One of these interacts strongly with a target proton (nucleus of the hydrogen atom) and produces many other particles.

and that

$$S(\Lambda^0) = S(\Sigma^+) = S(\Sigma^0) = S(\Sigma^-) = -1 \tag{9.21}$$

Similarly, from strong-production reactions such as

$$\begin{aligned} K^- + p &\longrightarrow \Xi^- + K^+ \\ \overline{K^0} + p &\longrightarrow \Xi^0 + K^+ \end{aligned} \tag{9.22}$$

we deduce that the cascade particles Ξ^0 and Ξ^- can be assigned the strangeness number $S = -2$, if the $\overline{K^0}$ and K^- have $S = -1$. The latter assignment is consistent with our identification of the K^- and $\overline{K^0}$ as antiparticles of the K^+ and K^0, respectively.

It is worth stressing, again, that the weak decays of the hadrons do not conserve strangeness. Therefore, if we assume strangeness to be conserved only in strong and electromagnetic interactions, then it follows that we cannot assign unique strangeness quantum numbers to leptons.

Isospin

The proton and the neutron are baryons with spin $\frac{1}{2}$, and are essentially degenerate in their mass. In fact, as we already indicated they are quite similar in their nuclear properties, except that the proton has a positive charge, whereas the neutron is electrically neutral. Correspondingly, their electromagnetic interactions are quite different, and, as we discussed earlier, even their magnetic dipole moments have opposite sign.

It has been realized for a long time that the strong force does not depend on the charge of a particle. In fact, studies of mirror nuclei (e.g., ^{3}H, ^{3}He) have demonstrated that the strong binding force between p–p, p–n, and n–n is essentially the same. Furthermore, scattering experiments have revealed that, if we correct for electromagnetic effects, the cross section for the scattering of two protons is the same as that for two neutrons. Thus, the strong interactions do not distinguish between a proton and a neutron. Consequently, if we imagine a world where only the strong force is present, and the weak and electromagnetic forces are turned off, then in such a world a proton would be indistinguishable from a neutron. (Our physical world is, of course, not like this. Because the strong force is so much stronger than the other forces, however, we can pretend that our world is close to this, and that the presence of the other forces will then simply provide small corrections to the simple picture.) In such a world, we can think of the proton and the neutron as two orthogonal states of the same particle that we can call the nucleon, and write the states for the neutron and proton as follows:

$$p = \begin{pmatrix} 1 \\ 0 \end{pmatrix}$$
$$n = \begin{pmatrix} 0 \\ 1 \end{pmatrix} \tag{9.23}$$

This language is very similar to that used in discussing the "spin up" and the "spin down" states of a spin-$\frac{1}{2}$ particle, which are also indistinguishable in the absence of any interaction that breaks rotational symmetry (e.g., a magnetic field). The two spin states will be degenerate in energy until we apply an external magnetic field, which picks out a preferred direction in space, and removes the degeneracy of the two states. In much the same way, we can think of the proton and the neutron as being degenerate in mass because of some symmetry of the strong force (or of the strong Hamiltonian), and we call this symmetry the isotopic-spin or isospin symmetry. In reality, the presence of the electromagnetic and weak forces breaks this symmetry, lifts the degeneracy in the masses, and allows us to distinguish between the neutron and the proton.

As indicated in Chapter II, just as with the proton and neutron, the three π mesons, namely π^+, π^-, and π^0, also have almost identical masses. And, just as in the case of the nucleons, the cross sections for scattering different pions on protons and neutrons are also found to be the same, once they are corrected for electromagnetic effects. Thus, it appears that the strong force does not distinguish between different kinds of π mesons. Therefore, in the absence of electromagnetic and weak forces, we can think of the three π mesons as corresponding to different states of one particle, the π meson, and we can represent the pion states as follows:

$$\pi^+ = \begin{pmatrix} 1 \\ 0 \\ 0 \end{pmatrix}$$

$$\pi^0 = \begin{pmatrix} 0 \\ 1 \\ 0 \end{pmatrix} \tag{9.24}$$

$$\pi^- = \begin{pmatrix} 0 \\ 0 \\ 1 \end{pmatrix}$$

These three states will, of course, be degenerate in mass in our hypothetical world. The analogy with spin, in this case, corresponds to the three spin projections of a $J = 1$ particle that are degenerate in energy for a rotationally invariant Hamiltonian.

Similarly, the (K^+, K^0) doublet, the $(\overline{K^0}, K^-)$ doublet, and the $(\Sigma^+, \Sigma^0, \Sigma^-)$ triplet correspond to states that can be considered as different manifestations of single particles, the K, $\overline{K}$, and Σ, respectively. In fact, this discussion can be extended to all the known hadrons, which can be classified into multiplets corresponding to some quantum number very much like the spin quantum number. We will refer to this quantum number as the strong isotopic spin or strong isospin, and its conservation suggests the invariance of the strong Hamiltonian under isospin transformations. These transformations correspond to rotations very much like those that occur for spin. In this case, however, the rotations are in an internal Hilbert space and not in space-time. The isospin quantum number (or I-spin) is observed to be conserved in strong interactions (it is a symmetry of the strong force). However, I-spin does not appear to be conserved in electromagnetic and weak processes.

Table 9.3 summarizes the strong isospin quantum numbers for different hadrons, as determined from different scattering experiments. The

Table 9.3 Isotopic spin assignments of a representative group of relatively long-lived hadrons

Hadron	*Mass* (MeV/c^2)	I	I_3
p	938.3	$1/2$	$1/2$
n	939.6	$1/2$	$-1/2$
π^+	139.6	1	1
π^0	135.0	1	0
π^-	139.6	1	-1
K^+	494.6	$1/2$	$1/2$
K^0	497.7	$1/2$	$-1/2$
$\overline{K}^0$	497.7	$1/2$	$1/2$
K^-	494.6	$1/2$	$-1/2$
η^0	548.8	0	0
Λ^0	1115.6	0	0
Σ^+	1189.4	1	1
Σ^0	1192.6	1	0
Σ^-	1197.4	1	-1
Ω^-	1672.4	0	0

assignment for the third-component, or projection of the isospin chosen in Table 9.3 is such that, in any given isospin multiplet, a particle with a larger positive charge has a higher value of the isospin projection. We have also denoted the projection as I_3 instead of the conventional notation I_z, in order to emphasize that isospin is not a space-time symmetry. We do not assign strong isospin quantum numbers to leptons or to the photon, because it is only the strong interactions that are invariant under isospin transformations and, therefore, these quantum numbers can be defined uniquely only in strong processes. The photon and the leptons do not participate in strong reactions, and hence one cannot uniquely assign strong isospin quantum numbers to them. As we will see in Chapter XIII, there also exists another symmetry called the *weak isospin symmetry*, which is fundamental to the Standard Model and which involves leptons and quarks.

As we have indicated, isospin is conserved in strong interactions, and this can be inferred from comparisons of different production and decay processes. We will leave these details to specific examples and problems in Chapter X.

GELL-MANN–NISHIJIMA RELATION

The assignment of the strangeness quantum number in Eq. 9.19, and the other choices we have made, may appear to be rather ad hoc. In fact, the assignments were made originally with the phenomenological observation in mind that the electric charge of a hadron can be related to its other quantum numbers through the Gell-Mann–Nishijima relation

$$Q = I_3 + Y/2 = I_3 + (B + S)/2 \tag{9.25}$$

where $Y = B+S$ is known as the strong hypercharge. (We will see later, in the context of the Standard Model, that there is a different relation involving the weak hypercharge, which holds for all fundamental particles.) We summarize the quantum numbers for several typical long-lived hadrons in Table 9.4. These are all consistent with Relation (9.25).

With the subsequent discovery of new particles with new "flavor" quantum numbers such as "charm" and "bottom," in addition to strangeness, the Gell-Mann–Nishijima relation has been generalized to include these as well. In the expanded relation, the hypercharge is defined to be the sum of the baryon number, strangeness number, and all the new flavor quantum numbers. With this modification, the original relation, namely,

$$Q = I_3 + \frac{Y}{2} \tag{9.26}$$

holds for all hadrons. Since charge and isospin are conserved in strong interactions, it follows that the generalized hypercharge is also conserved in such processes. In fact, strangeness and each of the flavors is conserved independently in strong interactions.

Table 9.4 Quantum numbers of a representative set of relatively long-lived hadrons

Hadron	Q	I_3	B	S	$Y = (B + S)$
π^+	1	1	0	0	0
π^0	0	0	0	0	0
π^-	−1	−1	0	0	0
K^+	1	$1/2$	0	1	1
K^0	0	$-1/2$	0	1	1
η^0	0	0	0	0	0
p	1	$1/2$	1	0	1
n	0	$-1/2$	1	0	1
Σ^+	1	1	1	−1	0
Λ^0	0	0	1	−1	0
Ξ^-	−1	$-1/2$	1	−2	−1
Ω^-	−1	0	1	−3	−2

PRODUCTION AND DECAYS OF RESONANCES

We already mentioned in Chapters II and IV that in the nuclear domain there exist resonances or excited states of the ground levels of nuclei. Similarly, it has been found that in the particle domain there are also excited states of hadrons, and that such resonances have typical lifetimes of the order of 10^{-23} sec. There are two ways to observe such short-lived particles. Let us first consider the $\Delta(1232)$, which is a π-N state that has $I = \frac{3}{2}$ (four different charge states). It was the first object of its kind to have been found, and it was discovered by Enrico Fermi and his collaborators in the study of π-N scattering as a function of energy. This direct way of searching for excited hadronic states is referred to as *formation* or *s-channel studies*. Using a pion beam, the probability of scattering from a nucleon target (i.e., the π-N cross section $\sigma_{\pi N}$) is measured as a function of the momentum of the pion, or, equivalently, the invariant mass $\sqrt{s}/c^2$ of the π-N system (see Eq. 1.64). Figure 9.2 sketches what is observed at low energies for the elastic scattering of π^+ on protons. The cross section rises from threshold (about 1080 MeV, corresponding to the sum of the rest masses of the π^+ and the proton) and reaches a maximum at $M_\Delta c^2 \sim 1230$ MeV; the peak has an observed full width at half maximum of $\Gamma_\Delta c^2 \sim 100$ MeV. The excitation spectrum can be characterized, essentially, by a Lorentzian, or "Breit-Wigner" form (after Gregory Breit and Eugene Wigner). This peak can be interpreted as a resonance in the π-N system, or as an excited state of the nucleon. There is an intrinsic uncertainty in the mass that is associated with the observed width of the line shape (corrected for any small effects due to the experimental resolution) that corresponds to a lifetime of the order of

$$\tau_\Delta \sim \frac{\hbar}{\Gamma_\Delta c^2} \simeq \frac{6.6 \times 10^{-22}}{100 \text{ MeV}} \text{ MeV sec} \simeq 10^{-23} \text{ sec} \tag{9.27}$$

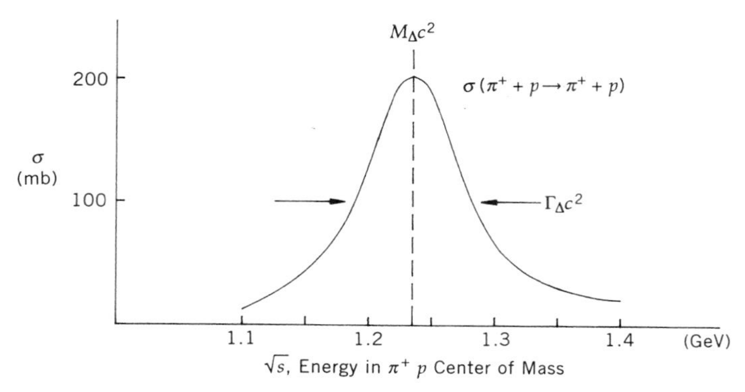

Figure 9.2 Sketch of the pion–nucleon elastic scattering cross section at low energies.

Clearly, not all excited states of hadrons can be found in this fashion. For example, resonances among pions cannot be produced in formation experiments because the fluxes needed to observe such systems would be forbiddingly high. Objects such as the ρ meson were discovered in final states involving multiple-pion production. The following reaction, when studied at a fixed energy, is a rich source of ρ^0 mesons:

$$\pi^- + p \longrightarrow \pi^+ + \pi^- + n \tag{9.28}$$

The way that the presence of a resonance can be detected in the final state is by plotting the invariant mass or the $\sqrt{s_{\pi\pi}}/c^2$ of the $\pi^+\pi^-$ system (see Eq. 1.64). If Reaction (9.28) proceeds through the intermediate step

$$\pi^- + p \longrightarrow \rho^0 + n$$

followed by (9.29)

$$\rho^0 \longrightarrow \pi^+\pi^-$$

then, after the ρ^0 decays, the π^+ and π^- will remain correlated. This is because energy-momentum conservation in the decay will assure that

$$\begin{aligned} E_\rho &= E_{\pi^+} + E_{\pi^-} \\ \vec{p}_\rho &= \vec{p}_{\pi^+} + \vec{p}_{\pi^-} \end{aligned} \tag{9.30}$$

and therefore the invariant mass of the two pions will maintain the mass of the ρ^0, namely,

$$M_\rho^2 c^4 = \left(E_\rho^2 - p_\rho^2 c^2\right) = (E_{\pi^+} + E_{\pi^-})^2 - (\vec{p}_{\pi^+} + \vec{p}_{\pi^-})^2\, c^2 = s_{\pi\pi} \tag{9.31}$$

Consequently, when we plot the distribution for the effective or invariant mass of two pions for many events corresponding to Reaction (9.28), namely, plot the number of events as a function of $\sqrt{s_{\pi\pi}}/c^2$, if there is a contribution from Reaction (9.29), we should then observe a peak in the distribution at $\sqrt{s_{\pi\pi}}/c^2 = M_\rho$. A typical result for Reaction (9.28) is shown in Fig. 9.3, and displays a peak at $M_\rho = 760$ MeV/c^2, with a width of $\Gamma_\rho \sim 150$ MeV/c^2, which characterizes the strong resonant interaction of the two pions.

The fact that resonances have Breit-Wigner form is an intrinsic consequence of the character of a decaying quantum state. The time dependence of the amplitude of any state with mean life $\hbar/\Gamma c^2$, and with a central value of mass of $M = M_0$, can be written in its own rest frame as (see the discussion on time development in Chapter XII)

$$\psi(t) \sim e^{-(ic^2/\hbar)(M_0 - i(\Gamma/2))t} \quad (\text{for } t > 0) \tag{9.32}$$

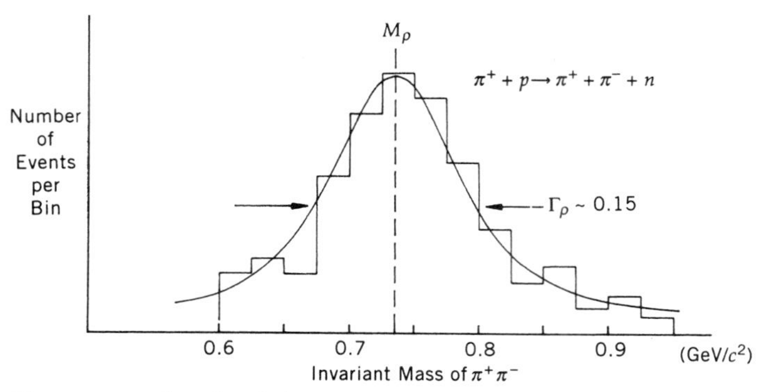

Figure 9.3 Sketch of the invariant mass of $\pi^+\pi^-$ pairs produced in Reaction (9.28).

This assures the exponential decay of the state with mean life $\hbar/\Gamma$:

$$|\psi(t)|^2 \sim e^{-\Gamma c^2 t/\hbar} \tag{9.33}$$

Taking the Fourier transform of Eq. 9.32 provides the amplitude in energy space (or in mass):

$$\psi(M) \sim \int_0^\infty dt\, \psi(t) e^{(i/\hbar)Mc^2 t} \tag{9.34}$$

This is simple to integrate, and, except for an overall normalization, yields

$$\psi(M) \simeq \frac{1}{(M - M_0) + i(\Gamma/2)} \tag{9.35}$$

which, upon squaring, yields the Lorentzian or the Breit-Wigner form for the resonant effect at $M = M_0$:

$$|\psi(M)|^2 \sim \frac{1}{(M - M_0)^2 + \Gamma^2/4} \tag{9.36}$$

DETERMINING SPINS

The spins of some of the stable elementary particles can in principle be obtained, for example, through a Stern-Gerlach type of experiment. Thus, from the splitting of a beam of particles in a magnetic field, we can deduce that the electron as well as the proton have spin angular momentum $\frac{1}{2}$. The neutrino, as we have seen earlier, was postulated to have spin $\frac{1}{2}$ in order for angular momentum to be conserved in β-decay. The spin of the photon can, of course, be determined from the classical properties of electromagnetic waves. We know that the electromagnetic field is described by a vector potential, which implies that the photon is a vector particle with spin-1, and that its wave function is proportional to its polarization vector $\vec{\epsilon}$. Normally, a spin-1 state has three angular momentum projections corresponding to $s_z = 1, 0, -1$. However, propagating electromagnetic waves are transverse, which means that the physical photon has no longitudinal degree of freedom. This is reflected in the fact that the electric ($\vec{E}$) and magnetic ($\vec{B}$) fields, and the polarization vector for the photon, are transverse to the direction of propagation ($\hat{k} = \vec{k}/k$):

$$\vec{E} = \vec{\epsilon} E_0 \, e^{i(\vec{k}\cdot\vec{r} - \omega t)}, \quad \vec{B} = \hat{k} \times \vec{E} \tag{9.37}$$

and obey the conditions

$$\vec{k} \cdot \vec{E} = 0, \quad \vec{k} \cdot \vec{B} = 0, \quad \vec{k} \cdot \vec{\epsilon} = 0 \tag{9.38}$$

These properties of the photon are related to its being massless, which is a feature that can be accommodated through the invariance of Maxwell's equations under what are traditionally referred to as "gauge" transformations of the electromagnetic potentials. (We discuss this in more detail in Chapter XIII.)

The spin of the π^0 meson can be deduced from the fact that it decays into two photons. In the rest frame of the π^0, the two photons must be emitted back-to-back with equal and opposite momenta (see Fig. 9.4). Let $\vec{k}$ denote the relative momentum of the two photons and let $\vec{\epsilon}_1$ and $\vec{\epsilon}_2$ represent their polarization vectors. The final-state in this decay consists of two identical bosons and, consequently, the final-state wave function, which corresponds to the product of the photon wave functions, must be symmetric under the exchange of the photons. As we have indicated, the photon wave function is proportional to its polarization vector. Now, the only quantities that we can construct from the vectors $\vec{k}$, $\vec{\epsilon}_1$, and $\vec{\epsilon}_2$ that are linear in both $\vec{\epsilon}_1$ and $\vec{\epsilon}_2$, and are symmetric under the exchange of the photon variables, are

$$\vec{k} \times (\vec{\epsilon}_1 \times \vec{\epsilon}_2), \quad \vec{k} \cdot (\vec{\epsilon}_1 \times \vec{\epsilon}_2), \quad \text{and} \quad \vec{\epsilon}_1 \cdot \vec{\epsilon}_2 \tag{9.39}$$

The first vanishes because the polarization vectors are transverse, namely,

$$\vec{k} \times (\vec{\epsilon}_1 \times \vec{\epsilon}_2) = (\vec{k} \cdot \vec{\epsilon}_2)\vec{\epsilon}_1 - (\vec{k} \cdot \vec{\epsilon}_1)\vec{\epsilon}_2 = 0$$

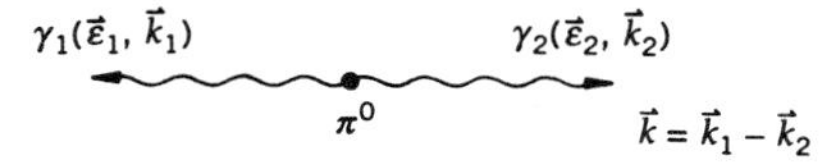

$\gamma_1(\vec{\epsilon}_1, \vec{k}_1)$ $\gamma_2(\vec{\epsilon}_2, \vec{k}_2)$

π^0 $\vec{k} = \vec{k}_1 - \vec{k}_2$

Figure 9.4 The decay of a π^0 into two photons, as viewed in its own rest frame.

Thus, we see that the simplest combinations satisfying all the symmetry properties are the scalar products in Eq. 9.39. Experimentally, one can measure the planes of polarization for the two photons, and they are found to be orthogonal. Thus, we conclude that the final-state wave function must be proportional to the scalar product

$$\vec{k} \cdot (\vec{\epsilon}_1 \times \vec{\epsilon}_2) \tag{9.40}$$

For the decay of the π^0 to take place (that is, for the transition amplitude to be nonzero), the pion wave function must have a component corresponding to the final-state wave function of the two photons. Therefore, we conclude that the pion wave function must also be a scalar under rotations. Namely, the pion must have spin zero.

We can similarly obtain the spin of the K^0 meson from the decay process $K^0 \longrightarrow 2\pi^0$. Once again, in the rest frame of the K^0 meson, the two π mesons must have equal and opposite momentum (see Fig. 9.5). The final state consists of two spin-zero particles, and consequently the total angular momentum of the final state (and the spin of the K^0) is the same as the relative orbital angular momentum of the two pions. Since the two π^0 mesons are identical bosons, the final-state wave function must be symmetric under the exchange of the two particles. If we let ℓ denote the orbital angular momentum of the final state, then the angular part of the wave function of the final state will be proportional to the spherical harmonics $Y_{\ell,m}(\theta, \phi)$. We have noted previously in Chapter III that, under the exchange of the two particles, these wave functions behave as follows:*

$$Y_{\ell,m}(\theta, \phi) \longrightarrow (-1)^{\ell} Y_{\ell,m}(\theta, \phi) \tag{9.41}$$

It follows, therefore, that for the final state to be symmetric under the exchange of the two particles, ℓ can only take on even values. Thus, we conclude that the spin of the K meson must be even, namely 0, 2, or 4, etc.

We can also argue, as we did in the case of the π^0, that, since pions are spinless, in the rest frame of the K meson, the only vector available for describing the final state is the relative momentum of the two π^0 mesons, or $\vec{k}$. Under an exchange of pions, the $\vec{k}$ vector changes sign, and we can conclude that the simplest wave function that can be constructed from this

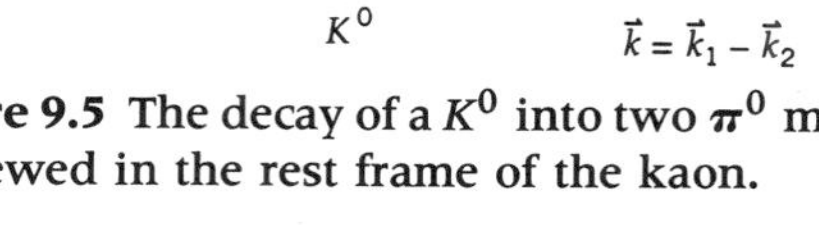

$\pi_1^0(\vec{k}_1)$ $\pi_2^0(\vec{k}_2)$

K^0 $\vec{k} = \vec{k}_1 - \vec{k}_2$

Figure 9.5 The decay of a K^0 into two π^0 mesons, as viewed in the rest frame of the kaon.

*Exchange of the two particles, in this case, corresponds to $\vec{r} \rightarrow -\vec{r}$, which is the same as reflection or the parity transformation (see Chapter XI).

vector, and still satisfy all the symmetry properties of the final state, is a scalar (e.g., any function of $\vec{k} \cdot \vec{k}$). Consequently, we deduce that the wave function for the K^0 meson must be a scalar, and that the K^0 is a spin-zero particle. In fact, the decay characteristics of the K mesons exclude the higher spin assignments of $J = 2, 4$, etc. That is, the angular distribution of the π^0 mesons in the K^0 rest frame shows no evidence for the presence of other than $\ell = 0$ contributions to the decay.

Let us next discuss how the spins of some of the baryons can be obtained. For example, let us analyze the following collision of a high-energy π^- with a proton at rest:

$$\pi^- + p \longrightarrow K^0 + Y_1^{*0} \tag{9.42}$$

where the hyperon (strange baryon) Y_1^{*0} subsequently decays through

$$Y_1^{*0} \longrightarrow \pi^0 + \Lambda^0 \tag{9.43}$$

Let us use the beam direction, namely the direction of the incident pion in the laboratory, as the axis of quantization for angular momentum. The component of the pion's orbital angular momentum along its direction of motion will clearly vanish ($\vec{L}_\pi \sim \vec{r} \times \vec{p}_\pi$, which is perpendicular to $\vec{p}_\pi$). Furthermore, the pion is spinless. For the initial state, therefore, the projection of the total angular momentum along the beam direction is given by the projection of the intrinsic spin of the proton, s_z, namely:

$$j_z = s_z(p) = \pm\tfrac{1}{2} \tag{9.44}$$

Now, let us restrict our study to events where the K^0 and Y_1^{*0} that are produced only along the beam axis, namely, close to the forward–backward direction in the center of mass of Reaction (9.42). Again, along this axis, the relative orbital momentum of the K^0–Y_1^{*0} system vanishes, and because the spin of the K^0 is zero, from the conservation of angular momentum we conclude that

$$s_z(Y_1^{*0}) = s_z(p) = \pm\tfrac{1}{2} \tag{9.45}$$

This allows us to conclude that the spin of the Y_1^{*0} hyperon is 1/2 or a higher half-integer value. To obtain the true spin, we would have to analyze the decay in Eq. 9.43 in substantial detail. Measuring the complexity of the angular decay distribution in the rest frame of the Y_1^{*0} would then determine its spin value, which is now known to be 3/2.

Similar analyses of other high-energy production processes, and the subsequent decays of the particles, have provided us with the spins of many hadrons. What has been found is that some of the particles can be related to each other and correlated into groups having similar quantum numbers.

VIOLATION OF QUANTUM NUMBERS

As we have seen, all quantum numbers appear to be conserved in strong processes. However, they may be violated in electromagnetic and weak interactions. We now discuss this with a few illustrative examples.

Weak Interactions

There are three kinds of weak processes in nature, which can be classified as follows: (a) hadronic decays, where only hadrons are present; (b) semileptonic processes, where both hadrons and leptons are present; and, finally, (c) leptonic processes, where only leptons are present. For example, the decays

$$\begin{aligned} \Lambda^0 &\to \pi^- + p \\ n &\to p + e^- + \overline{\nu}_e \\ \mu^- &\to e^- + \overline{\nu}_e + \nu_\mu \end{aligned} \tag{9.46}$$

represent the three different kinds of weak processes. Since most of the strong quantum numbers are not defined for leptons, it is not meaningful to discuss their violation in leptonic processes. Furthermore, even in the case of the semileptonic processes, we can only speak about the conservation or the violation of quantum numbers between the initial and the final hadronic states. Keeping this in mind, let us now examine some typical reactions.

HADRONIC WEAK-DECAYS

Consider the following decays of hadrons into other hadrons:

$$\begin{array}{llllll} \Lambda^0 & \longrightarrow & \pi^- & + & p & \\ I_3 = 0 & & I_3 = -1 & & I_3 = 1/2 & \\ S = -1 & & S = 0 & & S = 0 & \\ \Sigma^+ & \longrightarrow & p & + & \pi^0 & \\ I_3 = 1 & & I_3 = 1/2 & & I_3 = 0 & \\ S = -1 & & S = 0 & & S = 0 & \\ K^0 & \longrightarrow & \pi^+ & + & \pi^- & \\ I_3 = -1/2 & & I_3 = 1 & & I_3 = -1 & \\ S = 1 & & S = 0 & & S = 0 & \\ \Xi^- & \longrightarrow & \Lambda^0 & + & \pi^- & \\ I_3 = -1/2 & & I_3 = 0 & & I_3 = -1 & \\ S = -2 & & S = -1 & & S = 0 & \end{array} \tag{9.47}$$

We see experimentally that both isospin and strangeness are violated in these weak processes, and that a selection rule for such violations can be given by

$$|\Delta I_3| = 1/2 \qquad |\Delta S| = 1 \tag{9.48}$$

Also, we should add that, while both $\Delta I = 1/2$ and $\Delta I = 3/2$ appear to hold in these processes, experimentally, $\Delta I = 3/2$ transitions are found to be highly suppressed, and processes involving $|\Delta S| = 2$ are exceptionally rare.

Semileptonic Processes

Once again, we examine only a few examples to bring out the essential features of these decays. We emphasize, again, that we are considering only changes in the quantum numbers of the initial and the final-state hadrons.

$$\begin{array}{lllll}
n & \longrightarrow & p & + & e^- + \overline{\nu}_e \\
I_3 = -1/2 & & I_3 = 1/2 & & \\
S = 0 & & S = 0 & & \\
\\
\pi^- & \longrightarrow & \mu^- + \overline{\nu}_\mu & & \\
I_3 = -1 & & & & \\
S = 0 & & & & \\
\\
\pi^+ & \longrightarrow & \pi^0 & + & e^+ + \nu_e \\
I_3 = 1 & & I_3 = 0 & & \\
S = 0 & & S = 0 & & \\
\\
K^+ & \longrightarrow & \mu^+ + \nu_\mu & & \\
I_3 = 1/2 & & & & \\
S = 1 & & & & \\
\\
K^+ & \longrightarrow & \pi^0 & + & \mu^+ + \nu_\mu \\
I_3 = 1/2 & & I_3 = 0 & & \\
S = 1 & & S = 0 & & \\
\\
\Lambda^0 & \longrightarrow & p & + & e^- + \overline{\nu}_e \\
I_3 = 0 & & I_3 = 1/2 & & \\
S = -1 & & S = 0 & & \\
\\
\Sigma^- & \longrightarrow & n & + & e^- + \overline{\nu}_e \\
I_3 = -1 & & I_3 = -1/2 & & \\
S = -1 & & S = 0 & &
\end{array} \tag{9.49}$$

Thus, we see that the semileptonic decays can be classified into two types. The first kind has no change in the strangeness of the hadrons. These processes are known as *strangeness-preserving decays*, and are characterized, of course, by $|\Delta S| = 0$. In such processes we see that $|\Delta I_3| = 1$. Thus, the strangeness-conserving semileptonic processes satisfy

$$|\Delta S| = 0, \quad |\Delta I_3| = 1, \quad \Delta I = 1 \tag{9.50}$$

The second class of semileptonic decays do not conserve strangeness. Consequently, these decays are also known as *strangeness-changing processes,* and for these we find

$$|\Delta S| = 1, \quad |\Delta I_3| = 1/2, \quad \Delta I = 1/2 \quad \text{or} \quad 3/2 \tag{9.51}$$

Again, it is observed that the processes with $\Delta I = 3/2$ are highly suppressed, as are $|\Delta S| = 2$ transitions.

Electromagnetic Processes

We will also consider a few samples of electromagnetic decays, so as to bring out the essential features of these processes. Again, since the photon cannot always be assigned strong quantum numbers, the meaningful quantity to analyze is the change in the quantum numbers of the hadrons:

$$\begin{array}{lcl} \pi^0 & \longrightarrow & \gamma + \gamma \\ I_3 = 0 & & \\ S = 0 & & \\ \eta^0 & \longrightarrow & \gamma + \gamma \\ I_3 = 0 & & \\ S = 0 & & \\ \Sigma^0 & \longrightarrow & \Lambda^0 + \gamma \\ I_3 = 0 & & I_3 = 0 \\ S = -1 & & S = -1 \end{array} \tag{9.52}$$

This shows that strangeness is conserved in electromagnetic processes, while isospin is not. In fact, these processes are characterized by

$$|\Delta S| = 0, \quad |\Delta I_3| = 0, \quad \Delta I = 1 \quad \text{and} \quad 0 \tag{9.53}$$

In the following chapters we attempt to incorporate all these results into a theoretical framework, which is referred to as the Standard Model of particle physics.

Problems

IX.1 What quantum numbers, if any, are violated in the following reactions? Are the interactions strong, weak, electromagnetic, or none of the above?

(a) $\Omega^- \longrightarrow \Xi^0 + \pi^-$

(b) $\Sigma^+ \longrightarrow \pi^+ + \pi^0$

(c) $n \longrightarrow p + \pi^-$
(d) $\pi^0 \longrightarrow \mu^+ + e^- + \overline{\nu}_e$
(e) $K^0 \longrightarrow K^+ + e^- + \overline{\nu}_e$
(f) $\Lambda^0 \longrightarrow p + e^-$

IX.2 What quantum numbers, if any, are violated in the following processes. Would the reaction be strong, electromagnetic, weak, or unusually suppressed? Explain.
(a) $\Lambda^0 \longrightarrow p + e^- + \overline{\nu}_e$
(b) $K^- + p \longrightarrow K^+ + \Xi^-$
(c) $K^+ + p \longrightarrow K^+ + \Sigma^+ + \overline{K}^0$
(d) $p + p \longrightarrow K^+ + K^+ + n + n$
(e) $\Sigma^+(1385) \longrightarrow \Lambda^0 + \pi^+$
(f) $\overline{p} + n \longrightarrow \pi^- + \pi^0$ (See *CRC Handbook* for particle properties.)

IX.3 A π^0 meson with a momentum of 135 GeV/*c* decays into two photons. If the mean life of a π^0 is 8.5×10^{-17} sec, calculate to 10% accuracy how far the high-energy π^0 will travel prior to decay? What will be the approximate minimum value of the opening angle of the two photons in the laboratory?

IX.4 We will see in Chapter XIII that hadrons are composed of constituents known as quarks, and that mesons can be represented as quark–antiquark systems and baryons as three-quark systems. All quarks have baryon number $\frac{1}{3}$, and their other quantum numbers are listed in Table 9.5. Particles corresponding to all but the t-quark have now been detected. The antiquarks all have quantum numbers opposite in sign to the quarks. The isotopic spin of quarks can be inferred from the generalized Gell-Mann–Nishijima relation of Eq. 9.26.

Table 9.5 Properties of the quarks

Quark	*Symbol*	*Effective Mass (GeV/c²)*[a]	*Electric Charge (e)*	*"Flavor" Quantum Numbers*			
				Strangeness	*Charm*	*Bottom*	*Top*
Up	u	Small	$\frac{2}{3}$	0	0	0	0
Down	d	Small	$-\frac{1}{3}$	0	0	0	0
Strange	s	0.2	$-\frac{1}{3}$	−1	0	0	0
Charm	c	1.5	$\frac{2}{3}$	0	1	0	0
Bottom	b	4.7	$-\frac{1}{3}$	0	0	−1	0
Top[b]	t	>90	$\frac{2}{3}$	0	0	0	1

[a] Since quarks have not been observed as free particles, their masses cannot always be clearly defined.
[b] Still to be discovered.

The quark system uds can exist in more than one isospin state. What is the value of I_3 for this combination of quarks? What are the possible values of total I-spin for the uds states? Can you identify them with any known particles? (See, e.g., *CRC Handbook*)

IX.5 What is the baryon number, hypercharge, and isotopic spin of the following quark systems: (a) $u\bar{s}$, (b) $c\bar{d}$, (c) uud, (d) ddc, and (e) $s\bar{s}$. Using the *CRC Handbook*, can you identify these states with any known particles?

IX.6 Consider the following decays:

(a) $N^+(1535) \longrightarrow p + \eta^0$

(b) $\Sigma^+(1189) \longrightarrow p + \pi^0$

(c) $\rho^0(770) \longrightarrow \pi^0 + \gamma$

From the properties listed in the CRC tables, discuss the interactions involved in each case. What quantum numbers are violated? What are the possible values of orbital angular momentum in the final states?

Suggested Readings

Frauenfelder, H., and E. M. Henley. 1991. *Subatomic Physics*. Englewood Cliffs, N.J.: Prentice-Hall.

Griffiths, D. 1987. *Introduction to Elementary Particles*. New York: Wiley.

Williams, W. S. C. 1991. *Nuclear and Particle Physics*. London/New York: Oxford Univ. Press.

Chapter X

SYMMETRIES

INTRODUCTORY REMARKS

As we saw in the previous chapter, although several kinds of quantum numbers appeared to be conserved in strong high-energy processes, some of them were violated in weak and electromagnetic interactions. Such specific observations must reflect the inherent character of the underlying forces. Consequently, understanding the origin of conservation principles, and under what conditions they can be violated, would appear to be an important issue that could help us formulate a quantitative description of particle interactions. We will therefore first address the question of how conservation laws arise in physical theories. As we will see shortly, the surprisingly simple answer is that whenever there is an underlying symmetry in a physical system, namely if our system does not change when we make a change in some coordinate or other dynamical variable, then we can define a conserved "charge" (quantum number) associated with that symmetry. Conversely, if there is a conserved quantity associated with a physical system, then there exists an underlying invariance or symmetry principle responsible for its conservation. This observation, known as Noether's Theorem (after Emmy Noether), gives rise to powerful restrictions on the structure of physical theories. In this chapter, we discuss issues related to symmetries in physical systems.

SYMMETRIES IN THE LAGRANGIAN FORMALISM

In simple terms, any set of transformations that leaves the equations of motion of a system unchanged or invariant, defines what is known as a symmetry of that physical system. Symmetries can be discussed using either the Lagrangian or the Hamiltonian formalism, for both classical as well as quantum theories. In this section, we begin the discussion using the Lagrangian framework, which is most appropriate for studying relativistic systems—our ultimate goal.

Consider first an isolated nonrelativistic physical system consisting of two particles interacting through a potential that depends on the relative separation of the particles. The total kinetic energy and the potential energy for this system is given by

$$\begin{aligned} T &= \frac{1}{2}m_1\dot{\vec{r}}_1^{\,2} + \frac{1}{2}m_2\dot{\vec{r}}_2^{\,2} \\ V &= V(\vec{r}_1 - \vec{r}_2) \end{aligned} \tag{10.1}$$

where m_1, m_2 are the masses of the particles, and $\vec{r}_1, \vec{r}_2$ their coordinates measured with respect to some given origin. The equations of motion (Newton's equations), or the "dynamical equations," for this system take the form

$$\begin{aligned} m_1\ddot{\vec{r}}_1 &= -\vec{\nabla}_1 V(\vec{r}_1 - \vec{r}_2) = -\frac{\partial}{\partial \vec{r}_1}V(\vec{r}_1 - \vec{r}_2) \\ m_2\ddot{\vec{r}}_2 &= -\vec{\nabla}_2 V(\vec{r}_1 - \vec{r}_2) = -\frac{\partial}{\partial \vec{r}_2}V(\vec{r}_1 - \vec{r}_2) \end{aligned} \tag{10.2}$$

where by $(\partial/\partial\vec{r}_i)V(\vec{r}_1 - \vec{r}_2)$ we mean

$$\hat{x}\frac{\partial}{\partial x_i}V + \hat{y}\frac{\partial}{\partial y_i}V + \hat{z}\frac{\partial}{\partial z_i}V$$

with $i = 1, 2$, and $\hat{x}$, $\hat{y}$, and $\hat{z}$ are unit vectors along the x, y, and z axes of our fixed coordinate system.

Now, if we translate the origin of the coordinate system by a constant vector $-\vec{a}$, namely if we transform the coordinates as follows:

$$\begin{aligned} \vec{r}_1 &\to \vec{r}_1{}' = \vec{r}_1 + \vec{a} \\ \vec{r}_2 &\to \vec{r}_2{}' = \vec{r}_2 + \vec{a} \end{aligned} \tag{10.3}$$

then the dynamical equations for the system of particles in Eq. 10.2 do not change. This is simply a consequence of the fact that

$$V(\vec{r}_1 - \vec{r}_2) \rightarrow V(\vec{r}_1 + \vec{a} - \vec{r}_2 - \vec{a}) = V(\vec{r}_1 - \vec{r}_2) \tag{10.4}$$

Thus, a translation of the origin of the coordinate system defines a symmetry of the two-particle system, and we say that the physical system is invariant under spatial translations, namely, our physical system is not sensitive to any particular choice for the origin of our coordinate system. The consequence of this symmetry is quite interesting. We note from the form of the potential that the total force acting on the system vanishes, namely,

$$\vec{F}_{\mathrm{TOT}} = \vec{F}_1 + \vec{F}_2 = -\vec{\nabla}_1 V(\vec{r}_1 - \vec{r}_2) - \vec{\nabla}_2 V(\vec{r}_1 - \vec{r}_2) = 0 \tag{10.5}$$

(Equation 10.5 follows from the fact that $(\partial V/\partial \vec{r}_1) = -(\partial V/\partial \vec{r}_2)$.) Consequently, for the total momentum of the system, we obtain

$$\frac{d\vec{P}_{\mathrm{TOT}}}{dt} = \vec{F}_{\mathrm{TOT}} = 0 \tag{10.6}$$

In other words, the total momentum associated with the system is conserved—that is, it is a constant of the motion, and independent of time.

This kind of result may seem to be purely accidental, and correct only for our simple two-body example. In fact, however, it can be shown that for any symmetry associated with a physical system, there exists a conserved quantity. To see this, let us first rewrite the dynamical equations in Eq. 10.2 as follows:

$$\begin{aligned}
\frac{d}{dt}\frac{\partial T}{\partial \dot{\vec{r}}_1} &= -\frac{\partial V}{\partial \vec{r}_1} \\
\frac{d}{dt}\frac{\partial T}{\partial \dot{\vec{r}}_2} &= -\frac{\partial V}{\partial \vec{r}_2}
\end{aligned} \tag{10.7}$$

where the kinetic energy is defined in Eq. 10.1. (Again, we are using our shorthand notation in Eq. 10.7, and each equation represents, in fact, three separate relations, namely: $(d/dt)(\partial T/\partial \dot{q}) = -(\partial V/\partial q)$, for $q = x_i, y_i$, and z_i, and $i = 1, 2$). Furthermore, if we define the function

$$L = T - V \tag{10.8}$$

then, recognizing that the coordinate and the velocity of a particle are really independent variables, we can write the entire content of the dynamical equations in (10.2) or (10.7) schematically as

$$\frac{d}{dt}\frac{\partial L}{\partial \dot{\vec{r}}_i} - \frac{\partial L}{\partial \vec{r}_i} = 0 \qquad i = 1, 2 \tag{10.9}$$

The quantity $L(\vec{r}_i, \dot{\vec{r}}_i)$ is known as the Lagrangian of the system, and we note that by construction we have that

$$\frac{\partial L}{\partial \dot{\vec{r}}_i} = \frac{\partial T}{\partial \dot{\vec{r}}_i} = m_i \dot{\vec{r}}_i = \vec{p}_i \tag{10.10}$$

Consequently, the Hamiltonian ($H = T + V = 2T - L$) can now be obtained from the Lagrangian through the use of Eqs. 10.1 and 10.10, and written as

$$H = \sum_{i=1}^{2} [\vec{p}_i \cdot \dot{\vec{r}}_i - L(\vec{r}_i, \dot{\vec{r}}_i)] \tag{10.11}$$

All the preceding considerations carry over quite naturally to more complicated systems, and the Lagrangian for a general system with n-degrees of freedom (namely, n-coordinates and n-velocities) can be represented as

$$L = L(q_i, \dot{q}_i) \qquad i = 1, 2, \ldots, n \tag{10.12}$$

The momenta associated with, or "conjugate" to, the coordinates q_i can be defined as in Eq. 10.10 to be

$$p_i = \frac{\partial L}{\partial \dot{q}_i} \qquad i = 1, 2, \ldots, n \tag{10.13}$$

and the general dynamical equations of motion can be written in line with Eq. 10.9 as

$$\frac{d}{dt}\frac{\partial L}{\partial \dot{q}_i} - \frac{\partial L}{\partial q_i} = 0$$

or

$$\frac{dp_i}{dt} = \frac{\partial L}{\partial q_i} \qquad i = 1, 2, \ldots, n \tag{10.14}$$

Let us now suppose that the Lagrangian for a given physical system is independent of some particular coordinate, say q_m. Then

$$\frac{\partial L}{\partial q_m} = 0, \qquad \text{for the specified } m \tag{10.15}$$

As a result, the dynamical equation for $i = m$ in Eq. 10.14 will give

$$\frac{d}{dt} p_m = 0 \tag{10.16}$$

In other words, if the Lagrangian for a physical system does not depend explicitly on a given coordinate, then the corresponding conjugate momentum is conserved. Moreover, if a Lagrangian does not depend on some particular coordinate, it must be invariant under translations (redefinitions) of this coordinate. This therefore brings to focus the connection between the invariance of a theory and a corresponding conserved quantity.

As example, we saw in Chapter I that if we rewrite the two-body problem in terms of the relative coordinate $\vec{r} = \vec{r}_1 - \vec{r}_2$ and the center of mass coordinate $\vec{R}_{CM}$, then the potential energy, and therefore the Lagrangian, will be independent of $\vec{R}_{CM}$. Consequently, the corresponding momentum $\vec{P}_{CM}$, which we have seen to correspond to the total momentum of the system, will be a constant—a result we already recognized in Eq. 10.6.

As a second simple demonstration of these ideas, let us consider the motion of a free rotor. Here, in the absence of any force, the system has only kinetic energy, and therefore we can write

$$L = T = \frac{1}{2} I \dot{\theta}^2 \tag{10.17}$$

where I denotes the moment of inertia of the rotor and $\dot{\theta}$ its angular velocity. This Lagrangian is independent of the azimuthal coordinate θ of the rotor, and correspondingly we conclude (following our earlier argument) that

$$p_\theta = \frac{\partial L}{\partial \dot{\theta}} = I \dot{\theta} = \text{constant} \tag{10.18}$$

Thus, we see that the lack of dependence of the rotor Lagrangian on θ, which gives rise to a rotational invariance for the system, leads to a constant value for the angular momentum of the system. Such conclusions, as we have emphasized, are, in fact, quite general, and we summarize in Table 10.1 several common transformations and the associated quantities that are conserved when physical systems are invariant under these transformations.

The converse argument also holds, namely that for every quantity conserved in a physical system there exists an underlying invariance principle.

Table 10.1 Invariance of a system under a transformation and the corresponding conserved quantity

Transformation	*Conserved Quantity for System*
Space translation	Momentum
Time translation	Energy
Spatial rotation	Angular momentum
Rotation in isotopic-spin space	Isotopic spin

This, however, is much easier to see using the Hamiltonian formalism, to which we now turn.

SYMMETRIES IN THE HAMILTONIAN FORMALISM

The Hamiltonian formalism of classical mechanics goes over naturally to quantum mechanics, and therefore the discussion of symmetries in the context of the Hamiltonian formalism can be quite illuminating. Let us recall that a Hamiltonian $H(q_i, p_i)$ for a system with n-degrees of freedom is a function of n-coordinates and n-momenta. The equations of motion now comprise the first-order Hamilton relations, given by

$$\begin{aligned}\frac{dq_i}{dt} &= \dot{q}_i = \frac{\partial H}{\partial p_i}\\ \frac{dp_i}{dt} &= \dot{p}_i = -\frac{\partial H}{\partial q_i} \qquad i = 1, 2, \ldots, n\end{aligned} \tag{10.19}$$

Let us now introduce the bracket notation (Poisson bracket) between the independent coordinates and momenta. In general, the Poisson bracket for any two functions of q_i and p_i can be defined in terms of their partial derivatives with respect to these variables, as follows:

$$\{F(q_i, p_i), G(q_i, p_i)\} = \sum_{i=1}^{n}\left(\frac{\partial F}{\partial q_i}\frac{\partial G}{\partial p_i} - \frac{\partial F}{\partial p_i}\frac{\partial G}{\partial q_i}\right) = -\{G(q_i, p_i), F(q_i, p_i)\} \tag{10.20}$$

Thus, for the Poisson brackets of the coordinates and momenta, we obtain

$$\begin{aligned}\{q_i, q_j\} &= 0\\ \{p_i, p_j\} &= 0\\ \{q_i, p_j\} &= -\{p_j, q_i\} = \delta_{ij}\end{aligned} \tag{10.21}$$

where δ_{ij} is the Kronecker "delta," which equals unity for $i = j$ and zero for $i \neq j$. (It is worth pointing out that, in the transition to quantum mechanics, the Poisson brackets go over to commutators in a straightforward way.) Using the brackets, we note that

$$\begin{aligned}\{q_i, H\} &= \sum_j\left(\frac{\partial q_i}{\partial q_j}\frac{\partial H}{\partial p_j} - \frac{\partial q_i}{\partial p_j}\frac{\partial H}{\partial q_j}\right)\\ &= \sum_j \delta_{ij}\frac{\partial H}{\partial p_j} = \frac{\partial H}{\partial p_i}\end{aligned}$$

and

$$\{p_i, H\} = \sum_j \left(\frac{\partial p_i}{\partial q_j} \frac{\partial H}{\partial p_j} - \frac{\partial p_i}{\partial p_j} \frac{\partial H}{\partial q_j} \right)$$

$$= -\sum_j \delta_{ij} \frac{\partial H}{\partial q_j} = -\frac{\partial H}{\partial q_i} \tag{10.22}$$

Thus the dynamical equations in (10.19) can also be written as

$$\begin{aligned} \dot{q}_i &= \{q_i, H\} \\ \dot{p}_i &= \{p_i, H\} \end{aligned} \tag{10.23}$$

(To obtain the relations in Eq. 10.22 we have used the fact that the q_i and p_i are independent variables, and therefore the partial derivatives $\partial q_i / \partial q_j$ and $\partial p_i / \partial p_j$ vanish for $i \neq j$, and all $\partial q_i / \partial p_j$ and $\partial p_i / \partial q_j$ vanish as well.) In fact, if any physical observable $\omega(q_i, p_i)$ does not depend explicitly on time, then it follows from the chain rule of differentiation and use of Eq. 10.19 that its time evolution will be given by

$$\frac{d\omega(q_i, p_i)}{dt} = \{\omega(q_i, p_i), H\} \tag{10.24}$$

INFINITESIMAL TRANSLATIONS

Let us next consider an infinitesimal translation of the coordinates of the form

$$\begin{aligned} q_i \to q_i' &= q_i + \epsilon_i \\ p_i \to p_i' &= p_i \end{aligned} \tag{10.25}$$

where ϵ_i are infinitesimal constant (arbitrary) parameters defining the translation. Equivalently, we can write the infinitesimal changes of the dynamical variables as

$$\begin{aligned} \delta_\epsilon q_i &= q_i' - q_i = \epsilon_i \\ \delta_\epsilon p_i &= p_i' - p_i = 0 \end{aligned} \tag{10.26}$$

We note that if we define a function $g(q_i, p_i)$ as

$$g = \sum_j \epsilon_j p_j \tag{10.27}$$

we then obtain

$$\begin{aligned}\frac{\partial g}{\partial q_i} &= \frac{\partial(\sum_j \epsilon_j p_j)}{\partial q_i} = 0\\ \frac{\partial g}{\partial p_i} &= \frac{\partial(\sum_j \epsilon_j p_j)}{\partial p_i} = \sum_j \epsilon_j \delta_{ij} = \epsilon_i\end{aligned} \tag{10.28}$$

Thus, from the definition of the Poisson brackets in Eq. 10.20, we note that we can write

$$\begin{aligned}\{q_i, g\} &= \sum_j \left(\frac{\partial q_i}{\partial q_j}\frac{\partial g}{\partial p_j} - \frac{\partial q_i}{\partial p_j}\frac{\partial g}{\partial q_j}\right)\\ &= \sum_j \delta_{ij}\epsilon_j = \epsilon_i = \delta_\epsilon q_i\\ \{p_i, g\} &= \sum_j \left(\frac{\partial p_i}{\partial q_j}\frac{\partial g}{\partial p_j} - \frac{\partial p_i}{\partial p_j}\frac{\partial g}{\partial q_j}\right) = 0 = \delta_\epsilon p_i\end{aligned} \tag{10.29}$$

where we have used the results of Eq. 10.26 to relate the Poisson brackets to the infinitesimal changes of the dynamical variables. Now, using Eq. 10.20 or Eq. 10.21, we can also verify that the original and the transformed variables satisfy the same Poisson-bracket relations, namely,

$$\begin{aligned}\{q'_i, q'_j\} &= 0 = \{p'_i, p'_j\}\\ \{q'_i, p'_j\} &= \delta_{ij}\end{aligned} \tag{10.30}$$

In other words, the infinitesimal translations in Eq. 10.25 preserve the canonical Poisson-bracket structure, and correspondingly such transformations are known as canonical transformations.

Because the Hamiltonian is a function of the coordinates and momenta, its change under the transformation of Eq. 10.25 can be calculated using the chain rule of differentiation, as follows:

$$\begin{aligned}\delta_\epsilon H &= \sum_i \left(\frac{\partial H}{\partial q_i}\delta_\epsilon q_i + \frac{\partial H}{\partial p_i}\delta_\epsilon p_i\right)\\ &= \sum_i \frac{\partial H}{\partial q_i}\epsilon_i = \sum_i \left(\frac{\partial H}{\partial q_i}\frac{\partial g}{\partial p_i} - \frac{\partial H}{\partial p_i}\frac{\partial g}{\partial q_i}\right)\\ &= \{H, g\}\end{aligned} \tag{10.31}$$

where, in the middle step, we used the results of Eq. 10.28. We note that if the Hamiltonian does not change under our infinitesimal translation, that is, if

$$\delta_\epsilon H = \{H, g\} = 0 \tag{10.32}$$

then we can write

$$H(q_i', p_i') = H(q_i, p_i) \tag{10.33}$$

Furthermore, since the Poisson brackets between the q_i and the p_i do not change, it follows that the transformed dynamical equations coincide with the original equations of (10.23). That is,

$$\begin{aligned} \dot{q}_i' &= \{q_i', H(q_j', p_j')\} = \{q_i, H(q_j, p_j)\} \\ \dot{p}_i' &= \{p_i', H(q_j', p_j')\} = \{p_i, H(q_j, p_j)\} \end{aligned} \tag{10.34}$$

Namely, Eqs. 10.34 represent the same motion as Eqs. 10.23. This, in fact, reflects the very general result that when H does not change under an infinitesimal transformation, then such transformations define a symmetry of the dynamical equations of the system, or simply a symmetry of the physical system. In the present example, translations are a symmetry of the system if the Hamiltonian does not change under translations.

We see, from Eqs. 10.29 and 10.31, that the changes in q_i, p_i, and H under the infinitesimal translation of Eq. 10.25 can be obtained from their Poisson brackets with g. In fact, following the procedure in Eq. 10.31, it can be shown that the change in any observable can be obtained from its Poisson bracket with g. Thus, we can think of g as generating the infinitesimal translation, and, as a result, g is termed the *generator* of the transformations. We see from Eqs. 10.24, 10.32, and 10.27 that an infinitesimal translation will be a symmetry of a physical system if

$$\frac{dg}{dt} = \{g, H\} = 0$$

or

$$\frac{dp_i}{dt} = \{p_i, H\} = 0 \tag{10.35}$$

In other words, if translations are a symmetry of a system, then the momenta will be conserved, and, conversely, if the momenta are conserved, then translations will be a symmetry of the physical system. This is, of course, the same result that we obtained in Eq. 10.16 using the Lagrangian formalism.

INFINITESIMAL ROTATIONS

Let us next consider infinitesimal rotations for a two-dimensional system. We know that finite spatial rotations by an angle θ about the z-axis can be written as the set of transformations:

$$\begin{aligned} x' &= x\cos\theta - y\sin\theta \\ y' &= x\sin\theta + y\cos\theta \end{aligned} \tag{10.36}$$

For infinitesimal θ, we can replace $\cos\theta$ by $1 - \theta^2/2$, and $\sin\theta$ by θ, and, to first order in θ, we can write the transformations as

$$\begin{aligned} x' &= x - \theta y \\ y' &= \theta x + y \end{aligned} \tag{10.37}$$

Using matrix notation, this can be rewritten as follows:

$$\begin{pmatrix} x' \\ y' \end{pmatrix} = \begin{pmatrix} 1 & -\theta \\ \theta & 1 \end{pmatrix} \begin{pmatrix} x \\ y \end{pmatrix} \tag{10.38}$$

Defining $\delta_\theta x$ and $\delta_\theta y$ as the changes in the x and y coordinates, we can now write

$$\delta_\theta \begin{pmatrix} x \\ y \end{pmatrix} = \begin{pmatrix} x' - x \\ y' - y \end{pmatrix} = \theta \begin{pmatrix} -y \\ x \end{pmatrix} = \begin{pmatrix} 0 & -\theta \\ \theta & 0 \end{pmatrix} \begin{pmatrix} x \\ y \end{pmatrix} \tag{10.39}$$

Now, in terms of the generalized coordinates and momenta, we can write an infinitesimal rotation about the z-axis as

$$\begin{aligned} q_1 \to q_1' &= q_1 - \epsilon q_2 \\ q_2 \to q_2' &= q_2 + \epsilon q_1 \\ p_1 \to p_1' &= p_1 - \epsilon p_2 \\ p_2 \to p_2' &= p_2 + \epsilon p_1 \end{aligned} \tag{10.40}$$

(Here q_1, q_2 can be thought of as x, y, and p_1, p_2 as p_x, p_y.) Equivalently, we can write

$$\begin{aligned} \delta_\epsilon q_1 &= q_1' - q_1 = -\epsilon q_2 \\ \delta_\epsilon q_2 &= q_2' - q_2 = \epsilon q_1 \\ \delta_\epsilon p_1 &= p_1' - p_1 = -\epsilon p_2 \\ \delta_\epsilon p_2 &= p_2' - p_2 = \epsilon p_1 \end{aligned} \tag{10.41}$$

Now, going over to our matrix notation, and using column vectors for the coordinates and momenta, we get

$$\delta_\epsilon \begin{pmatrix} q_1 \\ q_2 \end{pmatrix} = \epsilon \begin{pmatrix} -q_2 \\ q_1 \end{pmatrix} = \begin{pmatrix} 0 & -\epsilon \\ \epsilon & 0 \end{pmatrix} \begin{pmatrix} q_1 \\ q_2 \end{pmatrix}$$

$$\delta_\epsilon \begin{pmatrix} p_1 \\ p_2 \end{pmatrix} = \epsilon \begin{pmatrix} -p_2 \\ p_1 \end{pmatrix} = \begin{pmatrix} 0 & -\epsilon \\ \epsilon & 0 \end{pmatrix} \begin{pmatrix} p_1 \\ p_2 \end{pmatrix}$$

which is similar to the transformations in Eq. 10.38. If we define a function $g(q_i, p_i)$ as proportional to the third or the z-component of orbital angular momentum, namely, proportional to $(\vec{r} \times \vec{p})_z$, as follows:

$$g = \epsilon(q_1 p_2 - q_2 p_1) = \epsilon \ell_z \tag{10.42}$$

then we have

$$\begin{aligned} \frac{\partial g}{\partial q_1} &= \epsilon p_2 \qquad & \frac{\partial g}{\partial q_2} &= -\epsilon p_1 \\ \frac{\partial g}{\partial p_1} &= -\epsilon q_2 \qquad & \frac{\partial g}{\partial p_2} &= \epsilon q_1 \end{aligned} \tag{10.43}$$

Therefore, using the definition for Poisson brackets, we obtain

$$\begin{aligned} \{q_1, g\} &= \frac{\partial g}{\partial p_1} = -\epsilon q_2 = \delta_\epsilon q_1 \\ \{q_2, g\} &= \frac{\partial g}{\partial p_2} = \epsilon q_1 = \delta_\epsilon q_2 \\ \{p_1, g\} &= -\frac{\partial g}{\partial q_1} = -\epsilon p_2 = \delta_\epsilon p_1 \\ \{p_2, g\} &= -\frac{\partial g}{\partial q_2} = \epsilon p_1 = \delta_\epsilon p_2 \end{aligned} \tag{10.44}$$

Once again, it can be shown that the Poisson brackets do not change under the transformation in Eq. 10.40, and the change in the Hamiltonian can be obtained, as before, to be

$$\begin{aligned} \delta_\epsilon H &= \sum_{i=1}^{2} \left(\frac{\partial H}{\partial q_i} \delta_\epsilon q_i + \frac{\partial H}{\partial p_i} \delta_\epsilon p_i \right) \\ &= \sum_{i=1}^{2} \left(\frac{\partial H}{\partial q_i} \frac{\partial g}{\partial p_i} - \frac{\partial H}{\partial p_i} \frac{\partial g}{\partial q_i} \right) \\ &= \{H, g\} = -\{g, H\} \end{aligned} \tag{10.45}$$

where we have used the results of Eqs. 10.44 and 10.20 in the preceding derivation. Thus, arguing as we did in Eqs. 10.32–10.34, we see again that rotations will be a symmetry of the dynamical equations if the Hamiltonian is invariant under such rotations, namely if

$$\delta_\epsilon H = -\{g, H\} = 0 \tag{10.46}$$

Through Eq. 10.24, this implies that

$$\{g, H\} = \frac{dg}{dt} = \epsilon \frac{d\ell_z}{dt} = 0 \tag{10.47}$$

We see therefore that, if rotations about the z-axis define a symmetry of the system, then the z-component of the orbital angular momentum will be conserved. Conversely, whenever the z-component of the orbital angular momentum is conserved, the physical system will be invariant under rotations about the z-axis.

We can similarly show that, for any general infinitesimal transformation, there exists a generator of the transformation. A physical system will be invariant under that transformation if the corresponding generator is conserved, and, conversely, if the generator of an infinitesimal transformation is conserved, the transformation will be a symmetry of the system.

SYMMETRIES IN QUANTUM MECHANICS

The transition from classical mechanics to quantum mechanics is best described within the framework of the Hamiltonian formalism. In quantum mechanics, classical observables are represented by Hermitian operators, and the Poisson brackets are replaced by appropriate commutation relations. The classical generators of infinitesimal transformations therefore become operators that define symmetry transformations for operators as well as for vectors in Hilbert space. In quantum theory, such symmetry transformations can be implemented in one of the two equivalent ways—namely, either by transforming the state vectors in the Hilbert space or by transforming the operators that act on them. This is quite similar to the two ways that a classical transformation can be implemented, namely, as a passive or an active transformation.

In quantum mechanics, any observable quantity corresponds to the expectation value of a Hermitian operator in a given quantum state, and its time evolution—if the operator does not depend explicitly on time—is given by Ehrenfest's theorem (compare with Eq. 10.24):

$$\frac{d}{dt}\langle Q\rangle = \frac{1}{i\hbar}\langle [Q, H]\rangle = \frac{1}{i\hbar}\langle (QH - HQ)\rangle \tag{10.48}$$

where we have denoted the expectation value of an operator Q in a state $|\psi\rangle$ as

$$\langle Q \rangle = \langle \psi | Q | \psi \rangle \tag{10.49}$$

It is clear, therefore, that an observable quantity will be conserved if and only if the corresponding quantum operator commutes with the Hamiltonian. That is, for any quantum state, we will obtain

$$\frac{d}{dt}\langle Q \rangle = 0$$

if and only if

$$[Q, H] = 0 \tag{10.50}$$

This is the quantum analogue of Eqs. 10.35 and 10.47, and we conclude that the infinitesimal transformations generated by an operator Q will define a symmetry of the theory if Eq. 10.50 holds; and as a consequence of the symmetry, the expectation value of Q in any quantum state will be independent of time (will be conserved). Conversely, if an observable or the expectation value of Q in any quantum state is conserved (is constant in time), then Q will generate a symmetry of the underlying physical system.

In quantum mechanics, when two operators commute, they can be simultaneously diagonalized, that is, they can have a complete set of common eigenfunctions. Thus, when the Hamiltonian has an underlying symmetry defined by the generator Q, the energy eigenstates will also be eigenfunctions of the operator Q, and will also be labeled by the quantum numbers corresponding to the eigenvalues of Q. Furthermore, these quantum numbers will be conserved in any physical process where the interaction Hamiltonian for a transition (e.g., decay or reaction) is invariant under the symmetry transformation. For transitions in which interaction Hamiltonians are not invariant under symmetry transformations, however, the corresponding quantum numbers do not have to be conserved. This result provides an understanding of why some quantum numbers are conserved, whereas others are violated in different interactions, and points to an essential first step in constructing physical theories of fundamental interactions.

As an example of quantum symmetries, let us again discuss translations. For simplicity, let us restrict ourselves to one dimension and consider an infinitesimal translation of the x-coordinate by a constant amount ϵ. We will implement the transformation on the state vectors and not on the operators, although the inverse is equally straightforward. Thus, for $x \to x + \epsilon$, with ϵ real, our wave function corresponding to a given state vector will change as follows:*

$$\psi(x) \to \psi(x-\epsilon) = \psi(x) - \epsilon \frac{d\psi(x)}{dx} + O(\epsilon^2) \tag{10.51}$$

Consequently, under such a transformation, the expectation value of the Hamiltonian will change as follows:

$$\begin{aligned}
\langle H \rangle &= \int_{-\infty}^{\infty} dx\, \psi^*(x) H(x) \psi(x) \\
\to \langle H' \rangle &= \int_{-\infty}^{\infty} dx\, \psi^*(x-\epsilon) H(x) \psi(x-\epsilon) \\
&= \int_{-\infty}^{\infty} dx\, \psi^*(x) H(x) \psi(x) - \epsilon \int_{-\infty}^{\infty} dx \frac{d\psi^*}{dx} H(x)\psi(x) \\
&\quad - \epsilon \int_{-\infty}^{\infty} dx\, \psi^*(x) H(x) \frac{d\psi(x)}{dx} + O(\epsilon^2)
\end{aligned}$$

We can integrate the middle term by parts to write

$$\int_{-\infty}^{\infty} dx \frac{d\psi^*(x)}{dx} H(x)\psi(x) = \int_{-\infty}^{\infty} dx \left[\frac{d}{dx}(\psi^* H \psi)\right] - \int_{-\infty}^{\infty} dx\, \psi^* \frac{d}{dx}(H\psi)$$

By assumption, the wave functions vanish at infinity, and therefore the first term on the right-hand side also vanishes, and we obtain

$$\langle H' \rangle = \langle H \rangle - \epsilon \int_{-\infty}^{\infty} dx\, \psi^*(x) \left(H \frac{d}{dx} - \frac{d}{dx} H\right) \psi(x) + O(\epsilon^2)$$

or

$$\langle H' \rangle = \langle H \rangle - \frac{i\epsilon}{\hbar} \langle [H, p_x] \rangle + O(\epsilon^2) \tag{10.52}$$

where, in the last step, we have identified the momentum operator with the spatial derivative

$$p_x \to -i\hbar \frac{d}{dx} \tag{10.53}$$

---

*Note that for the translation $x \to x + \epsilon$, the corresponding change in the wave function is $\psi(x) \to \psi(x - \epsilon)$. (See any standard book on quantum mechanics.)

Comparing this analysis to Eqs. 10.27 and 10.31 shows that, to first order in ϵ, the quantum generator (G) of infinitesimal space translations can be identified with the momentum operator, namely,

$$g = \epsilon G = -\frac{i\epsilon}{\hbar} p_x \tag{10.54}$$

and that the Hamiltonian will be invariant under translations of the x-coordinate if

$$[p_x, H] = 0 \tag{10.55}$$

Furthermore, if Eq. 10.55 holds, then, through Ehrenfest's theorem, $\langle p_x \rangle$ will be conserved. The Hamiltonian for a free particle, of mass m, in one dimension, clearly possesses this type of invariance:

$$H_{\text{free particle}} = \frac{p_x^2}{2m} \tag{10.56}$$

As we know, the energy eigenstates of the quantum mechanical free-particle Hamiltonian are plane waves, which are also the eigenstates of the momentum operator.

CONTINUOUS SYMMETRIES

Broadly speaking, all symmetry transformations of a theory can be classified into two categories: those that depend continuously on a set of parameters and those that correspond to some kind of reflection. Accordingly, they are known, respectively, as continuous and discrete transformations. All the examples of symmetry transformations that we have considered so far in this chapter can be identified with continuous transformations, since they depend on an arbitrary parameter of the transformation (e.g., ϵ). In the next chapter we turn to discrete transformations, but let us now proceed with our development of continuous symmetries.

It is only meaningful to speak about infinitesimal transformations when the transformations are continuous. In fact, for continuous transformations, the infinitesimal transformation has fundamental importance because any finite transformation can be described in terms of a series of successive infinitesimal transformations. This can be shown as follows. We note from Eq. 10.51 that the effect of an infinitesimal translation along the x-axis on a state $|\psi\rangle$ is given by the operator

$$U_x(\epsilon) = 1 - \frac{i\epsilon}{\hbar} p_x \tag{10.57}$$

acting on the state $|\psi\rangle$.

The operator corresponding to a finite translation along the x-axis, namely $U_x(\alpha)$, where α is no longer infinitesimal, can be obtained as follows. First, let us consider N successive infinitesimal translations by an amount ϵ. This will correspond to a total translation by an amount $N\epsilon$, and the operator representing such a transformation would merely correspond to the product of N infinitesimal translations in succession, namely,

$$U_x(N\epsilon) = \left(1 - \frac{i\epsilon}{\hbar}p_x\right)_1 \left(1 - \frac{i\epsilon}{\hbar}p_x\right)_2 \cdots \left(1 - \frac{i\epsilon}{\hbar}p_x\right)_N = \left(1 - \frac{i\epsilon}{\hbar}\, p_x\right)^N \tag{10.58}$$

Since ϵ is infinitesimal, $N\epsilon$ is also infinitesimal for any finite N. If N is very large, however, namely in the limit that $\epsilon \to 0$ and $N \to \infty$, the product can be finite. Thus, let us define the parameter of finite translation, α, as the limit $\alpha = N\epsilon$ when $\epsilon \to 0$ and $N \to \infty$. We see therefore that we can think of a finite translation as an infinitely large number of successive infinitesimal translations. It then follows from Eq. 10.58 that the operator corresponding to a finite translation is given by

$$U_x(\alpha) = \lim_{\substack{N\to\infty \\ \epsilon\to 0 \\ N\epsilon=\alpha}} \left(1 - \frac{i\epsilon}{\hbar}p_x\right)^N = \lim_{\substack{N\to\infty \\ \epsilon\to 0 \\ N\epsilon=\alpha}} \left(1 - \frac{i\alpha}{N\hbar}p_x\right)^N = e^{-(i\alpha/\hbar)p_x} \tag{10.59}$$

We see, therefore, that the operator for finite transformations is obtained simply by exponentiating the generators of infinitesimal transformations. (Clearly, similar expressions hold for finite translations along the other axes.)

Symmetry transformations normally define what is known as a "group." (See Appendix D for some basics on Group Theory.) Thus, for example, two successive translations can also be thought of as a single translation. Similarly, two successive rotations also define a rotation. The rules for combining two transformations (otherwise known as the group properties of the transformation) are completely determined by the commutation relations (or the "algebra") of the generators of the transformation. Thus, for example, for the case of translations along the x-axis, we saw that the generators are the commuting momentum operators:

$$[p_x, p_x] = 0 \tag{10.60}$$

In fact, all the momentum operators (along different axes) commute with one another:

$$[p_i, p_j] = 0, \quad i, j = x, y, \text{ or } z \tag{10.61}$$

For obvious reasons, such an algebra is known as a commutative or Abelian (after Niels Abel) algebra. A consequence of Eqs. 10.59–10.61 is that

$$
\begin{aligned}
U_j(\alpha)U_k(\beta) &= e^{-(i/\hbar)\alpha p_j} e^{-(i/\hbar)\beta p_k} \\
&= e^{-(i/\hbar)\beta p_k} e^{-(i/\hbar)\alpha p_j} \\
&= U_k(\beta)U_j(\alpha), \quad k, j = x, y, z
\end{aligned} \tag{10.62}
$$

and

$$
\begin{aligned}
U_x(\alpha)U_x(\beta) &= e^{-(i/\hbar)\alpha p_x} e^{-(i/\hbar)\beta p_x} \\
&= e^{-(i/\hbar)(\alpha+\beta)p_x} = U_x(\alpha + \beta) = U_x(\beta)U_x(\alpha)
\end{aligned} \tag{10.63}
$$

Namely, translations form what is referred to as a commutative or an Abelian group. The order of the two translations is not relevant. However, not all symmetry transformations have this property. As we know, infinitesimal rotations in quantum mechanics are generated by angular momentum operators (even classically, angular momenta generate rotations through Poisson bracket relations):

$$
\begin{aligned}
L_1 &= x_2p_3 - x_3p_2 \\
L_2 &= x_3p_1 - x_1p_3 \\
L_3 &= x_1p_2 - x_2p_1
\end{aligned} \tag{10.64}
$$

which satisfy the following quantum algebra (commutation relations):

$$
[L_j, L_k] = \sum_{\ell} i\hbar\epsilon_{jk\ell}L_\ell \qquad j, k, \ell = 1, 2, 3 \tag{10.65}
$$

where $\epsilon_{jk\ell}$ is the Levi-Civita symbol, which equals 1 if the j, k, ℓ combination is cyclical, -1 if it is not cyclical, and 0 if any two indices repeat. Equations 10.65 define the simplest noncommutative algebra (the generators do not commute), otherwise known as a non-Abelian algebra. A consequence of this noncommutative property is that the group of rotations behave quite differently from translations. In particular, unlike translations along two different directions, rotations about two different axes do not commute. The order of rotation is important.

The group of spatial rotations in three dimensions (known as $SO(3)$) has an algebraic structure very similar to that of the $SU(2)$ group, which is the group relevant to some internal symmetries, and is characterized by the properties of 2×2 unitary matrices that have determinants equal to unity. The $SU(2)$ group of transformations rotate state vectors in Hilbert space in a manner similar to that provided by spatial rotations, as we discuss next.

The states of a quantum mechanical system are defined by vectors in an abstract Hilbert space. And, just as normal vectors can be rotated in configuration (coordinate) space, so can the vectors corresponding to quantum mechanical states be rotated in an internal Hilbert space. Thus we see that continuous symmetries for a quantum system can be associated either with space-time symmetry transformations or with some internal symmetry transformations. We note that the transformations in the internal Hilbert space do not affect the space-time coordinates, and consequently in all such transformations, the space-time coordinates are kept fixed. Thus, if we consider a two-level system, where the two basic states are represented by the vectors

$$\begin{pmatrix} \psi_1(x) \\ 0 \end{pmatrix} \quad \text{and} \quad \begin{pmatrix} 0 \\ \psi_2(x) \end{pmatrix}$$

then a general rotation in the internal space of this two-dimensional system can be represented as

$$\delta \begin{pmatrix} \psi_1(x) \\ \psi_2(x) \end{pmatrix} = -\sum_{j=1}^{3} i\epsilon_j \frac{\sigma_j}{2} \begin{pmatrix} \psi_1(x) \\ \psi_2(x) \end{pmatrix} \tag{10.66}$$

Here σ_j are the 2×2 Pauli spin matrices, and the infinitesimal generators of $SU(2)$ are defined as

$$I_j = \frac{\sigma_j}{2}, \qquad j = 1, 2, 3$$
$$\sigma_1 = \begin{pmatrix} 0 & 1 \\ 1 & 0 \end{pmatrix}, \qquad \sigma_2 = \begin{pmatrix} 0 & -i \\ i & 0 \end{pmatrix}, \qquad \sigma_3 = \begin{pmatrix} 1 & 0 \\ 0 & -1 \end{pmatrix} \tag{10.67}$$

From the properties of the Pauli matrices, the I_j can be shown to satisfy the same algebra as satisfied by the angular momentum operators in Eq. 10.65.* If the Hamiltonian of the system is invariant under such an internal rotation, then there will be a conserved quantum number, and we can label the two states according to the eigenvalues of, say, the I_3 operator. In fact, the states $\binom{\psi_1(x)}{0}$ and $\binom{0}{\psi_2(x)}$ are eigenstates of I_3 with the eigenvalues $\pm\frac{1}{2}$. (Recall that, because the I_j do not commute, only one of them can be diagonal.) These two states will, of course, be degenerate in energy if such a rotation corresponds to a symmetry of the system. (Once again, this is

---

*Note that if we restrict the rotations to where $\epsilon_1 = 0$, $\epsilon_3 = 0$, and $\epsilon_2 = \epsilon$, then these internal rotations take the form

$$\delta \begin{pmatrix} \psi_1(x) \\ \psi_2(x) \end{pmatrix} = \frac{\epsilon}{2} \begin{pmatrix} -\psi_2(x) \\ \psi_1(x) \end{pmatrix} \tag{10.68}$$

which, when compared with Eq. 10.39, shows that the two components of the wave function can be thought of as coordinates in the internal space, and heightens the analogy with rotations in ordinary space.

very much like the situation with the spin-up and the spin-down states, which are degenerate in energy for a rotationally invariant system.) The strong isospin transformations that we discussed in the last chapter correspond to such internal rotations, and the degeneracy of the proton and the neutron mass can be thought of as a consequence of invariance of the strong-interaction Hamiltonian under such a symmetry transformation.

Example—Isotopic Spin

To bring the preceding formalism to a focus, we enlarge somewhat on the application of these ideas to isospin. If there is an isospin symmetry, then the implication is that our spin-up proton (p) with $I_3 = \frac{1}{2}$, and our spin-down neutron (n) with $I_3 = -\frac{1}{2}$, are indistinguishable. (In this section, we are denoting by I_3 the quantum number associated with the projection of isospin, namely the eigenvalue of the I_3-operator.) We can consequently define a new neutron and proton state as some linear superposition of the $|p\rangle$ and $|n\rangle$ vectors. We note from Eq. 10.68 (compare with Eqs. 10.36 and 10.39) that a finite rotation of our vectors in isospin space by an arbitrary angle θ about the I_2-axis leads to a set of transformed vectors $|p'\rangle$ and $|n'\rangle$:

$$
\begin{aligned}
|p'\rangle &= \cos\frac{\theta}{2}|p\rangle - \sin\frac{\theta}{2}|n\rangle \\
|n'\rangle &= \sin\frac{\theta}{2}|p\rangle + \cos\frac{\theta}{2}|n\rangle
\end{aligned}
\tag{10.69}
$$

Now, let us see what such an invariance implies about the nucleon–nucleon interaction. Our two-nucleon quantum states in the Hilbert space can be written in terms of the more fundamental states, which are either symmetric or antisymmetric under an exchange of particles. These correspond to the following four states:

$$
\begin{aligned}
|\psi_1\rangle &= |pp\rangle, \qquad & |\psi_2\rangle &= \frac{1}{\sqrt{2}}(|pn\rangle + |np\rangle) \\
|\psi_3\rangle &= |nn\rangle, \qquad & |\psi_4\rangle &= \frac{1}{\sqrt{2}}(|pn\rangle - |np\rangle)
\end{aligned}
\tag{10.70}
$$

Assuming, as in the case of normal spin, that I_3 is an additive quantum number, we can identify the isospin projections of $|\psi_1\rangle$ with $I_3 = +1$, $|\psi_2\rangle, |\psi_4\rangle$ with $I_3 = 0$, and $|\psi_3\rangle$ with $I_3 = -1$. Let us now see what the impact of the isospin transformation is, for example, on $|\psi_1\rangle$ and on $|\psi_4\rangle$. Under the rotation given in Eq. 10.69, these states will transform schematically as follows:

$$
\begin{aligned}
|\psi_1'\rangle &= |(\cos\frac{\theta}{2}p - \sin\frac{\theta}{2}n)(\cos\frac{\theta}{2}p - \sin\frac{\theta}{2}n)\rangle \\
&= \cos^2\frac{\theta}{2}\,|pp\rangle - \cos\frac{\theta}{2}\sin\frac{\theta}{2}\,(|pn\rangle + |np\rangle) + \sin^2\frac{\theta}{2}|nn\rangle \\
&= \cos^2\frac{\theta}{2}|\psi_1\rangle - \frac{1}{\sqrt{2}}\sin\theta|\psi_2\rangle + \sin^2\frac{\theta}{2}|\psi_3\rangle
\end{aligned}
\tag{10.71}
$$

$$
\begin{aligned}
|\psi_4'\rangle &= \frac{1}{\sqrt{2}}(|(\cos\frac{\theta}{2}p - \sin\frac{\theta}{2}n)(\sin\frac{\theta}{2}p + \cos\frac{\theta}{2}n)\rangle \\
&\quad - |(\sin\frac{\theta}{2}p + \cos\frac{\theta}{2}n)(\cos\frac{\theta}{2}p - \sin\frac{\theta}{2}n)\rangle) \\
&= \frac{1}{\sqrt{2}}(\cos^2\frac{\theta}{2} + \sin^2\frac{\theta}{2})(|pn\rangle - |np\rangle) = |\psi_4\rangle
\end{aligned}
\tag{10.72}
$$

We see, therefore, that $|\psi_4\rangle$ is totally insensitive to rotations. It must consequently correspond to a scalar (or a "singlet") combination, and thus should represent the $I = 0, I_3 = 0$ nucleon–nucleon system. We can also calculate the changes in the states $|\psi_2\rangle$ and $|\psi_3\rangle$ under a rotation, and we can show that the three remaining states transform into one another under the isospin rotation, just as the three components of a vector would transform into each other under a spatial rotation. If there is isospin invariance, then this must mean that the three states $|\psi_1\rangle$, $|\psi_2\rangle$, and $|\psi_3\rangle$, corresponding to $I_3 = 1, 0$, and -1, respectively, cannot be distinguished from each other in the sense that these states will be degenerate in energy. They are absolutely equivalent if there is isospin or n–p symmetry in the nucleon–nucleon interaction. Consequently, it appears that any two-nucleon system can be classified either as an $I = 0$ singlet or an $I = 1$ triplet in isotopic-spin space. The singlet and the three triplet states are independent of each other, and the three substates of $I = 1$ are indistinguishable if isospin is a symmetry of the system.

Similarly, we can form three-nucleon systems, that correspond to two doublets with $I = \frac{1}{2}$, and an independent quartet of states with $I = \frac{3}{2}$, etc. In this manner we can build multiplets in isotopic-spin space in the same way as we combine angular momentum states.

One important application of isospin invariance lies in the calculation of relative transition rates in decays or in interactions. As an example, let us examine how we might calculate the decay of the $\Delta(1232)$ into a pion and a nucleon. As we mentioned in Chapter IX, the $\Delta(1232)$ is a π-N resonance that was discovered by Fermi and his colleagues in the scattering of π mesons from nucleons. There are four members of the $\Delta(1232)$ family, which correspond to four charged states and to four I_3 projections of an $I = \frac{3}{2}$ isospin multiplet. To calculate the relative decay rates of the $\Delta(1232)$

into a pion and a nucleon, we must recognize that, if there is an isospin symmetry, then the total rate for $\Delta^{++}(1232)$, $\Delta^{+}(1232)$, $\Delta^{0}(1232)$, and for $\Delta^{-}(1232)$ must be identical because these members of the multiplet cannot be distinguished from each other on the basis of the strong interactions. In addition, under the transformation in Eq. 10.69, p and n, as well as π^+, π^0, and π^-, transform into one another, and cannot be told apart. We must therefore assume that the total rate for $\Delta(1232)$ to decay with a neutron in the final state must equal that for a proton in the final state, and similarly for the three pions. We can therefore form a table in which we list all the possible initial $\Delta(1232)$ members and all possible π and N combinations that do not violate charge conservation, and impose the previously given requirements of charge symmetry, or isospin invariance, for the strong decay. This is shown in Table 10.2. By requiring that the sum of the $p\pi^0$ and $n\pi^+$ rates in Δ^+ decay, as well as the sum of the $p\pi^-$ and $n\pi^0$ rates in Δ^0 decay, each add up to 1, we have assured that all the Δ members are equivalent in the sense that their total transition rates are the same, which we have for simplicity normalized to unity. Now, we impose the requirement that rates for decays involving a p or n in the final state are the same:

$$1 + x + y = (1 - x) + (1 - y) + 1 \tag{10.73}$$

In addition, we stipulate that the final rates for decays involving a π^+, π^0, or π^- be identical:

$$1 + (1 - x) = x + (1 - y) = y + 1 \tag{10.74}$$

We have more equations than unknowns, but there exists a set of consistent solutions, which is given in Table 10.2. The result indicates that, for example, the $\Delta^+(1232)$ will decay twice as often into $p + \pi^0$ as into $n + \pi^+$,

Table 10.2 Transition rates for $\Delta \to \pi N$, assuming isospin symmetry in the decay

Charge State of Δ	I_3	*Final State*	*Expected Rate*	*Solution*
Δ^{++}	$\frac{3}{2}$	$p\pi^+$	1	1
Δ^{+}	$\frac{1}{2}$	$p\pi^0$	x	$\frac{2}{3}$
		$n\pi^+$	$1 - x$	$\frac{1}{3}$
Δ^{0}	$-\frac{1}{2}$	$p\pi^-$	y	$\frac{1}{3}$
		$n\pi^0$	$1 - y$	$\frac{2}{3}$
Δ^{-}	$-\frac{3}{2}$	$n\pi^-$	1	1

and that the $\Delta^0(1232)$ will decay twice as often into $n + \pi^0$ as into $p + \pi^-$, and so forth. These relative rates are purely a consequence of isospin symmetry. The fact that these transition rates agree with data, suggests that isotopic spin is a symmetry of the strong interaction, and that both I and I_3 are conserved in strong processes. The solution we have just calculated could also have been obtained simply from tables of Clebsch-Gordan coefficients, which give the coupling of angular momenta. Our example, however, provides an instructive alternate procedure, one that has been emphasized by Robert Adair and Ilya Shmushkevich.

LOCAL SYMMETRIES

Continuous symmetries—whether space-time or internal—can again be of two kinds. First, the parameters of transformation can be constants, that is, global parameters, implying that the transformation is the same at all space-time points. In this case, the symmetry transformation is known as a global transformation. All of the continuous transformations that we have considered thus far fall into this category, and, as we have seen, invariance of a theory under such transformations provides conserved charges (quantum numbers). In contrast, if the parameters of transformation depend on the space-time coordinates—namely, if the magnitude of the transformation is different from point to point—then the symmetry transformation is known as a local transformation. In such a case, real physical forces must be introduced to maintain the symmetry. As an example, let us consider the time-independent Schrödinger equation

$$H\psi(\vec{r}) = \left(-\frac{\hbar^2}{2m}\vec{\nabla}^2 + V(\vec{r})\right)\psi(\vec{r}) = E\psi(\vec{r}) \tag{10.75}$$

Clearly, if $\psi(\vec{r})$ is a solution of this equation, then so is $e^{i\alpha}\psi(\vec{r})$, where α is a constant parameter. In other words, any quantum mechanical wave function can only be defined up to a constant phase, and therefore a constant phase transformation is a symmetry of a quantum mechanical system. This kind of transformation conserves the probability density of a quantum mechanical state, and, in fact, conservation of electric charge can be associated with just such a global phase transformation.

Consider next a local phase transformation

$$\psi(\vec{r}) \to e^{i\alpha(\vec{r})}\psi(\vec{r}) \tag{10.76}$$

where the phase depends explicitly on the space coordinate, namely every point in space has a different phase. (We wish to emphasize that there is no change in the space-time coordinates, but that the parameter of the phase transformation is different for different coordinate points.) Now, under the local phase transformation of Eq. 10.76, the gradient introduces an inhomogeneous term

$$\vec{\nabla}\left[e^{i\alpha(\vec{r})}\psi(\vec{r})\right] = e^{i\alpha(\vec{r})}\left[i(\vec{\nabla}\alpha(\vec{r}))\psi(\vec{r}) + \vec{\nabla}\psi(\vec{r})\right] \neq e^{i\alpha(\vec{r})}\vec{\nabla}\psi(\vec{r}) \tag{10.77}$$

Consequently, since the right-hand side of Eq. 10.75 remains homogeneous under the transformation in Eq. 10.76, we see that the Schrödinger equation cannot be invariant under a local phase transformation.

The transformation of Eq. 10.76 can, however, be made a symmetry of the Schrödinger equation if we arbitrarily introduce a modified gradient operator that contains a vector potential, as follows:

$$\vec{\nabla} \to \vec{\nabla} - i\vec{A}(\vec{r}) \tag{10.78}$$

and require that the vector potential $\vec{A}(\vec{r})$ change under the transformation of Eq. 10.76 as

$$\vec{A}(\vec{r}) \to \vec{A}(\vec{r}) + \vec{\nabla}\alpha(\vec{r}) \tag{10.79}$$

Now, the modified gradient operator will transform under the combined change as desired, namely,

$$\begin{aligned}\left(\vec{\nabla} - i\vec{A}(\vec{r})\right)\psi(\vec{r}) &\to \left(\vec{\nabla} - i\vec{A}(\vec{r}) - i(\vec{\nabla}\alpha(\vec{r}))\right)\left(e^{i\alpha(\vec{r})}\psi(\vec{r})\right) \\ &= e^{i\alpha(\vec{r})}\left(\vec{\nabla} - i\vec{A}(\vec{r})\right)\psi(\vec{r})\end{aligned} \tag{10.80}$$

which means that the local phase transformation in Eq. 10.76 will be a symmetry of the modified time-independent Schrödinger equation

$$\left(-\frac{\hbar^2}{2m}\left(\vec{\nabla} - i\vec{A}(\vec{r})\right)^2 + V(\vec{r})\right)\psi(\vec{r}) = E\psi(\vec{r}) \tag{10.81}$$

provided we require the added vector potential to transform as given in Eq. 10.79. We recognize Eq. 10.79 as a gauge transformation similar to that found in Maxwell's equations, and note, again, that invariance under a local phase transformation requires the introduction of additional fields. These fields are known as *gauge fields* (in the present case, $\vec{A}(\vec{r})$ can be interpreted as the electromagnetic vector potential), and lead to the introduction of definite physical forces. The symmetry group associated with the single-parameter phase transformations of Eq. 10.76 is Abelian (commuting symmetry), referred to as a $U(1)$ group. (These ideas are discussed further in Chapter XIII.)

Although we have discussed all these results in the context of a simple local phase symmetry, the general conclusions, namely that additional fields must be introduced to preserve a local symmetry, hold as well for more complicated symmetries. This observation—that additional forces must be introduced in order to implement a local symmetry—is particularly important in the construction of physical theories. As a starting point, we can turn the argument around, and suggest that the distinct fundamental forces in nature might originate only from such theories that have appropriate local invariances needed to generate those forces. This idea is referred to as the *gauge principle,* and such theories as *gauge theories,* and together they provide our current understanding of the fundamental interactions.

Problems

X.1 Using isotopic spin decomposition for the decays of the $I = 1$, ρ meson: $\rho^+ \to \pi^+\pi^0$, $\rho^- \to \pi^-\pi^0$, $\rho^0 \to \pi^+\pi^-$, and $\rho^0 \to \pi^0\pi^0$, prove that $\rho^0 \to \pi^0\pi^0$ is forbidden on the basis of isospin invariance (that is, use the Adair–Shmushkevich analysis).

X.2 Assuming invariance of strong interactions under rotations in isotopic-spin space, what would you predict for the ratios of transition rates in the following decays:

(a) For an $I = \frac{3}{2}$, K^* meson,

$$\frac{K^{*++} \to K^+\pi^+}{K^{*+} \to K^+\pi^0}, \qquad \frac{K^{*+} \to K^+\pi^0}{K^{*+} \to K^0\pi^+}, \qquad \frac{K^{*-} \to K^0\pi^-}{K^{*0} \to K^+\pi^-}$$

(b) What would you expect for the preceding processes if the K^* meson had $I = \frac{1}{2}$?

X.3 N^* baryons are $I = \frac{1}{2}$ excited states of the nucleon. On the basis of isospin invariance in strong interactions, compare the differences expected for N^* and Δ decays into the π-N systems discussed in Table 10.2.

X.4 What are the possible values of isotopic spin for the following systems? (a) A π^+ meson and an antiproton; (b) two neutrons; (c) a π^+ meson and a Λ^0; (d) a π^+ and a π^0 meson.

Suggested Readings

Frauenfelder, H., and E. M. Henley. 1991. *Subatomic Physics*. Englewood Cliffs, N.J.: Prentice-Hall.

Goldstein, H. 1980. *Classical Mechanics*. Reading, Mass.: Addison-Wesley.

Griffiths, D. 1987. *Introduction to Elementary Particles*. New York: Wiley.

Sakurai, J. J. 1964. *Invariance Principles and Elementary Particles*. Princeton, N.J.: Princeton Univ. Press.

Williams, W. S. C. 1991. *Nuclear and Particle Physics*. London/New York: Oxford Univ. Press.

Also, see standard texts on quantum mechanics, e.g., Das, A. and A. C. Melissinos. 1986. *Quantum Mechanics*. New York: Gordon & Breach.

Chapter XI

DISCRETE TRANSFORMATIONS

INTRODUCTORY REMARKS

Any set of transformations—either involving space-time or some internal space—can be best understood when described as a change in a reference frame. Continuous as well as discrete transformations can be discussed within this kind of framework. The previous chapter dealt with continuous symmetries, and we will now turn to discrete transformations.

PARITY

As mentioned in previous chapters, parity, otherwise known as space inversion, is a transformation that takes us from a right-handed coordinate frame to a left-handed one, or vice versa. Under this transformation, which we denote by the symbol P, the space-time four-vector changes as follows:

$$\begin{pmatrix} ct \\ x \\ y \\ z \end{pmatrix} \xrightarrow{P} \begin{pmatrix} ct \\ -x \\ -y \\ -z \end{pmatrix} \tag{11.1}$$

It is important to recognize that the parity operation is distinct from spatial rotations because a left-handed coordinate system cannot be obtained from a right-handed one through any combination of rotations. In fact, as we know, rotations define a set of continuous transformations, whereas the inversion of space coordinates does not. It is clear, therefore, that in a quantum theory, the quantum numbers corresponding to rotations and parity will be distinct.

Classically, we know that under an inversion of the coordinate system all components of a position and a momentum vector change sign, while their magnitudes are preserved:

$$\begin{aligned}
&\vec{r} \xrightarrow{P} -\vec{r} \\
&\vec{p} = m\dot{\vec{r}} \xrightarrow{P} -m\dot{\vec{r}} = -\vec{p} \\
&r = (\vec{r}\cdot\vec{r})^{1/2} \xrightarrow{P} [(-\vec{r})\cdot(-\vec{r})]^{1/2} = (\vec{r}\cdot\vec{r})^{1/2} = r \\
&p = (\vec{p}\cdot\vec{p})^{1/2} \xrightarrow{P} [(-\vec{p})\cdot(-\vec{p})]^{1/2} = (\vec{p}\cdot\vec{p})^{1/2} = p
\end{aligned} \tag{11.2}$$

This defines the behavior of normal scalar and vector quantities under space inversion. There are, however, scalar and vector quantities that do not transform under parity, as shown in Eq. 11.2. Thus, for example, the orbital angular momentum, which changes like a vector under a rotation, and which we therefore regard as a vector, behaves as follows:

$$\vec{L} = \vec{r}\times\vec{p} \xrightarrow{P} (-\vec{r})\times(-\vec{p}) = \vec{r}\times\vec{p} = \vec{L} \tag{11.3}$$

This is, in fact, just the opposite of how a normal vector transforms. Such vectors are consequently called *pseudovectors* or *axial vectors*. Similarly, there exists a class of scalars, for example, the volume of a parallelopiped, that transform oppositely from normal scalars:

$$\vec{a}\cdot(\vec{b}\times\vec{c}) \xrightarrow{P} (-\vec{a})\cdot(-\vec{b}\times-\vec{c}) = -\vec{a}\cdot(\vec{b}\times\vec{c}) \tag{11.4}$$

Such quantities are known as *pseudoscalars*. A vector can, of course, be labeled by one index (namely, by its components). There are more complex objects in physics that require more indices, and are known as *tensors*. The quadrupole moment, the Maxwell stress tensor, and the relativistic electromagnetic field strength $F_{\mu\nu}$, are examples of second-rank tensors (objects with two indices).

An important property of the parity operation is that two successive parity transformations leave the coordinate system unchanged, namely,

$$\vec{r} \xrightarrow{P} -\vec{r} \xrightarrow{P} \vec{r} \tag{11.5}$$

In quantum mechanics, if we think of P as representing the operator implementing the parity transformation, then from Eq. 11.5 we conclude that

$$P^2|\psi\rangle = +1|\psi\rangle \tag{11.6}$$

Therefore, the eigenvalues of the parity operator can be only ± 1. If we have a parity invariant theory, namely, a theory whose Hamiltonian H is invariant under inversion of coordinates, then, as discussed before, P commutes with H:

$$[P, H] = 0 \tag{11.7}$$

When P and H commute, the eigenstates of the Hamiltonian are also eigenstates of P, with eigenvalues of either $+1$ or -1. Because a wave function transforms under P as follows:

$$\psi(x) \xrightarrow{P} \psi(-x) \tag{11.8}$$

this implies that the stationary states of the Hamiltonian, which have a definite parity, can be classified as either even or odd functions. As an example, consider the one-dimensional harmonic oscillator. In this case the Hamiltonian is parity invariant:

$$H = \frac{p^2}{2m} + \frac{1}{2}m\omega^2 x^2 \xrightarrow{P} \frac{(-p)^2}{2m} + \frac{1}{2}m\omega^2(-x)^2 = H \tag{11.9}$$

and, as we know, the energy eigenstates, which are the Hermite polynomials, are either even or odd functions of x, but never a mixture of odd and even functions.

Consider next a rotationally invariant system in three dimensions. As we noted in Chapter X, the energy eigenstates in this case will also be eigenstates of the angular momentum operator. The wave function for the system can be written as

$$\psi_{n\ell m}(\vec{r}) = R_{n\ell}(r)Y_{\ell m}(\theta, \phi) \tag{11.10}$$

where the $Y_{\ell m}(\theta, \phi)$ are the spherical harmonics, discussed previously in Chapter III. The parity transformation in spherical coordinates takes the form

$$\begin{aligned} r &\xrightarrow{P} r \\ \theta &\xrightarrow{P} \pi - \theta \\ \phi &\xrightarrow{P} \pi + \phi \end{aligned} \tag{11.11}$$

and under this transformation the spherical harmonics behave as

$$Y_{\ell m}(\theta, \phi) \xrightarrow{P} Y_{\ell m}(\pi - \theta, \pi + \phi) = (-1)^{\ell} Y_{\ell m}(\theta, \phi) \tag{11.12}$$

Consequently, parity transforms any wave function that is an eigenstate of orbital angular momentum as follows:

$$\psi_{n\ell m}(\vec{r}) \xrightarrow{P} (-1)^{\ell} \psi_{n\ell m}(\vec{r}) \tag{11.13}$$

In general, a quantum mechanical wave function can have, in addition, an intrinsic parity or phase that is independent of its spatial transformation property in Eq. 11.13, and, correspondingly, a general quantum state that is described by eigenfunctions of orbital angular momentum will transform under parity as follows:

$$\psi_{n\ell m}(\vec{r}) \xrightarrow{P} \eta_{\psi}(-1)^{\ell} \psi_{n\ell m}(\vec{r}) \tag{11.14}$$

where η_{ψ} is known as the intrinsic parity of the quantum state. We can think of the intrinsic parity as the phase analog of intrinsic spin, which when added to the orbital angular momentum, yields the total angular momentum of a system. As a consequence of Eq. 11.6, the intrinsic parity satisfies the condition

$$\eta_{\psi}^2 = 1 \tag{11.15}$$

We can therefore define a total parity of any such quantum mechanical state as

$$\eta_{\text{TOT}} = \eta_{\psi}(-1)^{\ell} \tag{11.16}$$

A detailed analysis of relativistic quantum theories reveals that bosons have the same intrinsic parities as their antiparticles, whereas the relative intrinsic parity of fermions and their antiparticles is odd (opposite).

The classical Newton's equation of motion for a point particle has the form

$$m \frac{d^2 \vec{r}}{dt^2} = \vec{F} \tag{11.17}$$

If we assume the force $\vec{F}$ is either electromagnetic or gravitational, we can write

$$\vec{F} = \frac{C}{r^2} \hat{r} \tag{11.18}$$

where C is a constant. Clearly, since under inversion of coordinates both the left-hand side of Eq. 11.17 and the right-hand side of Eq. 11.18 change sign, Newton's equation for electromagnetic or gravitational interactions is invariant under space inversion. One can similarly show that Maxwell's equations are also invariant under parity.

When parity is a good symmetry, then the intrinsic parities of different particles can be determined by analyzing different decay or production processes, as is shown in the examples. It should be recognized, however, that it is not possible to determine an absolute parity of any system because, starting with some set of assignments, we could change the parities of all states without observing a physical consequence of that change. This is similar to the situation that exists, for example, in defining the absolute sign of a charge or any other quantum number. A convention is needed to define intrinsic parities of objects that differ in some fundamental way—either through the presence of an electric charge, strangeness, or some other characteristic. The accepted convention is to choose the intrinsic parities of the proton, the neutron, and the Λ hyperon as $+1$. The parities of other particles can then be obtained relative to these assignments through the analysis of various experiments.

When parity is conserved, it then restricts the kind of decay processes that may be observed experimentally. Let us consider, for example, particle A decaying in its rest frame into particles B and C:

$$A \longrightarrow B + C \tag{11.19}$$

If J denotes the spin of the decaying particle, then conservation of angular momentum requires that the total angular momentum of the final state must also be J. In particular, if the two decay products are spinless, then their relative orbital angular momentum (ℓ) must equal the spin of A,

$$\ell = J \tag{11.20}$$

Conservation of parity in the decay then implies that

$$\eta_A = \eta_B \eta_C (-1)^\ell = \eta_B \eta_C (-1)^J \tag{11.21}$$

If the decaying particle also has spin-zero, then for the process in Eq. 11.19 to take place we must have

$$\eta_A = \eta_B \eta_C \tag{11.22}$$

In this case, the allowed decays can be represented as follows:

$$\begin{aligned} 0^+ &\longrightarrow 0^+ + 0^+ \\ 0^+ &\longrightarrow 0^- + 0^- \\ 0^- &\longrightarrow 0^+ + 0^- \end{aligned} \tag{11.23}$$

where by $J^P = 0^+(0^-)$ we mean our normal convention of labeling a spin-zero particle with intrinsic parity $+1(-1)$. It also follows that, if parity is conserved, the following decays cannot take place:

$$\begin{aligned} 0^+ &\not\rightarrow 0^+ + 0^- \\ 0^- &\not\rightarrow 0^+ + 0^+ \\ 0^- &\not\rightarrow 0^- + 0^- \end{aligned} \tag{11.24}$$

Example 1—Parity of π^- Meson

Consider the absorption of a very low-energy (or, as is usually termed, a "stopping") π^- on deuterium nuclei:

$$\pi^- + d \longrightarrow n + n \tag{11.25}$$

If ℓ_i and ℓ_f denote the orbital angular momenta in the initial and final states, respectively, then conservation of parity in the reaction would require

$$\eta_\pi \eta_d (-1)^{\ell_i} = \eta_n \eta_n (-1)^{\ell_f} \tag{11.26}$$

where η_π, η_d, andη_n represent the intrinsic parities of the three particles. Because the intrinsic parity of the deuteron is $+1$, and $\eta_n^2 = +1$, it follows that

$$\eta_\pi = (-1)^{\ell_f - \ell_i} = (-1)^{\ell_f + \ell_i} \tag{11.27}$$

The capture process is known to proceed from an $\ell_i = 0$ state, and consequently, we get that

$$\eta_\pi = (-1)^{\ell_f} \tag{11.28}$$

Now, the spin of the deuteron is $J_d = 1$, and this leaves the following possibilities for the state of the two neutrons:

$$\begin{aligned} |\psi_{nn}^{(1)}\rangle &= |J = 1, S = 1, \ell_f = 0 \text{ or } 2\rangle \\ |\psi_{nn}^{(2)}\rangle &= |J = 1, S = 1, \ell_f = 1\rangle \\ |\psi_{nn}^{(3)}\rangle &= |J = 1, S = 0, \ell_f = 1\rangle \end{aligned} \tag{11.29}$$

where the state with $S = 0$ corresponds to the antisymmetric spin state $(\uparrow\downarrow - \downarrow\uparrow)$, while the state with $S = 1$ is given by the symmetric spin state $(\uparrow\downarrow + \downarrow\uparrow)$ of two neutrons. Because the two neutrons are identical fermions, their overall wave function must be antisymmetric, which excludes all but $|\psi_{nn}^{(2)}\rangle$ from consideration, and specifies that the pion is a pseudoscalar, or has an intrinsic parity of $\eta_\pi = -1$.

Example 2—Parity of Δ(1232)

As we discussed in Chapter IX, the Δ(1232) is a π-N resonance that decays strongly into a pion and a nucleon:

$$\Delta(1232) \longrightarrow \pi + N \tag{11.30}$$

The parity of the Δ can therefore be written as

$$\eta_{\Delta} = \eta_{\pi}\eta_{N}(-1)^{\ell} \tag{11.31}$$

where ℓ is the relative orbital angular momentum in the final state. We have just seen that $\eta_\pi = -1$ and η_N is defined as $+1$. Consequently, we have that the parity of the Δ is given uniquely by the orbital wave found in the final state. From the observed angular distribution of the π and N in the rest frame of the Δ, it is found that $\ell = 1$, and consequently the parity of the Δ relative to the nucleon is $\eta_\Delta = +1$. (The spin of the Δ is also known to be $J = \frac{3}{2}$.)

VIOLATION OF PARITY

Until the late 1950s, it was believed that parity was a symmetry of all fundamental interactions. Namely, physics was believed to be the same whether described in a right-handed coordinate system or in a left-handed one. In the early 1950s, however, two weak-decays were observed that were very puzzling. Let us denote these as

$$\begin{aligned} \theta^+ &\longrightarrow \pi^+ + \pi^0 \\ \tau^+ &\longrightarrow \pi^+ + \pi^+ + \pi^- \end{aligned} \tag{11.32}$$

These decays were interesting because the decaying particles were observed to have the same masses and lifetimes. (Subsequently, both were also found to have spin-zero, a fact we will use here to simplify our argument.) Naively, one would therefore conclude that the θ^+ and τ^+ were one and the same particle. However, this assumption presented a conflict with parity conservation. To see this, note that in the rest frame of the decaying particles, the total angular momentum of the initial state is zero (i.e., assuming both θ^+ and τ^+ have spin-zero). The final states in the two processes involve only π mesons, which also have spin-zero. Thus, conservation of angular momentum requires that the relative orbital angular momentum in the $\pi^+\pi^0$ final state must vanish ($\ell_f = 0$). For the $\pi^+\pi^+\pi^-$ final state, the situation is somewhat more complicated because there are two relative angular momenta in the final state (namely, the relative angular momentum of the $2\pi^+$ and that of the π^- with respect to this system), but from experiment we know that both of them correspond to $\ell = 0$. Consequently, the intrinsic parities of the θ^+ and τ^+ can be obtained from the intrinsic parities of the π mesons, which are known to be pseudoscalars ($\eta_{\pi^+} = \eta_{\pi^-} = \eta_{\pi^0} = -1$):

$$\begin{aligned} \eta_{\theta^+} &= \eta_{\pi^+}\eta_{\pi^0} = 1 \\ \eta_{\tau^+} &= \eta_{\pi^+}\eta_{\pi^+}\eta_{\pi^-} = -1 \end{aligned} \tag{11.33}$$

We see, therefore, that if parity is conserved in these decays, the θ^+ and τ^+ have opposite intrinsic parities, and therefore cannot correspond to different decay modes of one object. Alternatively, we can assume that the θ^+ and τ^+ are the same particle, but, of course, only if parity is not conserved in their decays. Tsung-Dao Lee and Chen-Ning Yang, in fact, undertook a systematic study of all the experimentally known weak-decays, and concluded that there was no evidence supporting conservation of parity in weak processes. They postulated that the weak interactions violate parity, and suggested experiments to test their conjecture, which was clearly and rapidly confirmed to be correct. The two decays in Eq. 11.32 are understood now as the two weak-decay modes of the K^+ meson

$$\begin{aligned} K^+ &\longrightarrow \pi^+ + \pi^0 \\ K^+ &\longrightarrow \pi^+ + \pi^+ + \pi^- \end{aligned} \tag{11.34}$$

in which parity is not conserved.*

The experiment that showed conclusively that parity is violated in weak interactions involved a study of β-decay of polarized ^{60}Co. The experimental techniques were simple, yet quite powerful, and we will sketch out the procedures. Consider the decay

$$^{60}\text{Co} \longrightarrow {}^{60}\text{Ni} + e^- + \overline{\nu}_e \tag{11.35}$$

which is equivalent to the β-decay of a neutron. The experiment used a crystal of cobalt salt, whose nuclear spins were polarized by applying a strong external magnetic field. The temperature of the salt was lowered to about 0.01 K in order to minimize thermal motion that leads to depolarization. The angular distribution (θ_e) of the emitted electrons was measured relative to the direction of the applied magnetic field, and the electrons were found to be emitted preferentially in a direction opposite to the field, and therefore opposite to the spin direction of the cobalt nuclei. Namely,

---

*We should point out that the antiparticle of the K^+, namely the K^- meson, has decay channels analogous to those of Eqs. 11.32 and 11.34, and these are

$$K^- \longrightarrow \pi^- + \pi^0$$

and

$$K^- \longrightarrow \pi^- + \pi^+ + \pi^- \tag{11.34'}$$

if $\vec{s}$ denotes the spin of ^{60}Co and $\vec{p}$ the momentum of the emitted electron, then the experiment showed that the expectation value of $\cos\theta_e$ was finite and negative:

$$\langle \cos\theta_e \rangle = \langle \frac{\vec{s}\cdot\vec{p}}{|\vec{s}||\vec{p}|} \rangle = \langle \psi | \frac{\vec{s}\cdot\vec{p}}{|\vec{s}||\vec{p}|} | \psi \rangle \quad < 0 \tag{11.36}$$

Because spin is an angular momentum, and therefore an axial vector, under a parity transformation, our observable $\langle \cos\theta_e \rangle$ changes sign:

$$\begin{aligned} \langle \cos\theta_e \rangle &= \langle \frac{\vec{s}\cdot\vec{p}}{|\vec{s}||\vec{p}|} \rangle \xrightarrow{P} \langle \frac{\vec{s}\cdot(-\vec{p})}{|\vec{s}||\vec{p}|} \rangle \\ &= -\langle \frac{\vec{s}\cdot\vec{p}}{|\vec{s}||\vec{p}|} \rangle = -\langle \cos\theta_e \rangle \end{aligned} \tag{11.37}$$

Therefore, if a right-handed and a left-handed coordinate system were physically equivalent, we would expect to observe the same value in the two frames, and consequently from Eq. 11.37 we conclude that, if parity is conserved, $\langle \cos\theta_e \rangle$ must equal its negative value, and must therefore vanish, namely,

$$\langle \cos\theta_e \rangle \sim \langle \vec{s}\cdot\vec{p} \rangle = 0 \tag{11.38}$$

implying that the electrons would be emitted with equal probability for $\cos\theta_e > 0$ and for $\cos\theta_e < 0$ in the decay. The finite negative value observed for this quantity therefore shows that the two coordinate systems are not equivalent, and that parity is violated in weak interactions. In fact, we now know that parity is violated maximally in weak processes. Since the original experiment by Mme. C. S. Wu and her collaborators, others have confirmed this result. The basic principle in these kinds of experiments is similar, namely, they all try to measure the expectation value of a quantity that should vanish if parity is conserved.

TIME REVERSAL

In simple terms, time reversal corresponds to inverting the time axis, or the direction of the flow of time. In classical mechanics this transformation can be represented as

$$\begin{aligned} & t \xrightarrow{T} -t \\ & \vec{r} \xrightarrow{T} \vec{r} \\ & \vec{p} = m\dot{\vec{r}} \xrightarrow{T} -m\dot{\vec{r}} = -\vec{p} \\ & \vec{L} = \vec{r}\times\vec{p} \xrightarrow{T} \vec{r}\times(-\vec{p}) = -\vec{L} \end{aligned} \tag{11.39}$$

Newton's equations of motion given in Eq. 11.17, being second order in the time derivative, are invariant under time reversal for both electromagnetic and gravitational interactions. In fact, we can also show that Maxwell's equations are invariant under time inversion. However, not all macroscopic systems are time-reversal invariant. In fact, for macroscopic systems, statistical mechanics defines a unique direction for the flow of time, as the one for which entropy (disorder) increases. Microscopic systems, on the other hand, appear to respect time-reversal invariance. The implementation of time-reversal symmetry into a theoretical formalism, however, is not entirely as straightforward as for the other symmetries.

Let us consider the time-dependent Schrödinger equation

$$i\hbar \frac{\partial \psi}{\partial t} = H\psi \tag{11.40}$$

Being a first-order equation in the time derivative, it therefore cannot be invariant under the simplest time inversion:

$$\psi(\vec{r}, t) \xrightarrow{T} \psi(\vec{r}, -t) \tag{11.41}$$

If, however, we require the wave function to transform under time reversal as

$$\psi(\vec{r}, t) \xrightarrow{T} \psi^*(\vec{r}, -t) \tag{11.42}$$

then, assuming H to be real, from the complex conjugate of Eq. 11.40, we also obtain

$$-i\hbar \frac{\partial \psi^*}{\partial t}(\vec{r}, t) = H\psi^*(\vec{r}, t) \tag{11.43}$$

and now letting $t \to -t$, we obtain

$$i\hbar \frac{\partial \psi^*}{\partial t}(\vec{r}, -t) = H\psi^*(\vec{r}, -t) \tag{11.44}$$

Hence, the Schrödinger equation can be made invariant under time inversion, namely, both ψ and its time-reversed solution can obey the same equation, provided that time reversal for quantum mechanical wave functions is defined by Eq. 11.42.

Consequently, the operator representing time reversal in quantum mechanics is quite unconventional in that it takes a wave function to its complex conjugate. (Technically, such operators are called *antilinear*.) Because time-dependent wave functions are necessarily complex, it follows that quantum mechanical wave functions cannot be eigenfunctions of the time-reversal operator. Consequently, there is no simple quantum number that can be associated with time-reversal invariance. Physically, however, invariance under time reversal implies that the transition amplitudes for the process $i \to f$ and the time-reversed one $f \to i$ will have the same magnitude, namely,

$$|M_{i\to f}| = |M_{f\to i}| \tag{11.45}$$

Here $M_{i\to f}$ denotes the matrix element for the transition from an initial state $|i\rangle$ to a final state $|f\rangle$. Equation 11.45 is referred to conventionally as the principle of detailed balance. It states that the quantum mechanical probability for a forward reaction to take place is the same as for the time-reversed process. We wish to emphasize, however, that the transition rates for the two processes can be quite different. From Fermi's Golden Rule, we know that the rates are given by

$$\begin{aligned} W_{i\to f} &= \frac{2\pi}{\hbar}|M_{i\to f}|^2\rho_f \\ W_{f\to i} &= \frac{2\pi}{\hbar}|M_{f\to i}|^2\rho_i \end{aligned} \tag{11.46}$$

where ρ_f and ρ_i represent the density of states for the end products in the two reactions. These can be quite different, depending on the masses of the particles involved, and, correspondingly, the rates can be different, even if detailed balance is valid. The principle of detailed balance has been verified for many processes, and in fact, has been used to determine the spins of particles by comparing the rates for the forward and backward reactions.

Experimentally, time-reversal invariance appears to be valid in almost all known fundamental processes. The most spectacular test of this invariance principle for electromagnetic interactions comes from a search for an electric dipole moment of the neutron. As we have already discussed, although the neutron has no electric charge, it has a magnetic dipole moment, suggesting the presence of an extended charge distribution inside the neutron. If the centers of the positive and the negative charge distributions do not coincide, then the neutron can also have an electric dipole moment. We can estimate the magnitude of such a dipole moment from simple dimensional arguments to be

$$\mu_{e\ell} \lesssim ed \simeq e \times 10^{-13}\text{ cm} \simeq 10^{-13}e\text{–cm} \tag{11.47}$$

where we have used the fact that the typical size of the neutron, and therefore the maximum separation of the charge centers, is about $d \simeq 10^{-13}$ cm. Since, for the neutron, the only preferred spatial direction is its spin axis, if the neutron had a nonvanishing electric dipole moment, it could only point along that axis. Experimentally, therefore, one searches for a finite value of $\langle \vec{\mu}_{e\ell} \cdot \vec{s}\rangle$. The most precise experiments to date yield an upper limit of

$$\mu_{e\ell} \lesssim 10^{-25} e\text{–cm} \tag{11.48}$$

This is, of course, consistent with no electric dipole moment, and is $\simeq 12$ orders of magnitude smaller than our naive estimate of Eq. 11.47.

A finite value of $\mu_{e\ell}$, and therefore a finite expectation value for $\langle \vec{\mu}_{e\ell} \cdot \vec{s} \rangle$, would imply violation of T-invariance. To see this, note that under time reversal the operator for the projection of the electric dipole moment along the spin direction transforms as follows:

$$\vec{\mu}_{e\ell} \cdot \vec{s} \xrightarrow{T} \vec{\mu}_{e\ell} \cdot (-\vec{s}) = -\vec{\mu}_{e\ell} \cdot \vec{s} \tag{11.49}$$

Here we used the fact that $\vec{\mu}_{e\ell}$ transforms as $\sim e\vec{r}$, which does not change under time reversal. The spin, on the other hand, being an angular momentum changes sign (see Eq. 11.39). Consequently,

$$\langle \vec{\mu}_{e\ell} \cdot \vec{s} \rangle \xrightarrow{T} -\langle \vec{\mu}_{e\ell} \cdot \vec{s} \rangle \tag{11.50}$$

and, if time reversal is a symmetry of the system, then this quantity must vanish. The result in Eq. 11.48 can therefore be regarded as an impressive upper limit on T-violation in electromagnetic interactions. There is, however, a question of interpretation of this result, because under the parity transformation we also obtain

$$\langle \vec{\mu}_{e\ell} \cdot \vec{s} \rangle \xrightarrow{P} \langle (-\vec{\mu}_{e\ell}) \cdot \vec{s} \rangle = -\langle \vec{\mu}_{e\ell} \cdot \vec{s} \rangle \tag{11.51}$$

This means that a finite electric dipole moment could also arise as a consequence of parity violation. We know from experiments, however, that parity is conserved in electromagnetic interactions, but violated in weak processes. Consequently, the presence of an electric dipole moment could arise from an interplay of electromagnetic and weak interactions. In fact, a small contribution to the electric dipole moment is expected from the weak interaction. Anything beyond that could be attributed to T-violation in electromagnetic processes or to other new physical mechanisms. The upper limit on the value of the electric dipole moment of the neutron (which is comparable to the upper limit obtained for the electric dipole moment of the electron) can therefore be interpreted as providing limits on T-violation in electromagnetic and P-violation in weak interactions.

CHARGE CONJUGATION

Both parity and time reversal are discrete space-time symmetry transformations. One can naturally ask whether there also exist discrete transformations in the internal Hilbert space of a quantum mechanical system.

Charge conjugation defines such a transformation. Here the space-time coordinates are, again, unchanged and the discrete transformation only affects the internal properties of the quantum mechanical state.

Let us recall that the classification of the electron as particle and positron as antiparticle is arbitrary. In fact, the definition of positive and negative charge, positive and negative strangeness, the assignment of baryon number, etc., as we have stated before, are all a matter of convention. Once a choice is made, however, we can measure the quantum numbers of other particles relative to the defined assignments. The charge conjugation operation inverts the internal quantum numbers of states, and thereby relates particles to their antiparticles. Classically, one can represent charge conjugation as the following transformation on electric charge Q:

$$Q \xrightarrow{C} -Q \tag{11.52}$$

Since the electric charge is the source of electric and magnetic fields, it follows that under such a transformation

$$\begin{aligned} \vec{E} &\xrightarrow{C} -\vec{E} \\ \vec{B} &\xrightarrow{C} -\vec{B} \end{aligned} \tag{11.53}$$

(This is simply because both $\vec{E}$ and $\vec{B}$ are linear in electric charge.) It is straightforward to show that Maxwell's equations are invariant under such a transformation.

For a quantum mechanical state $|\psi(Q, \vec{r}, t)\rangle$, where Q represents all the internal quantum numbers such as charge, lepton number, baryon number, and strangeness, charge conjugation reverses all the charges,

$$|\psi(Q, \vec{r}, t)\rangle \xrightarrow{C} |\psi(-Q, \vec{r}, t)\rangle \tag{11.54}$$

Consequently, a state can be an eigenstate of the charge conjugation operator C if, at the very least, it is electrically neutral. Thus, for example, the photon (γ), the positronium atom ($e^- - e^+$), the π^0 meson, etc. can be eigenstates of C. Not all charge–neutral states are eigenstates of C, however (they may carry other internal quantum numbers that are nonzero). For example, the following are, clearly, not eigenstates of C:

$$\begin{aligned} |n\rangle &\xrightarrow{C} |\overline{n}\rangle \\ |\pi^- p\rangle &\xrightarrow{C} |\pi^+ \overline{p}\rangle \\ |K^0\rangle &\xrightarrow{C} |\overline{K^0}\rangle \end{aligned} \tag{11.55}$$

Because two consecutive charge conjugation transformations will leave a state unchanged, it follows that the eigenvalues of C, or the charge parities of an eigenstate, can be only ± 1. Thus, for example, from Eq. 11.53 we conclude that the photon, or the quantum of electromagnetic interactions, must have a charge parity of -1,

$$\eta_C(\gamma) = -1 \tag{11.56}$$

If charge conjugation is a symmetry of the theory, namely if

$$[C, H] = 0 \tag{11.57}$$

then the charge parities in any given process must be conserved. Electromagnetic interactions are invariant under charge conjugation (remember that Maxwell's equations do not change under C). Consequently, from the decay of the π^0,

$$\pi^0 \longrightarrow \gamma + \gamma \tag{11.58}$$

we conclude that, for charge parity to be conserved, the π^0 must be even under C:

$$\eta_C(\pi^0) = \eta_C(\gamma)\eta_C(\gamma) = (-1)^2 = +1 \tag{11.59}$$

Invariance under charge conjugation therefore leads to restrictions on the kinds of interactions or decays that can take place. For example, a π^0 cannot decay to an odd number of photons, because that would violate conservation of C-parity:

$$\pi^0 \not\rightarrow n\gamma \qquad n \text{ odd} \tag{11.60}$$

While charge conjugation is known to be a symmetry of the electromagnetic and of the strong interactions, we can argue, as follows, that it must be violated in weak interactions. As we have emphasized, charge conjugation does not change space-time properties, and therefore the handedness of a quantum state is insensitive to such a transformation. Thus, under charge conjugation, we obtain

$$\begin{aligned} |\nu_L\rangle &\xrightarrow{C} |\overline{\nu}_L\rangle \\ |\overline{\nu}_R\rangle &\xrightarrow{C} |\nu_R\rangle \end{aligned} \tag{11.61}$$

where the subscripts L and R refer to left- and right-handed neutrinos (or antineutrinos), respectively. We pointed out earlier that there is no evidence for the existence of right-handed neutrinos or left-handed antineutrinos. Consequently, the charge conjugate process of β-decay cannot take place, and charge conjugation therefore cannot be a symmetry of such interactions. Nevertheless, although both P and C symmetry are violated in β decay, the combined transformation of CP appears to be a symmetry of such processes. One can see this heuristically as follows:

$$\begin{aligned} |\nu_L\rangle &\xrightarrow{P} |\nu_R\rangle \xrightarrow{C} |\overline{\nu}_R\rangle \\ |\overline{\nu}_R\rangle &\xrightarrow{P} |\overline{\nu}_L\rangle \xrightarrow{C} |\nu_L\rangle \end{aligned} \tag{11.62}$$

that is, the combined operation of CP takes a physical state to another physical state, which is not what the C or P operations do individually. Nevertheless, the CP operation is not a symmetry of all weak interactions, as we will discover in the following chapter.

CPT Theorem

We have seen that the discrete symmetries P, T, and C appear to be violated in some processes. However, Georg Lüders, Wolfgang Pauli, and Julian Schwinger showed independently, that the combined operation of CPT must be a symmetry of essentially any theory that is invariant under Lorentz transformations. That is, even if the individual transformations do not represent symmetries of any given theory, the product transformation will be a symmetry. This is known as the CPT theorem, and it leads to certain very interesting conclusions, which we summarize below.

1. As a consequence of CPT invariance, one can show that, if particles have normal statistics (either Bose–Einstein or Fermi–Dirac), then those with integer spin satisfy Bose-Einstein statistics, whereas those with half-integer spin obey Fermi-Dirac statistics. This consequence has additional implications for relativistic theories, in that it requires an operator with integer spin to be quantized using commutation relations, while anticommutation relations are to be used for operators with half-integer spin.
2. The fact that particles and antiparticles have identical masses and lifetimes is a consequence of CPT invariance of physical theories.
3. Invariance under CPT also leads to the conclusion that all the internal quantum numbers of antiparticles are opposite to those of the particles.

The CPT theorem is consistent with all known experimental observations, and CPT appears to be a true symmetry of all interactions.

Problems

XI.1 The $\rho^0(770)$ has $J^P = 1^-$, and it decays strongly into $\pi^+\pi^-$ pairs. From symmetry and angular momentum considerations, explain why the decay $\rho^0(770) \to \pi^0\pi^0$ is forbidden.

XI.2 What is the charge conjugate reaction to $K^- + p \to \overline{K^0} + n$? Can a K^-p system be an eigenstate of the charge conjugation operator? Similarly, discuss the reaction $\overline{p} + p \to \pi^+ + \pi^-$.

XI.3 If ρ^0 mesons are produced in states with spin projection $J_z = 0$ along their line of flight, what would you expect for the angular distribution of $\rho^0 \to \pi^+ + \pi^-$ decay products in the ρ^0 rest frame? (See Appendix B for the appropriate $Y_{\ell,m}(\theta, \phi)$ functions.) What would be your answer if the initial ρ^0 had spin projection $J_z = +1$?

XI.4 The Ξ^- has $J^P = \frac{1}{2}^+$. It decays through the weak interaction into a Λ^0 and a π^-. If $J_\Lambda^P = \frac{1}{2}^+$ and $J_\pi^P = 0^-$, what are the allowed relative orbital angular momenta for the Λ–π^- system?

XI.5 Which of the following decays are forbidden by C-invariance: (a) $\omega^0 \to \pi^0 + \gamma$; (b) $\eta' \to \rho^0 + \gamma$; (c) $\pi^0 \to \gamma + \gamma + \gamma$; (d) $J/\psi \to \overline{p} + p$; (e) $\rho^0 \to \gamma + \gamma$. (Check the *CRC* tables to see if these decays take place.)

XI.6 Although the orbital wave for any strong $\pi - N$ state determines the parity of that state, different ℓ-values do not necessarily yield different decay angular distributions. In particular, show that a $J = \frac{1}{2}, J_Z = +\frac{1}{2}, \pi - N$ resonance decays the same way whether it has $\ell = 0$ or $\ell = 1$. Similarly, show that a $J = \frac{3}{2}, J_Z = +\frac{1}{2}, \pi - N$ system has the same decay angular distribution for $\ell = 1$ as for $\ell = 2$. (*Hint:* Expand the wave function for the state in terms of the products of $S = \frac{1}{2}$ spin-states and the appropriate $Y_{\ell,m}(\theta, \phi)$.)

Suggested Readings

Frauenfelder, H., and E. M. Henley. 1991. *Subatomic Physics*. Englewood Cliffs, N.J.: Prentice-Hall.

Goldstein, H. 1980. *Classical Mechanics*. Reading, Mass.: Addison-Wesley.

Griffiths, D. 1987. *Introduction to Elementary Particles*. New York: Wiley.

Sakurai, J. J. 1964. *Invariance Principles and Elementary Particles*. Princeton, N.J.: Princeton Univ. Press.

Williams, W. S. C. 1991. *Nuclear and Particle Physics*. London/New York: Oxford Univ. Press.

Also, see standard texts on quantum mechanics, e.g., Das, A., and A. C. Melissinos. 1986. *Quantum Mechanics*. New York: Gordon & Breach.

Chapter XII

NEUTRAL KAONS AND *CP* VIOLATION

INTRODUCTORY REMARKS

As we saw in the previous chapter, weak interactions violate, separately, both the C and P symmetries. Nevertheless, it was thought until the early 1960s that the combined operation of CP might hold for all interactions. However, as we discuss below, there are processes in which even the combined operation of CP is violated. Because the CP transformation takes a physical particle state to a physical antiparticle state (compare, for example, Eqs. 11.61 and 11.62), invariance under CP is equivalent to having a particle–antiparticle symmetry in nature. The universe, however, is known to be dominated by matter, with essentially no antimatter present, which is tantamount to saying that there is, in fact, a particle–antiparticle asymmetry in the universe. This would suggest that CP may not be a symmetry of the fundamental interactions. In this chapter we discuss the violation of CP in weak processes. It should be recognized that, if CPT is a symmetry of all physical systems, then a violation of CP automatically implies that T must also be violated. CP violation therefore implies that there exist microscopic (subatomic) systems where time has a unique direction of flow.

NEUTRAL KAONS

We have already discussed the τ-θ puzzle, where we concluded that the decays of the $\theta^+(\theta^-)$ and the $\tau^+(\tau^-)$ can be identified as two decay channels of the $K^+(K^-)$. There are analogous decays of the neutral kaon partners of the K^+ and K^- that we consider in this chapter. In particular, we concentrate on the hadronic final states:

$$\begin{aligned} \theta^0 &\longrightarrow \pi^0 + \pi^0 \\ \theta^0 &\longrightarrow \pi^+ + \pi^- \\ \tau^0 &\longrightarrow \pi^0 + \pi^0 + \pi^0 \\ \tau^0 &\longrightarrow \pi^+ + \pi^- + \pi^0 \end{aligned} \tag{12.1}$$

The first question that naturally comes to mind is how the θ^0 and τ^0 are related to the K^0 and $\overline{K^0}$. To make this connection, we first discuss the production and decay characteristics of the neutral kaons.

Both K^0 and $\overline{K^0}$ mesons can be produced in strong-interaction processes such as

$$\begin{aligned} K^- + p &\longrightarrow \overline{K^0} + n \\ K^+ + n &\longrightarrow K^0 + p \\ \pi^- + p &\longrightarrow \Lambda^0 + K^0 \end{aligned} \tag{12.2}$$

In these reactions the kaons are produced in states of unique strangeness, namely $S = +1$ for the K^0, and $S = -1$ for the $\overline{K^0}$. As we know, the K^0 can be identified as the $I_3 = -\frac{1}{2}$ isospin partner of the K^+, and the $\overline{K^0}$ as the $I_3 = +\frac{1}{2}$ isospin partner of the K^-. The $\overline{K^0}$ is the antiparticle of the K^0, and can be distinguished from it due to the difference in their strangeness. The neutral kaons produced in the preceding collisions are unstable, and decay through the weak interaction after traveling some distance ℓ (in a time t_{lab}) in the laboratory. The distance traveled prior to decay is related to the proper time through the velocity, v, of the kaon,

$$\ell = v t_{\text{lab}} = v\gamma t_{\text{proper}}, \qquad \gamma = \left(1 - \frac{v^2}{c^2}\right)^{-1/2} \tag{12.3}$$

and the mean proper time is just the lifetime, τ_{K^0}, of the K^0,

$$\tau_{K^0} = \langle t_{\text{proper}} \rangle \tag{12.4}$$

By measuring the velocity and the decay length (ℓ) of a K^0, we can determine its proper time, and from a sample of such events extract the lifetime. Because the $\overline{K^0}$ is the antiparticle of the K^0, it follows from the CPT theorem that the two particles must have identical masses and lifetimes.

The results of experiments that study τ_{K^0} are sketched in Fig. 12.1, and are remarkable indeed. Instead of observing a single characteristic decay time (exponential drop-off) that would be expected for any unique eigenstate of the free-particle Hamiltonian, the data indicates that there are two distinct lifetimes associated with both the K^0 and the $\overline{K^0}$. This can only be understood if we assume that the K^0 and the $\overline{K^0}$ states consist of a superposition of two distinct states with different lifetimes: a short-lived one, originally labeled K_1^0, and a longer lived one, labeled K_2^0. The events corresponding to K_1^0-decays are of the θ^0 variety (namely, two-pion channels), while those corresponding to the K_2^0-decays are of the τ^0 variety (that is, three-pion channels). The results found for K^0- and $\overline{K^0}$-decays are completely consistent with each other in the sense that the decay modes and lifetimes observed for the K_1^0 and K_2^0 components in both cases are the same:

$$\begin{aligned} \tau_1 &\simeq 0.9 \times 10^{-10} \text{ sec} \\ \tau_2 &\simeq 5 \times 10^{-8} \text{ sec} \end{aligned} \tag{12.5}$$

The fact that the K^0 and the $\overline{K^0}$ can decay through common channels

$$\begin{aligned} K^0 &\longrightarrow \pi^0 + \pi^0 \\ \overline{K^0} &\longrightarrow \pi^0 + \pi^0 \end{aligned} \tag{12.6}$$

suggests that these particles can mix through higher orders in the weak interaction. That is, although the K^0 and $\overline{K^0}$ are distinguishable because of their strangeness quantum number, and are represented by orthogonal states, they do not remain orthogonal as time evolves and weak interactions set in. That is, in the presence of weak interactions the two kaons share the same decay channels. This is a consequence of the fact that weak interactions do not conserve strangeness. It is therefore possible to have

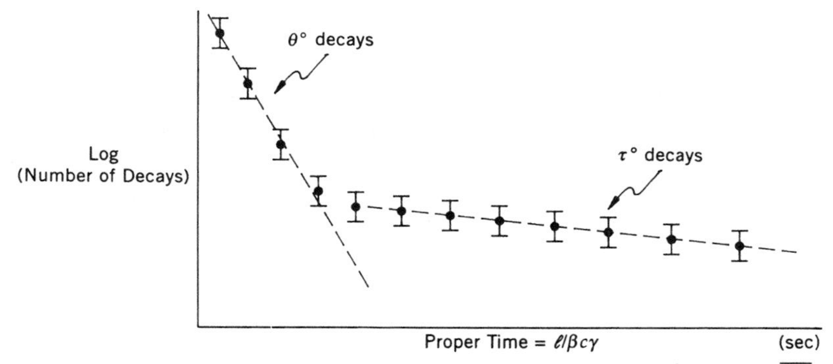

Figure 12.1 Sketch of the distribution in proper time for K^0s, or for $\overline{K^0}$s, calculated from their velocities and distances traveled prior to decay.

transitions between a K^0 and a $\overline{K^0}$, for example, through $2\pi^0$ intermediate states (see Fig. 12.2):

$$K^0 \xrightarrow{H_{\rm wk}} \pi^0 + \pi^0 \xrightarrow{H_{\rm wk}} \overline{K^0} \tag{12.7}$$

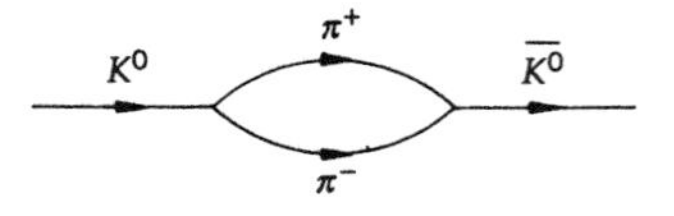

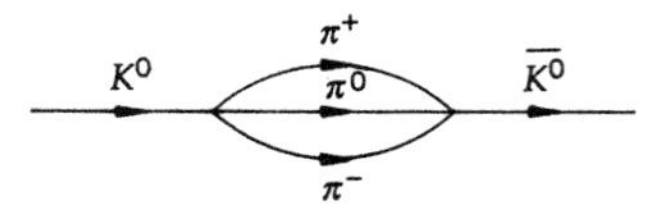

Figure 12.2 Possible transformations of K^0 to $\overline{K^0}$.

Thus, the K^0 and $\overline{K^0}$ particle states, although eigenstates of the strong-interaction Hamiltonian ($H_{\rm st}$), cannot be eigenstates of the weak interaction Hamiltonian ($H_{\rm wk}$). That is, schematically, we have for the strong interactions:

$$\langle \overline{K^0} | K^0 \rangle = 0$$
$$\langle \overline{K^0} | H_{\rm st} | K^0 \rangle = 0$$

with $H_{\rm st}|K^0\rangle = m_{K^0}c^2|K^0\rangle$, $H_{\rm st}|\overline{K^0}\rangle = m_{\overline{K^0}}c^2|\overline{K^0}\rangle$, and $m_{K^0} = m_{\overline{K^0}} \simeq$ 498 MeV/c^2

$$\begin{aligned} S|K^0\rangle &= +1|K^0\rangle, \quad S|\overline{K^0}\rangle = -1|\overline{K^0}\rangle \\ I_3|K^0\rangle &= -\tfrac{1}{2}|K^0\rangle, \quad I_3|\overline{K^0}\rangle = \tfrac{1}{2}|\overline{K^0}\rangle \end{aligned} \tag{12.8}$$

while for the weak interactions we have

$$\langle \overline{K^0} | H_{\rm wk} | K^0 \rangle \neq 0 \tag{12.9}$$

Since the decay of K mesons is a weak process, the observed K_1^0 and K_2^0 particles, with unique lifetimes, can be thought of as corresponding to the eigenstates of $H_{\rm wk}$. Furthermore, because both the K^0 and the $\overline{K^0}$ appear to be superpositions of the K_1^0 and K_2^0 states, it follows that the K_1^0 and the K_2^0 must also be superpositions of K^0 and $\overline{K^0}$.

CP EIGENSTATES OF NEUTRAL KAONS

To determine which linear superposition of states will correspond to the eigenstates of the weak Hamiltonian, let us assume for simplicity that CP

is a symmetry of the weak interactions. Consequently, we will construct linear superpositions of K^0 and $\overline{K^0}$ that are eigenstates of CP. We will choose the phases for the K^0 and $\overline{K^0}$ states as follows:

$$\begin{aligned} CP|K^0\rangle &= -C|K^0\rangle = -|\overline{K^0}\rangle \\ CP|\overline{K^0}\rangle &= -C|\overline{K^0}\rangle = -|K^0\rangle \end{aligned} \tag{12.10}$$

where we have used the fact that K mesons are pseudoscalars, and consequently have odd intrinsic parities. Using Eq. 12.10, we can define two linear orthonormal combinations of K^0 and $\overline{K^0}$ that will be eigenstates of the CP operator, namely,

$$\begin{aligned} |K_1^0\rangle &= \frac{1}{\sqrt{2}}\left(|K^0\rangle - |\overline{K^0}\rangle\right) \\ |K_2^0\rangle &= \frac{1}{\sqrt{2}}\left(|K^0\rangle + |\overline{K^0}\rangle\right) \end{aligned} \tag{12.11}$$

Applying the CP operator to the K_1^0 and K_2^0 states gives

$$\begin{aligned} CP|K_1^0\rangle &= \frac{1}{\sqrt{2}}\left(CP|K^0\rangle - CP|\overline{K^0}\rangle\right) \\ &= \frac{1}{\sqrt{2}}\left(-|\overline{K^0}\rangle + |K^0\rangle\right) = \frac{1}{\sqrt{2}}\left(|K^0\rangle - |\overline{K^0}\rangle\right) = |K_1^0\rangle \\ CP|K_2^0\rangle &= \frac{1}{\sqrt{2}}\left(CP|K^0\rangle + CP|\overline{K^0}\rangle\right) \\ &= \frac{1}{\sqrt{2}}\left(-|\overline{K^0}\rangle - |K^0\rangle\right) = -\frac{1}{\sqrt{2}}\left(|K^0\rangle + |\overline{K^0}\rangle\right) = -|K_2^0\rangle \end{aligned} \tag{12.12}$$

Thus, the two states $|K_1^0\rangle$ and $|K_2^0\rangle$, which do not carry unique strangeness, can be defined as eigenstates of CP with eigenvalues $+1$ and -1, respectively. If CP is conserved in weak processes, we can then identify K_1^0 and K_2^0 with θ^0 and τ^0, respectively. In fact, we see that in the rest frame of θ^0, the two π^0 mesons must have zero orbital angular momentum ($\ell = 0$), and the final-state $\pi^0\pi^0$ system will be in an eigenstate of CP with eigenvalue $+1$. This is consistent with our identification of K_1^0 with the θ^0 decay mode:

$$\theta^0 = K_1^0 \longrightarrow \pi^0 + \pi^o \tag{12.13}$$

Recalling that pions are pseudoscalar mesons, a similar analysis of the $3\pi^0$-decay mode of the τ^0 shows that the final state, in this case, is an eigenstate of CP with eigenvalue -1, which confirms the identification of K_2^0 with the τ^0 decay:

$$\tau^0 = K_2^0 \longrightarrow \pi^0 + \pi^0 + \pi^o \tag{12.14}$$

Note that the momentum (and therefore the phase space or the density of states) available for the two-body decay in Eq. 12.13 will be substantially larger than the phase space for the three body decay in Eq. 12.14. Consequently, if our analysis is valid, then we can predict that the rate for decay of the K_1^0 will be much greater than that for K_2^0, leading to the expectation that the two particles will have different lifetimes—K_1^0 being short-lived compared to K_2^0. This prediction of two lifetimes was, in fact, a principal result of an analysis of the problem by Murray Gell-Mann and Abraham Pais prior to the discovery of the K_2^0.

STRANGENESS OSCILLATION

We can also invert the relations in Eq. 12.11 to obtain

$$\begin{aligned} |K^0\rangle &= \frac{1}{\sqrt{2}}(|K_1^0\rangle + |K_2^0\rangle) \\ |\overline{K^0}\rangle &= -\frac{1}{\sqrt{2}}(|K_1^0\rangle - |K_2^0\rangle) \end{aligned} \tag{12.15}$$

Thus the interactions involving the K^0 and $\overline{K^0}$ mesons can be understood as follows. In the strong production processes, such as given in Eqs. 12.2, only the eigenstates of the strong Hamiltonian are produced, namely $|K^0\rangle$ and $|\overline{K^0}\rangle$. As we see from Eq. 12.15, however, these states are superpositions of the $|K_1^0\rangle$ and $|K_2^0\rangle$, which are eigenstates of the weak Hamiltonian. (The mass and lifetime eigenvalues of the K_1^0 and K_2^0 are discussed below.) At the time of their production, the K^0 or $\overline{K^0}$ correspond to the specific superpositions of K_1^0 and K_2^0 given in Eq. 12.15. As these specific mixtures of the $|K_1^0\rangle$ and $|K_2^0\rangle$ states start propagating in vacuum, however, both the $|K_1^0\rangle$ and $|K_2^0\rangle$ components decay away. However, the state $|K_1^0\rangle$ decays much faster than $|K_2^0\rangle$, and, as a result, after some time, the initial $|K^0\rangle$ or $|\overline{K^0}\rangle$ will have primarily the $|K_2^0\rangle$ component left over. We know from Eq. 12.11, however, that K_2^0 has an equal admixture of K^0 and $\overline{K^0}$, which means that, starting out either as a pure K^0 or as a pure $\overline{K^0}$ state, any neutral kaon will evolve into a state of mixed strangeness. This phenomenon is known as $K^0 - \overline{K^0}$ or strangeness oscillation, and it can be observed experimentally. For example, to detect the presence of $\overline{K^0}$ resulting from the time evolution of initially pure K^0, we can examine the interactions of neutral kaons produced in any reaction as a function of the distance from their point of origin. At the point of production, the system is in a purely $S = +1$ state, but as the K_1^0 component decays away, the $\overline{K^0}$ component starts developing, and can interact strongly with the medium (say protons) to produce hyperons of $S = -1$, as follows:

$$\begin{aligned} \overline{K^0} + p &\to \Sigma^+ + \pi^+ + \pi^- \\ \overline{K^0} + p &\to \Lambda^0 + \pi^+ + \pi^0 \end{aligned} \tag{12.16}$$

On the other hand, conservation of strangeness does not allow production of hyperons by K^0s, namely,

$$\begin{aligned} K^0 + p &\not\to \Sigma^+ + \pi^+ + \pi^- \\ K^0 + p &\not\to \Lambda^0 + \pi^+ + \pi^0 \end{aligned}$$

Consequently detecting hyperon production in the medium will signal the presence of $\overline{K^0}$. This is, in fact, exactly what is observed. Close to where the K^0 are produced, there are no $\overline{K^0}$, and no secondary interactions of the kind given in Eq. 12.16 are found. Further downstream, however, there is evidence for the emergence of $\overline{K^0}$, as can be inferred from the observation of hyperon production.

The phenomenon of $K^0 - \overline{K^0}$ oscillations, as we will see shortly, has been used to measure the small mass difference between the K_1^0 and the K_2^0. It is similar to the technique that we alluded to earlier that is used to search for finite values of the masses of neutrinos. There have also been searches for oscillations between neutrons and antineutrons. However, only the neutral K-meson system has been observed to have these interesting quantum effects.

K_1^0 REGENERATION

Another interesting process involving the $K^0 - \overline{K^0}$ system is what is known as K_1^0 regeneration. This regeneration was originally proposed to occur by Abraham Pais and Oreste Piccioni. The idea relies on the fact that the cross section for $\overline{K^0}$ interactions with nucleons is different from (greater than) the cross section for K^0 interactions with nucleons. (We might expect $\sigma(\overline{K^0}N) > \sigma(K^0N)$, since strong $\overline{K^0}N$ collisions can produce everything that strong K^0N collisions can yield, and, in addition, can produce hyperons, as indicated in Eqs. 12.16.) Let us consider a beam of K^0 that is allowed to evolve in vacuum (through K_1^0 decay) into essentially a pure K_2^0 beam. If we now let the K_2^0 interact with some target material, then because the absorption of the $\overline{K^0}$ component is greater than that of the K^0 component of the K_2^0, the admixture of K^0 and $\overline{K^0}$ in the K_2^0 beam will change; in particular, if all the $\overline{K^0}$ is preferentially removed in strong interactions with the material, then we will be left with a transmitted beam that is pure K^0, and is consequently an equal mixture of K_1^0 and K_2^0. Hence, starting with a K_2^0 beam, we can regenerate the K_1^0 by passing the K_2^0 through some medium. This effect has also been observed and confirmed in many experiments.

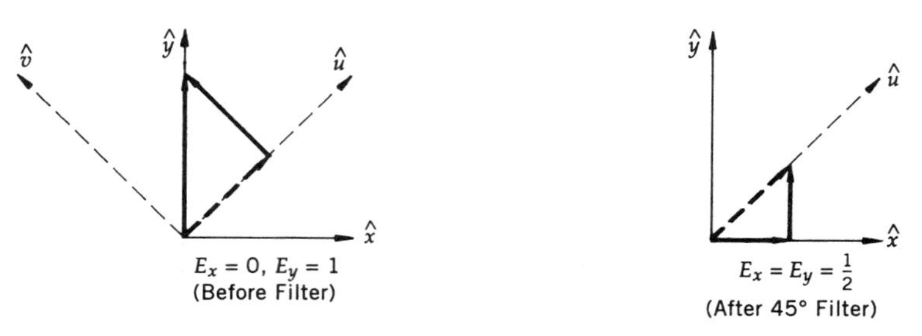

Figure 12.3 Polarization vector of an electric field before and after a filter that transmits light along 45° ($\hat{u}$).

Although the K_1^0 regeneration process may appear to be somewhat exotic, it has, in fact, a simple analogy in optics in the absorption of linearly polarized light. Just as the K^0 and $\overline{K^0}$ can be expressed using the K_1^0 and K_2^0 basis vectors, and vice versa, so can light polarized along $\hat{x}$ or $\hat{y}$ be expressed in terms of polarization vectors $\hat{u}$ and $\hat{v}$ that are rotated by 45° relative to $\hat{x}$ and $\hat{y}$ (see Fig. 12.3), and vice versa. If we pass light polarized along $\hat{y}$ through a filter that absorbs light along $\hat{v}$ (that is, a filter rotated by 45° relative to $\hat{y}$), then the partially transmitted light will be polarized along $\hat{u}$. This transmitted light can be decomposed along $\hat{x}$ and $\hat{y}$, to yield an equal mixture of components along both directions. Thus, starting off with light that has the electric field polarized along $\hat{y}$ (or with a K_2^0 beam), we absorb away the component along $\hat{v}$ (selectively deplete the $\overline{K^0}$ fraction of K_2^0), and thereby generate a component that is orthogonal to the incident polarization, namely along $\hat{x}$ (regenerate the orthogonal K_1^0 state).

VIOLATION OF *CP* INVARIANCE

Since $|K_2^0\rangle$ is a state with CP eigenvalue -1, if CP is conserved in weak interactions, then we would conclude that such a state could not decay into two pions. Namely the following transitions would not take place:

$$K_2^0 \not\to \pi^0 + \pi^0$$
$$K_2^0 \not\to \pi^+ + \pi^- \tag{12.17}$$

Nevertheless, an experiment performed in 1963 by James Christenson, James Cronin, Val Fitch, and René Turlay revealed that the long-lived component of K^0 did, in fact, decay into two pions. Because this implies that the long-lived and the short-lived components of K^0 need not be the K_2^0

and K_1^0 states that are the eigenstates of CP, we will refer to them as K_L^0 and K_S^0 (for "K-long and K-short") from now on. The current measured branching rates of the K_L^0 into $\pi^+\pi^-$ and into $\pi^0\pi^0$ are of the order of 0.1% of all the decays:

$$\frac{K_L^0 \to \pi^+\pi^-}{K_L^0 \to \text{ ALL}} = 2 \times 10^{-3}, \qquad \frac{K_L^0 \to \pi^0\pi^0}{K_L^0 \to \text{ ALL}} = 9 \times 10^{-4} \tag{12.18}$$

From the lifetimes of the K_L^0 and K_S^0 (Eq. 12.5), we can deduce that the decay rate of $K_L^0 \to 2\pi$ is about 5×10^{-6} that of the decay rate of the short-lived K_S^0 into two pions.

The experiment of Cronin, Fitch and their colleagues involved passing a pure K^0 beam of momentum $\sim$ 1 GeV/c through an evacuated tube that was about 15 meters long. The short-lived component in the beam was expected to have decayed completely by the time the beam reached the end of the tube ($\langle \ell_{K_S^0} \rangle = \gamma\beta c\tau_{K_S^0} \sim 6$ cm). The goal of the experiment was to search for 2π-decays at the end of the tube, and thereby establish a more stringent upper limit on the 2π-decay mode of K_2^0, but, instead, the experiment found 2π-decays representing the first clear evidence for CP violation.

As we mentioned in the introduction, the violation of CP is important in gaining an understanding of the matter–antimatter asymmetry in the universe. However, the character of CP violation is somewhat unusual. In particular, parity violation in weak interactions is maximal, while CP is violated only infinitesimally. In fact, for all practical purposes, we can think of CP as being conserved. It is only in a few isolated systems, such as the K^0 mesons, that the CP violation manifests itself. Thus, we can regard CP as almost a symmetry of physical systems.

Unlike parity violation, CP violation is particularly difficult to incorporate into theories. There are two ways one can try to do this:

1. We can assume that the weak Hamiltonian is not invariant under CP, and therefore the eigenstates of the weak Hamiltonian are not eigenstates of the CP operator. In such a case, the physical states will be superpositions of CP-odd and CP-even components.
2. The alternative is to assume that the weak Hamiltonian is CP invariant, and that its eigenstates are therefore eigenstates of CP. But that there exists an additional interaction Hamiltonian that operates only on systems such as the K^0 mesons. This new Hamiltonian is not CP invariant and is therefore the source of CP violation.

Let us elaborate on the two possibilities in some detail.

In the first scenario, the weak Hamiltonian is not invariant under CP. However, the violation of CP is very small. Thus, we can expect the eigenstates of the physical particles to differ from the eigenstates of CP only infinitesimally. Let us define the two eigenstates of the weak Hamiltonian in terms of the eigenstates of H_{st} (i.e., $|K^0>$ and $|\overline{K^0}>$), as follows (for a more formal discussion, see next section):

$$\begin{aligned}|K_S^0\rangle &= \frac{1}{\sqrt{2(1+|\epsilon|^2)}}\big((1+\epsilon)|K^0\rangle - (1-\epsilon)|\overline{K^0}\rangle\big)\\ &= \frac{1}{\sqrt{2(1+|\epsilon|^2)}}\Big[\big(|K^0\rangle - |\overline{K^0}\rangle\big) + \epsilon\big(|K^0\rangle + |\overline{K^0}\rangle\big)\Big]\\ &= \frac{1}{\sqrt{(1+|\epsilon|^2)}}\big(|K_1^0\rangle + \epsilon|K_2^0\rangle\big) \end{aligned} \tag{12.19}$$

$$\begin{aligned}|K_L^0\rangle &= \frac{1}{\sqrt{2(1+|\epsilon|^2)}}\big((1+\epsilon)|K^0\rangle + (1-\epsilon)|\overline{K^0}\rangle\big)\\ &= \frac{1}{\sqrt{2(1+|\epsilon|^2)}}\Big[\big(|K^0\rangle + |\overline{K^0}\rangle\big) + \epsilon\big(|K^0\rangle - |\overline{K^0}\rangle\big)\Big]\\ &= \frac{1}{\sqrt{(1+|\epsilon|^2)}}\big(|K_2^0\rangle + \epsilon|K_1^0\rangle\big) \end{aligned} \tag{12.20}$$

Here ϵ is an infinitesimal, complex parameter, reflecting a strangeness imbalance, and the deviation of the K_L^0 and K_S^0 states from true CP eigenstates, and, therefore, the degree of violation of CP in the system. In other words, we have constructed the physical short-lived and long-lived neutral K mesons as the admixtures of the CP eigenstates. These new states, therefore, cannot be eigenstates of CP, as can be checked:

$$\begin{aligned}CP|K_S^0\rangle &= \frac{1}{\sqrt{(1+|\epsilon|^2)}}\big(CP|K_1^0\rangle + \epsilon CP|K_2^0\rangle\big)\\ &= \frac{1}{\sqrt{(1+|\epsilon|^2)}}\big(|K_1^0\rangle - \epsilon|K_2^0\rangle\big) \neq |K_S^0\rangle\\ CP|K_L^0\rangle &= \frac{1}{\sqrt{(1+|\epsilon|^2)}}\big(CP|K_2^0\rangle + \epsilon CP|K_1^0\rangle\big)\\ &= \frac{1}{\sqrt{(1+|\epsilon|^2)}}\big(-|K_2^0\rangle + \epsilon|K_1^0\rangle\big) \neq -|K_L^0\rangle \end{aligned} \tag{12.21}$$

More interestingly, these new physical states are not even orthogonal:

$$\begin{aligned}\langle K_L^0|K_S^0\rangle &= \frac{1}{1+|\epsilon|^2}\big(\langle K_2^0| + \epsilon^*\langle K_1^0|\big)\big(|K_1^0\rangle + \epsilon|K_2^0\rangle\big)\\ &= \frac{1}{1+|\epsilon|^2}\big(\epsilon\langle K_2^0|K_2^0\rangle + \epsilon^*\langle K_1^0|K_1^0\rangle\big)\\ &= \frac{\epsilon+\epsilon^*}{1+|\epsilon|^2} = \frac{2\,\mathrm{Re}\,\epsilon}{1+|\epsilon|^2} = \langle K_S^0|K_L^0\rangle \end{aligned} \tag{12.22}$$

The lack of orthogonality of the two states is, in some sense, expected, since both have the same decay channels (such as the 2π and 3π modes), and this lack of orthogonality, in fact, is also a measure of the degree of CP violation. In other words, in this scenario, it is still the state $|K_1^0\rangle$ which decays into two pions. Since $|K_L^0\rangle$ contains a small admixture of $|K_1^0\rangle$, however, it also has a small probability of decaying into two pions.

Let us next parameterize the ratios of the K_L^0 and K_S^0 decay amplitudes by the complex numbers

$$\begin{aligned} \eta_{+-} &= \frac{\text{Amp}\,(K_L^0 \to \pi^+ + \pi^-)}{\text{Amp}\,(K_S^0 \to \pi^+ + \pi^-)} \\ \eta_{00} &= \frac{\text{Amp}\,(K_L^0 \to \pi^0 + \pi^0)}{\text{Amp}\,(K_S^0 \to \pi^0 + \pi^0)} \end{aligned} \tag{12.23}$$

We note that while these decays can take place through $\Delta I = \frac{1}{2}$ or $\frac{3}{2}$ channels, experimentally, as we mentioned earlier, the amplitude for $\Delta I = \frac{3}{2}$ is highly suppressed. Thus, for simplicity, we will take these decays as proceeding through $\Delta I = \frac{1}{2}$, which implies that the 2π systems are in the $I = 0$ state of isotopic spin. Then, from the definition of $|K_S^0\rangle$ and $|K_L^0\rangle$ in Eqs. 12.19 and 12.20 we can conclude that

$$\eta_{+-} = \eta_{00} = \epsilon \tag{12.24}$$

Thus, in this scenario, we would expect the ratios for the two decay modes in Eq. 12.23 to be the same. In fact, the experimental measurements are consistent with this expectation:

$$\begin{aligned} |\eta_{+-}| &= (2.27 \pm 0.02) \times 10^{-3} \\ \phi_{+-} &= (47 \pm 1)^\circ \end{aligned} \tag{12.25}$$

$$\begin{aligned} |\eta_{00}| &= (2.25 \pm 0.02) \times 10^{-3} \\ \phi_{00} &= (47 \pm 2)^\circ \end{aligned} \tag{12.26}$$

where we have parameterized

$$\eta_{+-} = |\eta_{+-}|e^{i\phi_{+-}} \tag{12.27}$$

and

$$\eta_{00} = |\eta_{00}|e^{i\phi_{00}} \tag{12.28}$$

In the alternate scenario, where we assume the physical states to be CP eigenstates and ascribe CP violation to the existence of an additional weaker Hamiltonian that contributes directly in the decays of kaons, there are many possible models. These are commonly referred to as milliweak or superweak theories, depending on the expected strengths of such interactions. In these theories, CP violation is purely a property of the kaon system, and is not expected in other regimes. The simplest of these phenomenological theories that is consistent with experiment is the superweak theory of Lincoln Wolfenstein, wherein the CP violating Hamiltonian changes strangeness of the K^0 by two units ($|\Delta S| = 2$). In this scenario, CP violation is a first-order process, and, correspondingly, the strength of such a Hamiltonian is much weaker. In fact, the superweak Hamiltonian has a strength about 10^{-8} times that of the conventional weak theory. (Hence the name superweak.) This theory also leads to the experimentally verified prediction that $\eta_{+-} = \eta_{00}$, and is identical to the earlier scenario, at least, when the $\Delta I = \frac{3}{2}$ channel is ignored. Including the $\Delta I = \frac{3}{2}$ admixture does not affect the prediction of the superweak theory, but does provide small contributions expected from the Standard Model for the case of the first scenario. Although, at this stage of experimental study the superweak scenario is still in agreement with the latest measurements, it is generally believed that CP violation is primarily due to the CP noninvariance of the weak Hamiltonian. The issue of whether there is a direct decay of the K_2^0 component into two pions is still unresolved.

TIME DEVELOPMENT AND ANALYSIS OF THE K^0–$\overline{K^0}$ SYSTEM

In the absence of weak interactions, we know that the states $|K^0\rangle$ and $|\overline{K^0}\rangle$ are eigenstates of the strong Hamiltonian, and describe distinct particle and antiparticle states. These are stationary states in a two-dimensional Hilbert space, and can be identified with the basis vectors

$$
\begin{aligned}
|K^0\rangle &\longrightarrow \begin{pmatrix} 1 \\ 0 \end{pmatrix} \\
|\overline{K^0}\rangle &\longrightarrow \begin{pmatrix} 0 \\ 1 \end{pmatrix}
\end{aligned}
\tag{12.29}
$$

Any normalized general state, in this space, can, of course, be written as a linear superposition of the two of the form

$$
\begin{aligned}
|\psi\rangle &= \frac{1}{\left(|a|^2 + |b|^2\right)^{1/2}}\left(a|K^0\rangle + b|\overline{K^0}\rangle\right) \\
&\longrightarrow \frac{1}{\left(|a|^2 + |b|^2\right)^{1/2}}\begin{pmatrix} a \\ b \end{pmatrix}
\end{aligned}
\tag{12.30}
$$

In the presence of weak interactions, however, the states in Eq. 12.29 will not be stationary. In fact, they can decay through many available weak channels. To describe the K^0–$\overline{K^0}$ system, in this case, requires an expansion of our Hilbert space through the incorporation of the other final states. Alternately, we can keep the analysis simple by restricting ourselves to the two-dimensional Hilbert space, and incorporating the effect of decays into some effective Hamiltonian. Because we will be describing a system whose states decay in time, that is, whose probability is not conserved, the effective Hamiltonian will no longer be Hermitian (see for example, Eq. 9.32). Nevertheless, the time evolution of a general two-dimensional vector in this space will still be governed by the time-dependent Schrödinger equation

$$i\hbar \frac{\partial|\psi(t)\rangle}{\partial t} = H_{\text{eff}}|\psi(t)\rangle \tag{12.31}$$

where H_{eff} is a 2×2 complex (non-Hermitian) matrix operator, which can be written in general as

$$H_{\text{eff}} = M - \frac{i}{2}\Gamma \tag{12.32}$$

with

$$\begin{aligned} M &= \frac{1}{2}\left(H_{\text{eff}} + H_{\text{eff}}^{\dagger}\right) \\ \Gamma &= i\left(H_{\text{eff}} - H_{\text{eff}}^{\dagger}\right) \end{aligned}$$

so that (12.33)

$$\begin{aligned} M^{\dagger} &= M \qquad \text{or} \qquad M_{jk}^{*} = M_{kj} \\ \Gamma^{\dagger} &= \Gamma \qquad \text{or} \qquad \Gamma_{jk}^{*} = \Gamma_{kj}, \qquad j,k = 1,2 \end{aligned}$$

where Γ and M are Hermitian 2×2 matrices. Note that, if Γ does not vanish, then

$$H_{\text{eff}}^{\dagger} \neq H_{\text{eff}} \tag{12.34}$$

In fact, as we have just emphasized, since H_{eff} has to incorporate decays, it cannot be Hermitian. We presume therefore that Γ is related to the lifetimes of the states.

For the time evolution of $|\psi\rangle$ in Eq. 12.31 we can write

$$\begin{aligned} i\hbar \frac{\partial|\psi(t)\rangle}{\partial t} &= H_{\text{eff}}|\psi(t)\rangle = \left(M - \frac{i}{2}\Gamma\right)|\psi(t)\rangle \\ -i\hbar \frac{\partial\langle\psi(t)|}{\partial t} &= \langle\psi(t)|H_{\text{eff}}^{\dagger} = \langle\psi(t)|\left(M + \frac{i}{2}\Gamma\right) \end{aligned} \tag{12.35}$$

From these equations, it is straightforward to show that

$$\frac{\partial\langle\psi(t)|\psi(t)\rangle}{\partial t} = -\frac{1}{\hbar}\langle\psi(t)|\Gamma|\psi(t)\rangle \tag{12.36}$$

Thus, we conclude that, since decays reduce probability, the matrix Γ, for any state in this two-dimensional space, must satisfy the requirement

$$\langle\psi(t)|\Gamma|\psi(t)\rangle \geq 0 \tag{12.37}$$

In other words, the matrix Γ must have positive or vanishing eigenvalues. Therefore, as anticipated, Γ represents the decay characteristics of the system. The eigenvalues of matrix M, which correspond to the real parts of the energy levels of the system, define the masses of the states in their own rest frames ($\vec{p} = 0$). For this reason, M is known as the *mass matrix* of the system, while Γ is commonly referred to as the *decay matrix*.

Let us further note that if we represent H_{eff} by a general 2×2 matrix of the form

$$H_{\text{eff}} = \begin{pmatrix} A & B \\ C & D \end{pmatrix} \tag{12.38}$$

then from Eq. 12.29 we obtain

$$\begin{aligned} \langle K^0|H_{\text{eff}}|K^0\rangle &= A \\ \langle\overline{K^0}|H_{\text{eff}}|\overline{K^0}\rangle &= D \end{aligned} \tag{12.39}$$

Invariance of H_{eff} under CPT then leads to the requirement that

$$A = D \tag{12.40}$$

We can therefore write the general form of H_{eff}, consistent with CPT invariance, as follows:

$$H_{\text{eff}} = \begin{pmatrix} A & B \\ C & A \end{pmatrix} \tag{12.41}$$

Let us next construct the eigenstates of H_{eff}. Let us assume the following parametrization for the two eigenstates:

$$\begin{aligned} |K_S^0\rangle &= \frac{1}{\left(|p|^2 + |q|^2\right)^{1/2}}\left(p|K^0\rangle + q|\overline{K^0}\rangle\right) \\ &\rightarrow \frac{1}{\left(|p|^2 + |q|^2\right)^{1/2}}\begin{pmatrix} p \\ q \end{pmatrix} \end{aligned}$$

$$|K_L^0\rangle = \frac{1}{\left(|r|^2 + |s|^2\right)^{1/2}}\left(r|K^0\rangle + s|\overline{K^0}\rangle\right)$$
$$\rightarrow \frac{1}{\left(|r|^2 + |s|^2\right)^{1/2}}\begin{pmatrix} r \\ s \end{pmatrix} \tag{12.42}$$

where p, q, r, s are complex numbers that define the K_S^0 and K_L^0 eigenstates of H_{eff}, which in the particle's rest frame ($\vec{p} = 0$) have eigenvalues $m_S - (i/2)\gamma_S$ and $m_L - (i/2)\gamma_L$, respectively. Namely,

$$H_{\text{eff}}|K_S^0\rangle = \left(m_S - \frac{i}{2}\gamma_S\right)|K_S^0\rangle$$
$$H_{\text{eff}}|K_L^0\rangle = \left(m_L - \frac{i}{2}\gamma_L\right)|K_L^0\rangle \tag{12.43}$$

where m_S and m_L, and γ_S and γ_L, correspond to the masses and widths of the two eigenstates. (Note that throughout this section we have set $C = 1$.) In the basis of the K_L^0, K_S^0 eigenstates, the diagonal elements of H_{eff} are, of course, the two eigenvalues of Eq. 12.43. In this basis, the sum of the eigenvalues equals the trace (Tr) of H_{eff}. Since the trace of a matrix is the same in any basis, however, Tr H_{eff} always equals the sum of the two eigenvalues. Thus, from Eq. 12.41, we obtain

$$\text{Tr}H_{\text{eff}} = 2A = \left(m_S - \frac{i}{2}\gamma_S\right) + \left(m_L - \frac{i}{2}\gamma_L\right)$$

or

$$A = \frac{1}{2}(m_S + m_L) - \frac{i}{4}(\gamma_S + \gamma_L) \tag{12.44}$$

Writing out the first equation in 12.43, we obtain

$$\begin{pmatrix} A & B \\ C & A \end{pmatrix}\begin{pmatrix} p \\ q \end{pmatrix} = \left(m_S - \frac{i}{2}\gamma_S\right)\begin{pmatrix} p \\ q \end{pmatrix}$$

or

$$\begin{pmatrix} A - m_S + \frac{i}{2}\gamma_S & B \\ C & A - m_S + \frac{i}{2}\gamma_S \end{pmatrix}\begin{pmatrix} p \\ q \end{pmatrix} = 0 \tag{12.45}$$

Equation 12.45 defines a set of coupled linear, homogeneous equations in the unknowns p and q, and a nontrivial solution in such a case exists only if the determinant of the coefficient matrix vanishes. In other words, for a nontrivial solution of Eq. 12.45 to exist we must have

$$\det\begin{pmatrix} A - m_S + \frac{i}{2}\gamma_S & B \\ C & A - m_S + \frac{i}{2}\gamma_S \end{pmatrix} = 0 \tag{12.46}$$

or

$$BC = \left(A - m_S + \frac{i}{2}\gamma_S\right)^2 = \left[\frac{1}{2}(m_L - m_S) - \frac{i}{4}(\gamma_L - \gamma_S)\right]^2$$

or

$$\frac{1}{2}(m_L - m_S) - \frac{i}{4}(\gamma_L - \gamma_S) = \pm\sqrt{BC} \tag{12.47}$$

Substituting this back into Eq. 12.45, leads to the conclusion that the coefficients p and q satisfy

$$\frac{p}{q} = \pm\sqrt{\frac{B}{C}} \tag{12.48}$$

Similarly, from the second equation in 12.43, we can obtain

$$\frac{r}{s} = \mp\sqrt{\frac{B}{C}} = -\frac{p}{q} \tag{12.49}$$

Thus, if we choose $r = p$ and $s = -q$, we can write

$$\begin{aligned} |K_S^0\rangle &= \frac{1}{\left(|p|^2 + |q|^2\right)^{1/2}}\left(p|K^0\rangle + q|\overline{K^0}\rangle\right) \\ |K_L^0\rangle &= \frac{1}{\left(|p|^2 + |q|^2\right)^{1/2}}\left(p|K^0\rangle - q|\overline{K^0}\rangle\right) \end{aligned} \tag{12.50}$$

The choice made in Eqs. 12.19 and 12.20 would then correspond to

$$p = 1 + \epsilon \qquad q = -(1 - \epsilon) \tag{12.51}$$

Inverting the relations in Eq. 12.50, we can now write

$$\begin{aligned} |K^0\rangle &= \frac{\left(|p|^2 + |q|^2\right)^{1/2}}{2p}\left(|K_S^0\rangle + |K_L^0\rangle\right) \\ |\overline{K^0}\rangle &= \frac{\left(|p|^2 + |q|^2\right)^{1/2}}{2q}\left(|K_S^0\rangle - |K_L^0\rangle\right) \end{aligned} \tag{12.52}$$

Since $|K_S^0\rangle$ and $|K_L^0\rangle$ are eigenstates of H_{eff}, we note from Eqs. 12.31 and 12.43 that we can write

$$\begin{aligned} |K_S^0(t)\rangle &= e^{-(i/\hbar)(m_S - (i/2)\gamma_S)t}|K_S^0\rangle \\ |K_L^0(t)\rangle &= e^{-(i/\hbar)(m_L - (i/2)\gamma_L)t}|K_L^0\rangle \end{aligned} \tag{12.53}$$

Recall, again, that we have set $C = 1$ in this section, (cf. Eq. 9.32).

The states in Eq. 12.53 clearly decay in time, with lifetimes given by

$$\tau_S = \frac{\hbar}{\gamma_S}$$
$$\tau_L = \frac{\hbar}{\gamma_L} \qquad (12.54)$$

which correspond, of course, to the previously cited lifetimes of $\tau_S \simeq 0.9\times 10^{-10}$ sec and $\tau_L \simeq 5\times 10^{-8}$ sec. Also, as we stated earlier, m_L and m_S can be identified with the masses of the long-lived and the short-lived particles, respectively. It is worth emphasizing that, because the K_L^0 and K_S^0 are not each other's antiparticles (unlike the K^0 and $\overline{K^0}$), their decays and masses do not have to be identical.

Let us now assume that we start with an initially pure K^0 beam. The evolution of such a beam can be obtained from Eqs. 12.52 and 12.53 as follows:

$$\begin{aligned}
|K^0(t)\rangle &= \frac{\left(|p|^2+|q|^2\right)^{1/2}}{2p}\left(|K_S^0(t)\rangle + |K_L^0(t)\rangle\right) \\
&= \frac{\left(|p|^2+|q|^2\right)^{1/2}}{2p}\left[e^{-(i/\hbar)(m_S-(i/2)\gamma_S)t}|K_S^0\rangle + e^{-(i/\hbar)(m_L-(i/2)\gamma_L)t}|K_L^0\rangle\right] \\
&= \frac{\left(|p|^2+|q|^2\right)^{1/2}}{2p}\left[e^{-(i/\hbar)(m_S-(i/2)\gamma_S)t}\frac{1}{\left(|p|^2+|q|^2\right)^{1/2}}\left(p|K^0\rangle + q|\overline{K^0}\rangle\right)\right. \\
&\quad \left. + e^{-(i/\hbar)(m_L-(i/2)\gamma_L)t}\frac{1}{\left(|p|^2+|q|^2\right)^{1/2}}\left(p|K^0\rangle - q|\overline{K^0}\rangle\right)\right] \\
&= \frac{1}{2p}\left[p\left(e^{-(i/\hbar)(m_S-(i/2)\gamma_S)t} + e^{-(i/\hbar)(m_L-(i/2)\gamma_L)t}\right)|K^0\rangle\right. \\
&\quad \left. + q\left(e^{-(i/\hbar)(m_S-(i/2)\gamma_S)t} - e^{-(i/\hbar)(m_L-(i/2)\gamma_L)t}\right)|\overline{K^0}\rangle\right] \qquad (12.55)
\end{aligned}$$

Thus, the probability of finding the state $|K^0>$ in the beam at a later time t is given by

$$\begin{aligned}
P(K^0,t) &= |\langle K^0|K^0(t)\rangle|^2 \\
&= \frac{1}{4}\left|\left(e^{-(i/\hbar)(m_S-(i/2)\gamma_S)t} + e^{-(i/\hbar)(m_L-(i/2)\gamma_L)t}\right)\right|^2 \\
&= \frac{1}{4}\left(e^{-(\gamma_S t/\hbar)} + e^{-(\gamma_L t/\hbar)} + e^{-(1/2\hbar)(\gamma_S+\gamma_L)t}\cdot 2\cos(m_L-m_S)t/\hbar\right) \\
&= \frac{1}{4}e^{-t/\tau_S} + \frac{1}{4}e^{-t/\tau_L} + \frac{1}{2}e^{-1/2(1/\tau_S+1/\tau_L)t}\cos\frac{\Delta m t}{\hbar} \qquad (12.56)
\end{aligned}$$

Here we have defined the mass difference:

$$\Delta m = m_L - m_S \tag{12.57}$$

Similarly, we can obtain the probability of finding the state $|\overline{K^0}\rangle$ at a time t in the original $|K^0\rangle$ beam:

$$\begin{aligned} P(\overline{K^0}, t) &= |\langle \overline{K^0}|K^0(t)\rangle|^2 \\ &= \left|\frac{q}{p}\right|^2 \left[\frac{1}{4}e^{-t/\tau_S} + \frac{1}{4}e^{-t/\tau_L} - \frac{1}{2}e^{-1/2(1/\tau_S + 1/\tau_L)t} \cos\frac{\Delta m t}{\hbar}\right] \end{aligned} \tag{12.58}$$

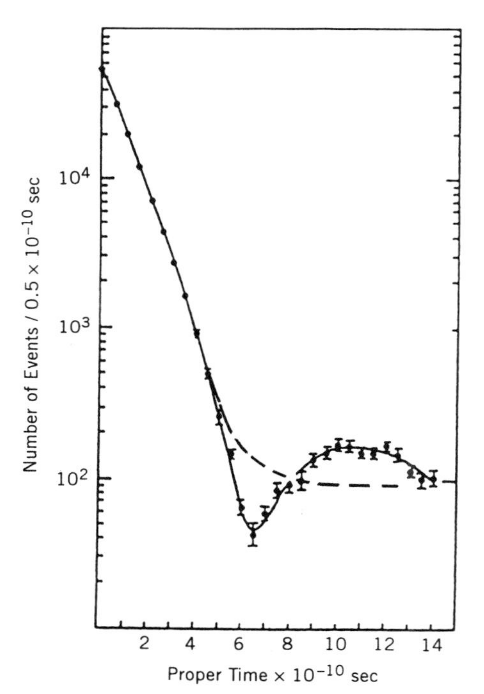

Figure 12.4 Data for $K^0_{L,S} \to \pi^+\pi^-$ as a function of the proper time, after passing a K^0_L beam through a carbon regenerator. The dashed curve shows the shape expected in the absence of $K^0_L - K^0_S$ interference. The solid curve shows a fit including interference, which is used to determine ϕ_{+-}.

(After W. C. Carithers et al., 1975. Phys. Rev. Lett. 34: 1244.)

We note from Eqs. 12.57 and 12.58 that, if the two states $|K^0_S\rangle$ and $|K^0_L\rangle$ had identical masses, namely, for $\Delta m = 0$, the beam intensities would only show exponential fall-off with two characteristic time scales. Experimentally, however, an oscillatory behavior (strangeness oscillation) is also observed, implying a finite mass difference for the two particles. This mass splitting can, in fact, be measured from the period of oscillation, and has the value

$$\Delta m = m_L - m_S \simeq 3.5 \times 10^{-12} \text{ MeV}/c^2 \tag{12.59}$$

The mass splitting is indeed small (remember $m_K \sim 500$ MeV/c^2), and, when related to a possible K^0–$\overline{K^0}$ mass difference, implies that CPT invariance holds to high accuracy in decays of neutral kaons. It also suggests that the K^0–$\overline{K^0}$ mixing is a second-order effect in the weak Hamiltonian, satisfying $|\Delta S| = 1$ for nonleptonic weak interactions. (Note that a direct K^0–$\overline{K^0}$ transition would correspond to $|\Delta S| = 2$.)

Starting off with Eq. 12.55, we can also calculate the probability of finding a $|K_L^0\rangle$ or $|K_S^0\rangle$ in the beam as a function of time. Because both $|K_L^0\rangle$ and $|K_S^0\rangle$ decay into $\pi^+\pi^-$ pairs, by studying the number of $\pi^+\pi^-$ (or $\pi^0\pi^0$) decays as a function of proper time, we can observe quantum-mechanical interference in the two-pion decay modes of the $|K_L^0\rangle$ and $|K_S^0\rangle$. That is, if we measure the square of the sum of the amplitudes

$$\left| |K_L^0 \to 2\pi\rangle + |K_S^0 \to 2\pi\rangle \right|^2 \tag{12.60}$$

such data can provide values of the relative phases ϕ_{+-} and ϕ_{00} from the interference term in Eq. 12.60. The result of this type of measurement is shown in Fig. 12.4.

SEMILEPTONIC K^0 DECAYS

From studies of K^0 and $\overline{K^0}$ produced in reactions such as given in Eq. 12.2, we know that when a K^0 decays semileptonically there is a positron in the final state, while for the decay of the $\overline{K^0}$ there is an electron:

$$\begin{aligned} K^0 &\longrightarrow \pi^- + e^+ + \nu_e \\ \overline{K^0} &\longrightarrow \pi^+ + e^- + \overline{\nu_e} \end{aligned} \tag{12.61}$$

Under the CP operation, all particles, including neutrinos, transform to their physical antiparticles. Consequently, the decays just given can, in principle, provide additional insight into CP violation in the kaon system.

One interesting possibility is to start off with a beam that is composed primarily of either K^0 or $\overline{K^0}$, and use the fact that there is strangeness oscillation in time to study the variation in the number of decays involving e^+(denoted by N^+) and the number involving e^-(denoted by N^-). (See Eqs. 12.56 and 12.58.) As the K_S^0 component decays away, the original beam will turn completely into pure K_L^0. Now, if K_L^0 is an eigenstate of CP, then it will have equal admixtures of K^0 and $\overline{K^0}$, and therefore an

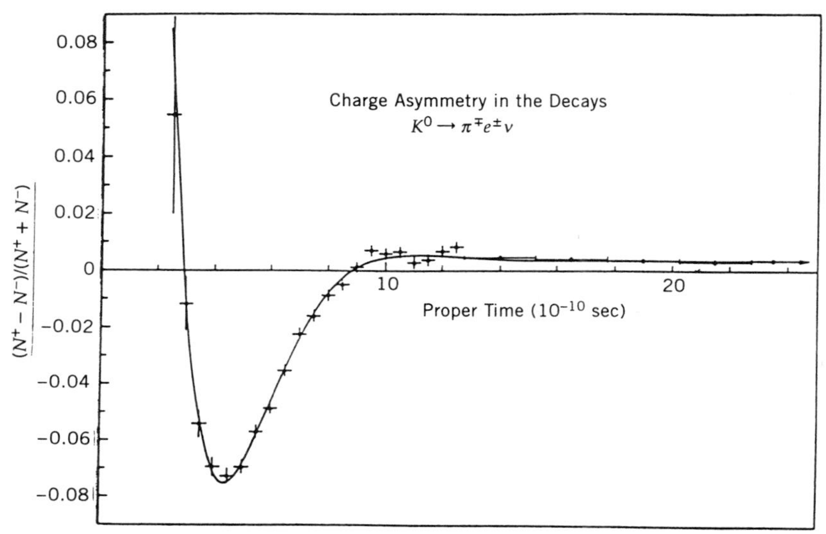

Figure 12.5 The charge asymmetry observed for $K^0 \rightarrow \pi^- e^+ \nu$ and $\overline{K^0} \rightarrow \pi^+ e^- \overline{\nu}$ as a function of proper time, when the starting beam is predominantly K^0. The observed interference effect is sensitive to the $K_L^0 - K_S^0$ mass difference. For large values of proper time, the asymmetry represents a *CP* violating effect and determines the strangeness imbalance in K_L^0.
(After S. Gjesdal et al., 1974. Phys. Lett. **52B**: 113.)

identical number of e^+ and e^- decays. If, however, the K_L^0 is not an eigenstate of CP, namely if CP is violated in the neutral kaon system, then there will be an asymmetry in the number of e^+ and e^- decays. That is, as time goes by, and oscillations have died down (recall that $\tau_S \ll \tau_L$), there will be an asymmetry observed in N^+ and N^-, which depends on the relative strengths of the K^0 and $\overline{K^0}$ components in K_L^0. This strangeness asymmetry, which appears explicitly in our definition in Eq. 12.20, also determines the value of $|q/p|^2$ in Eq. 12.58, which is observed to be $\sim 3.3 \times 10^{-3}$ (and corresponds to 2 Re ϵ). This result is shown in Fig. 12.5.

Problems

XII.1 Ignoring CP violation, to 10% accuracy, plot the probability of observing $\overline{K^0}$ as a function of time in a beam that is initially ($t = 0$) pure K^0.

XII.2 Using the parameters η_{+-} and ϕ_{+-} (of Eq. 12.27), derive an expression for the rate of $K^0 \rightarrow \pi^+\pi^-$ decay as a function of time. Assume that you start with a pure K^0 beam that develops according to Eq. 12.55. You may ignore the overall normalization of the decay rate.

Suggested Readings

Frauenfelder, H., and E. M. Henley. 1991. *Subatomic Physics*. Englewood Cliffs, N.J.: Prentice-Hall.

Griffiths, D. 1987. *Introduction to Elementary Particles*. New York: Wiley.

Kabir, P. K. 1968. *The CP Puzzle*. New York: Academic Press.

Williams, W. S. C. 1991. *Nuclear and Particle Physics*. London/New York: Oxford Univ. Press.

Chapter XIII

THE STANDARD MODEL

INTRODUCTORY REMARKS

In Chapter IX, we discussed the properties of only a few low-mass hadrons. As the energies available in accelerators increased, however, it was possible to produce many more excited states of those particles, with larger masses and higher spins. In fact, by the mid-1960s, a whole host of new particles were found to exist, and it was questioned whether they could all be regarded as fundamental constituents of matter. As we argued earlier, even the lightest baryons, namely the proton and the neutron, show indirect evidence of substructure. For example, the large anomalous magnetic moments observed for these particles clearly imply a complex internal distribution of currents, which is a conclusion that has particularly dramatic impact for the case of the neutron. In 1964, from the pattern of the observed spectrum of hadrons, Murray Gell-Mann and George Zweig suggested independently that all such particles could be understood as composed of quark constituents. These constituents had rather unusual properties (see Table 9.5), and were initially regarded as primarily calculational tools rather than as true physical objects.

A series of measurements performed in the late 1960s at the Stanford Linear Accelerator Center (SLAC) on electron scattering from hydrogen and deuterium revealed that the data could be most easily understood if protons and neutrons were composed of pointlike objects that had charges of $-\frac{1}{3}e$ and $\frac{2}{3}e$. These experiments, led by Jerome Friedman, Henry Kendall, and Richard Taylor, corresponded to a modern parallel of the original work of the Rutherford group, where, instead of finding "pointlike" nuclei within atoms, the presence of pointlike "quarks" or "partons" was deduced from the characteristics of the inelastically scattered electrons. (Recall that the nuclei in the original experiments of the Rutherford group were not probed very deeply and therefore did not break apart in the collisions with α-particles; on the other hand, the scattering of electrons at SLAC involved sufficient momentum transfers to break apart the neutrons and protons.)

Thus, by the early 1970s, it became increasingly clear that hadrons were not fundamental pointlike particles. In contrast, leptons still do not show any evidence of structure, even for the highest available momentum transfers. It is natural therefore to regard leptons as elementary particles, but to regard hadrons as being composed of more fundamental constituents. This line of thought—completely phenomenological in the beginning—merged the observations from electron scattering with those from particle spectroscopy and the quark model, and culminated in the present Standard Model. The Standard Model incorporates all the known fundamental particles, namely, the quarks, leptons, and the gauge bosons, and it provides a theory describing three of the basic forces of nature—the strong, the weak, and the electromagnetic.

QUARKS AND LEPTONS

As we saw earlier, each charged lepton has its own neutrino, and there are three families (or flavors) of such leptons, namely,

$$\begin{pmatrix} \nu_e \\ e^- \end{pmatrix} \quad \begin{pmatrix} \nu_\mu \\ \mu^- \end{pmatrix} \quad \text{and} \quad \begin{pmatrix} \nu_\tau \\ \tau^- \end{pmatrix} \tag{13.1}$$

In writing this, we have used the convention introduced previously in connection with the strong isospin symmetry, namely, the higher member of a given multiplet carries a higher electric charge. We now believe that the constituents of the hadrons—the quarks—also come in three families (see Problem IX.4):

$$\begin{pmatrix} u \\ d \end{pmatrix} \quad \begin{pmatrix} c \\ s \end{pmatrix} \quad \text{and} \quad \begin{pmatrix} t \\ b \end{pmatrix} \tag{13.2}$$

The top quark (t) has yet to be found, and several experiments are currently searching for particles that would contain this quark. However, the experimental evidence for the existence of the other quarks, namely, the up (u), down (d), charm (c), strange (s), and the bottom (b) quarks, is extensive. The assumed charges and baryon content of the different quarks were given in Table 9.5. The baryon numbers are all $B = \frac{1}{3}$, and the charges are

$$\begin{aligned} Q[u] &= Q[c] = Q[t] = +\tfrac{2}{3}e \\ Q[d] &= Q[s] = Q[b] = -\tfrac{1}{3}e \end{aligned} \tag{13.3}$$

Although the fractional nature of the electric charges has been deduced indirectly for only the u and d quarks from electron scattering, phenomenologically, such charge assignments also provide a natural way for classifying the existing hadrons as bound states of quarks. Quarks also carry other quantum numbers (see Table 9.5). For example, because we defined the strangeness of the K^+ as $+1$, we will see shortly that the strange quark will have to be assigned a strangeness of -1. The charm, top, and the bottom quarks, correspondingly, carry their own flavor quantum numbers. Of course, each quark has its own antiquark, which has opposite electric charge and other internal quantum numbers, such as strangeness and charm.

QUARK CONTENT OF MESONS

The quarks, like the leptons, are pointlike fermionic particles. In other words, they have spin angular momentum of $\frac{1}{2}$. This suggests that, since mesons have integer spin, then, if they are bound states of quarks, they can only consist of an even number of these particles. In fact, detailed analysis reveals that every known meson can be described as a bound state of a quark and an antiquark. Thus, for example, a π^+ meson, which has spin-zero and electric charge $+1$, can be described as the bound state

$$\pi^+ = u\overline{d} \tag{13.4}$$

It follows, therefore, that the π^- meson, which is the antiparticle of the π^+, can be described as the bound state

$$\pi^- = \overline{u}d \tag{13.5}$$

The π^0 meson, which is charge neutral, can, in principle, be described as a bound state of any quark and an antiquark of the same kind. However, other considerations, such as the fact that all three π mesons belong to an isospin multiplet, and should therefore have the same internal structure, lead to a description of the π^0 meson as

$$\pi^0 = \frac{1}{\sqrt{2}}(u\overline{u} - d\overline{d}) \tag{13.6}$$

The strange mesons can similarly be described as bound states of a quark and an antiquark, where one of the constituents is strange. Thus, we can identify the following systems:

$$\begin{aligned} K^+ &= u\bar{s} \\ K^- &= \bar{u}s \\ K^0 &= d\bar{s} \\ \overline{K}^0 &= \bar{d}s \end{aligned} \tag{13.7}$$

It is quite easy to check that not only are the charge assignments right, but even the strangeness quantum numbers work out to be correct if we assign a strangeness quantum number $S = -1$ to the s quark. We note that, since there are other quarks with new quantum numbers, phenomenologically, on the basis of the quark model we would also expect new kinds of mesons. Many such mesons have already been found. For example, the charge–neutral J/ψ meson whose discovery in 1974, by independent groups headed by Samuel Ting and by Burton Richter, suggested the first evidence for the existence of the charm quark, can be described as a bound state of charmonium (named in analogy with positronium):

$$J/\psi = c\bar{c} \tag{13.8}$$

This is a "normal" meson, in the sense that the quantum numbers of charm add up to zero, but its properties (decays) cannot be explained using just u, d, and s quarks. There are, of course, mesons that contain "open" charm, such as

$$\begin{aligned} D^+ &= c\bar{d} \\ D^- &= \bar{c}d \\ D^0 &= c\bar{u} \\ \overline{D}^0 &= \bar{c}u \end{aligned} \tag{13.9}$$

We can think of these mesons as the charm analogues of the K mesons, and the properties of these mesons have by now been studied in great detail. In analogy with the K^+ meson, the D^+ meson is defined to have charm flavor of $+1$, which then defines the charm quantum number for the c quark to be $+1$. Experimentally, we also know of mesons that carry both strangeness and charm quantum numbers, two of these are denoted as

$$\begin{aligned} D_s^+ &= c\bar{s} \\ D_s^- &= \bar{c}s \end{aligned} \tag{13.10}$$

Finally, we also have extensive experimental evidence for mesons where one of the constituents is a bottom quark. These are, again, analogous to the K mesons, and have structure of the form

$$\begin{aligned} B^+ &= u\overline{b} \\ B^- &= \overline{u}b \\ B_d^0 &= d\overline{b} \\ \overline{B}_d^0 &= \overline{d}b \end{aligned} \tag{13.11}$$

The charge–neutral states involving b and s quarks are particularly interesting because, just like the K^0–$\overline{K}^0$ system, they are expected to exhibit CP violation in their decays:

$$\begin{aligned} B_s^0 &= s\overline{b} \\ \overline{B}_s^0 &= \overline{s}b \end{aligned} \tag{13.12}$$

Recent proposals to build e^+e^- colliders to be used as "B-factories," are intended primarily to study the properties of these neutral B mesons in the clean environment of e^+e^- collisions:

$$e^+ + e^- \longrightarrow B + \overline{B} \tag{13.13}$$

QUARK CONTENT OF BARYONS

Just as mesons can be thought of as bound states of quarks and antiquarks, so can baryons be considered as constructed out of these constituents. Since baryons carry half-integral spin angular momenta (they are fermions), however, they can be formed from only an odd number of quarks. Properties of baryons are most consistent with being composed of only three quarks. Thus, we can think of the proton and the neutron as corresponding to the bound states

$$\begin{aligned} p &= uud \\ n &= udd \end{aligned} \tag{13.14}$$

Similarly, the hyperons, which carry a strangeness quantum number, can be described by

$$\begin{aligned} \Lambda^0 &= uds \\ \Sigma^+ &= uus \\ \Sigma^0 &= uds \\ \Sigma^- &= dds \end{aligned} \tag{13.15}$$

Also, the cascade particles, which carry two units of strangeness, can be described as

$$\begin{aligned} \Xi^0 &= uss \\ \Xi^- &= dss \end{aligned} \tag{13.16}$$

Since all baryons have baryon number of unity, it follows, therefore, that each quark must carry a baryon number of $\frac{1}{3}$. Furthermore, since a meson consists of a quark and an antiquark, and since an antiquark would have a baryon number $-\frac{1}{3}$, we conclude that mesons do not carry baryon number, which is consistent with our previous discussion.

NEED FOR COLOR

In trying to extend the quark model to all the baryons, however, one soon runs into a difficulty. We have already discussed the Δ^{++} baryon, which is nonstrange, carries two units of positive charge, and has spin angular momentum of $\frac{3}{2}$. Thus, naively, we can conclude that the Δ^{++} can be described by three up quarks:

$$\Delta^{++} = uuu \tag{13.17}$$

This substructure satisfies all the known quantum numbers and, in the ground state (namely, when all relative orbital waves should vanish), the three up quarks can have parallel spins, in order to provide a resultant value of $J = \frac{3}{2}$. We must conclude, however, that the wave function for this final state, which consists of three identical fermions, would therefore be symmetric under the exchange of any two quarks. This is, of course, incompatible with the Pauli principle, which requires a wave function containing identical fermions to be totally antisymmetric. It would appear, therefore, that the quark model cannot adequately describe the Δ^{++}. On the other hand, the model works so well for other hadrons that it would seem unwise to give it up entirely. An interesting resolution can be attained if it is assumed that all quarks carry an additional internal quantum number, and that the final state in Eq. 13.17 is, in fact, antisymmetric in the space corresponding to this quantum number.

This additional degree of freedom is referred to as *color*, and it is believed that each of the quarks comes in three different colors. Namely, the quark multiplets take the form

$$\begin{pmatrix} u^a \\ d^a \end{pmatrix} \quad \begin{pmatrix} c^a \\ s^a \end{pmatrix} \quad \begin{pmatrix} t^a \\ b^a \end{pmatrix} \qquad a = \text{red, blue, green} \tag{13.18}$$

At this point of our development, color can be regarded as merely a new quantum number needed for phenomenological reasons for understanding the substructure of hadrons. However, we will soon see that, in fact, color is to the strong interaction what charge is to the electromagnetic force, namely, the source of the respective fields.

All the known hadrons do not appear to carry any net color, and therefore correspond to bound states of quarks and of antiquarks of zero total color quantum number, or, simply stated, hadrons are color–neutral bound states of quarks. Under the interchange of any two quarks, the color singlet wave function of three quarks changes sign, while that of a quark–antiquark color singlet does not. This hypothesis leads to an excellent description of all known baryons as bound states of three quarks, and of mesons as bound states of quark–antiquark pairs. In particular, it also explains the structure of the Ω^- baryon, which, as we know, has a strangeness of -3 and spin angular momentum of $\frac{3}{2}$, and simply corresponds to the ground state of three strange quarks:

$$\Omega^- = sss \tag{13.19}$$

We see once again how the symmetry property in the color space plays a crucial role in assuring the overall antisymmetry of the fermionic wave function for this state.

The theoretical postulate of color seems rather ad hoc, especially since the observable hadrons do not carry a color quantum number. The existence of color can, however, be established experimentally as follows. Consider the annihilation of an electron and positron, leading to the creation of a muon and antimuon pair or a quark and antiquark pair. The process can be thought of as going through the production of an intermediate virtual photon, as shown in Fig. 13.1. The cross section for the production of hadrons in this process depends on the number of ways a photon can produce a quark–antiquark pair. This must therefore be proportional to the number of available quark colors. Namely, the ratio of production cross sections

$$R = \frac{\sigma(e^- e^+ \to \text{hadrons})}{\sigma(e^- e^+ \to \mu^- \mu^+)} \tag{13.20}$$

is proportional to the number of colors. A detailed analysis of this quantity at any energy shows that there are exactly three quark colors. Because

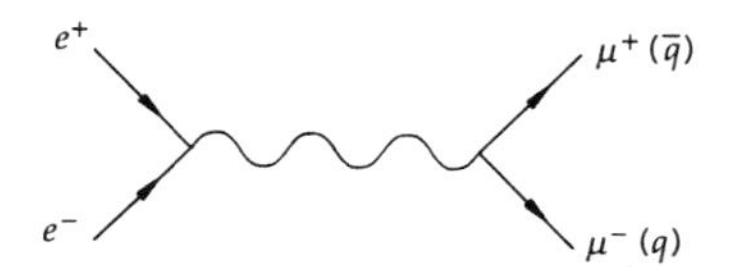

Figure 13.1 The annihilation of e^+e^- through a virtual photon into a $\mu^+\mu^-$ or $q\bar{q}$ pair.

the production of hadrons depends, in addition, on the electric charges of the quarks, such data also confirm the fractional nature of electric charge carried by quarks. In ending this section, we wish to point out that electron–positron annihilation at high energies is one of the cleanest ways to establish the presence of new quark flavors. That is, when the energy of the e^+e^- system exceeds the threshold for production of hadrons containing new quarks, the ratio in Eq. 13.29 must increase and display a step at that energy. Of course, in addition, beyond that threshold, the new hadrons can be observed in the final states of such collisions.

Example—Quark Model for Mesons

We will now apply the symmetry requirements of the strong interactions to $q\overline{q}$ wave functions, and thereby deduce the quantum numbers that we would expect for the spectrum of charge–neutral meson states in a simple nonrelativistic quark model. Specifically, we will establish the spin (J), parity (P), and charge conjugation (C) quantum-number restrictions that apply to such systems. The $q\overline{q}$ wave function is a product of separate wave functions, each of which has a unique symmetry of its own under the exchange of the two particles:

$$\Psi = \psi_{\text{space}}\psi_{\text{spin}}\psi_{\text{charge}} \tag{13.21}$$

where ψ_{space} denotes the space-time part of the $q\overline{q}$ wave function, ψ_{spin} represents the intrinsic-spin, and ψ_{charge} the charge conjugation properties. We have ignored the part of the wave function that is associated with the color degree of freedom because we know that this will always have even symmetry for mesons.

The symmetry of ψ_{space} under the exchange of the q and $\overline{q}$ is, as usual, determined by the spherical harmonics and the relative orbital angular momentum of the q and $\overline{q}$. If we call the exchange operation X, then, schematically, we have

$$X\psi_{\text{space}} = XY_{\ell m}(\theta, \phi) = (-1)^{\ell}\psi_{\text{space}} \tag{13.22}$$

Therefore, if Ψ is a state of definite parity, the spatial part of the wave function will be either symmetric or antisymmetric under an exchange, depending on whether ℓ is even or odd.

The effect of the exchange operation on ψ_{spin} will depend on whether the two quark spins are in a spin state $s = 0$ or $s = 1$. Considering the states with $s_z = 0$, we obtain

$$\begin{aligned} s = 0: \quad & X[|\uparrow\downarrow\rangle - |\downarrow\uparrow\rangle] = -[|\uparrow\downarrow\rangle - |\downarrow\uparrow\rangle] \\ s = 1: \quad & X[|\uparrow\downarrow\rangle + |\downarrow\uparrow\rangle] = +[|\uparrow\downarrow\rangle + |\downarrow\uparrow\rangle] \end{aligned} \tag{13.23}$$

Thus, we have

$$X\psi_{\text{spin}} = (-1)^{s+1}\psi_{\text{spin}} \tag{13.24}$$

Under the action of the exchange operator, q and $\overline{q}$ become interchanged, and consequently, we can think of this as the operation of charge conjugation in the space of ψ_{charge}. To determine the charge conjugation properties of such a state, let us impose the Pauli principle on our two-fermion system, namely, let us require that the overall wave function change sign under an interchange of q and $\overline{q}$. Note that, in this discussion, we are using a generalized form of the Pauli principle and are treating q and $\overline{q}$ as identical fermions corresponding to spin-up and spin-down states in the space of ψ_{charge}. Thus, we require

$$X\Psi = -\Psi \tag{13.25}$$

Now, using the results of Eqs. 13.22, 13.24, and 13.25, we can write

$$\begin{aligned} X\Psi &= X\psi_{\text{space}}X\psi_{\text{spin}}X\psi_{\text{charge}} = -\Psi \\ &= (-1)^{\ell}\psi_{\text{space}}(-1)^{s+1}\psi_{\text{spin}}C\psi_{\text{charge}} = -\Psi \end{aligned} \tag{13.26}$$

Consequently, for Eq. 13.26 to hold, we conclude that the meson state must be an eigenstate of charge conjugation with charge parity

$$\eta_C = (-1)^{\ell+s} \tag{13.27}$$

Thus, for the meson states that are eigenstates of charge conjugation, Eq. 13.27 establishes a relationship between the orbital wave, the intrinsic-spin value, and the C quantum number of the $q\overline{q}$ system.

The only relevant quantum number that is still missing in our discussion is the parity of the allowed states. The parity of Ψ is given by the product of the intrinsic parities of the constituents and the effect of inversion of spatial coordinates. As we discussed previously in Chapter XI, the relative intrinsic parity of a particle and an antiparticle with spin $\frac{1}{2}$ is odd. Consequently, the total parity of our state Ψ is

$$P\Psi = -(-1)^{\ell}\Psi = (-1)^{\ell+1}\Psi$$

or the total parity quantum number is

$$\eta_P = (-1)^{\ell+1} \tag{13.28}$$

Since the spins of the mesons are obtained from the addition of the orbital and intrinsic angular momenta of the $q\overline{q}$ pair

$$\vec{J} = \vec{L} + \vec{S} \tag{13.29}$$

Table 13.1 Lowest lying meson states expected in the quark model

ℓ	s	j	η_P	η_C	*Meson*[a]
0	0	0	−	+	π^0, η
0	1	1	−	−	$\rho^0, \omega, \phi, J/\psi$
1	0	1	+	−	$b_1^0(1235)$
1	1	0	+	+	$a_0(1980), f_0(975)$
1	1	1	+	+	$a_1^0(1260), f_1(1285)$
1	1	2	+	+	$a_2^0(1320), f_2(1270)$

[a] For other properties of these mesons, see the *CRC Handbook*.

we now have all the ingredients for forming an allowed spectrum of mesons. Table 13.1 lists all the possible lowest lying states, all of which correspond to known mesons.

WEAK ISOSPIN AND COLOR SYMMETRY

As we have seen, leptons and quarks come as doublets, or in pairs, and quarks, in addition, carry a color quantum number. The existence of such groupings, and the color degrees of freedom, suggest the presence of new underlying symmetries for this overall structure. From our discussion of spin and isospin, we can associate the doublet structure with a noncommuting (non-Abelian) symmetry group $SU(2)$. We will continue to refer to this underlying symmetry group as isospin, since it is an internal symmetry. Unlike strong isospin, however, which is used only to classify hadrons, the isospin in the present case also classifies leptons. Leptons, on the other hand, interact weakly and, therefore, this symmetry must be related to the weak interaction. Correspondingly, the isospin symmetry associated with quarks and leptons is referred to as *weak isospin*. This symmetry is quite distinct from that of the strong isospin symmetry we discussed previously. However, just as with strong isospin, where the symmetry is discernible only when the electromagnetic interaction (electric charge) can be ignored, similarly, the essential character of the weak isospin symmetry can be fully discerned only when the electromagnetic force is "turned off." Under such circumstances, the up and down states of Eqs. 13.1 and 13.2 would be equivalent and could not be distinguished.

For the case of weak isospin symmetry, we can define a weak hypercharge for each quark and lepton, based on a general form of the Gell-Mann–Nishijima relation of Eq. 9.25, namely,

$$Q = I_3 + \frac{Y}{2}$$

or

$$Y = 2(Q - I_3) \tag{13.30}$$

where Q is the charge of the particle and I_3 the projection of its weak isospin quantum number. Thus, for the (ν, e^-) doublet we obtain

$$\begin{aligned} Y(\nu) &= 2\left(0 - \tfrac{1}{2}\right) = -1 \\ Y(e^-) &= 2\left(-1 + \tfrac{1}{2}\right) = -1 \end{aligned} \tag{13.31}$$

Similarly, for the (u, d) quark doublet, we have

$$\begin{aligned} Y(u) &= 2\left(\tfrac{2}{3} - \tfrac{1}{2}\right) = 2 \times \tfrac{1}{6} = \tfrac{1}{3} \\ Y(d) &= 2\left(-\tfrac{1}{3} + \tfrac{1}{2}\right) = 2 \times \tfrac{1}{6} = \tfrac{1}{3} \end{aligned} \tag{13.32}$$

The weak hypercharge quantum number for other quark and lepton doublets can be obtained in the same manner. (In fact, in the Standard Model, only left-handed particles have a doublet structure. The right-handed quarks and the right-handed charged leptons are all singlets with $I = 0$, and there are no right-handed neutrinos. We also wish to point out that, as can be seen from Eq. 13.30, the weak hypercharge quantum number is the same for both members of any doublet. This equality is required if weak hypercharge is to be regarded as a $U(1)$ symmetry of the type specified in Eq. 10.76.)

Again, the color symmetry of quarks is an internal symmetry. It can be shown that it is similar to isospin in that it involves rotations—however, the rotations are in an internal space of three dimensions—corresponding to the three distinct colors of the quarks. The relevant symmetry group is known as $SU(3)$. The interactions of quarks are assumed to be invariant under such $SU(3)$ rotations in color space, leading to an equivalence of quarks of different color. (This condition is needed in order to have consistency with experimental observations.) Because the color quantum number is carried only by quarks, namely, by the constituents of hadrons, we expect this symmetry to be associated only with the strong interaction.

GAUGE BOSONS

As we saw previously, the presence of a global symmetry is useful for providing a classification of particle states according to some quantum number (e.g., strong isospin). A local symmetry, on the other hand, leads to the introduction of physical forces. Since weak isospin and the color symmetry are associated with rather distinct interactions, it is interesting to ask whether the corresponding physical forces—namely, the strong (color) and the weak forces—might arise purely from the requirement that these symmetries be local. Years of painstaking theoretical development, coupled with detailed experimental verification, has led to the conclusion that this is indeed very likely. It is the current understanding that the local symmetries underlying the electromagnetic, weak, and the strong interactions are obtained from the $U_Y(1)$, $SU_L(2)$, and $SU_{\text{color}}(3)$ symmetry groups, respectively. The group corresponding to the weak hypercharge symmetry, $U_Y(1)$, corresponds to a local Abelian symmetry group, while $SU_L(2)$ and $SU_{\text{color}}(3)$ are the non-Abelian groups corresponding to weak isospin and color symmetries.* From the Gell-Mann–Nishijima formula of Eq. 13.30, we see that the electric charge is related to the weak hypercharge and the weak isospin, from which it follows that the electromagnetic $U_Q(1)$ symmetry can be regarded as a particular combination of the weak isospin and weak hypercharge symmetries.

In Chapter X, we showed in a simple example how local invariance necessarily leads to the introduction of gauge potentials, such as the vector potential in electromagnetic interactions. When these potentials are quantized, they provide the carriers of the force, otherwise known as gauge particles. Thus, for example, the carrier of the electromagnetic interaction, or its gauge boson, is the photon. There are three gauge bosons associated with the weak interactions, and they are known as the W^+, W^-, and Z^0 bosons. (These bosons were first observed by Carlo Rubbia and collaborators at the antiproton–proton collider at the CERN Laboratory outside of Geneva, Switzerland.) For the strong (color) interactions, there are eight gauge bosons, and they are known as *gluons* (g). (The number of gauge bosons associated with a symmetry is a characteristic of the symmetry group.) All the gauge bosons have spin $J = 1$. Furthermore, the gluons, or the gauge bosons of color symmetry, are electrically neutral, but carry the color quantum number. This is in contrast to the photon, which is the carrier of the force between charged particles, but does not itself carry electric charge. This difference can be attributed to the Abelian nature of the $U_Q(1)$ symmetry that describes the photon and the non-Abelian nature of $SU_{\text{color}}(3)$ that describes gluons. Figure 13.2 displays several examples of how the different gauge bosons can induce transitions between different fermions, and, in the case of the color force, between gluons as well.

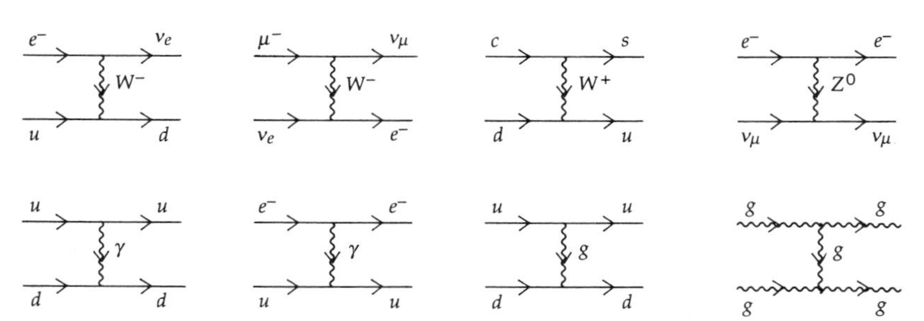

Figure 13.2 Typical interactions among leptons and quarks and between gluons, mediated by different gauge bosons.

*Because the doublet structure of quarks and leptons involves only left-handed particles, the weak isospin symmetry group is also conventionally denoted by $SU_L(2)$. This kind of structure is essential for incorporating the properties of neutrinos and of parity violation in weak interactions.

DYNAMICS OF THE GAUGE PARTICLES

In this section, we try to bring out some basic features of the dynamics of these gauge particles. For simplicity, let us consider Maxwell's equations, which describe the dynamics of gauge particles associated with electromagnetic interactions. For the other interactions, the dynamical equations of the gauge bosons, although similar in form, are more complicated. The basic conclusions regarding the properties of such particles, however, can be obtained from the simpler case. Consider Maxwell's equations in vacuum:

$$\begin{aligned}
\vec{\nabla}\cdot\vec{E} &= 0 \\
\vec{\nabla}\cdot\vec{B} &= 0 \\
\vec{\nabla}\times\vec{E} &= -\frac{1}{c}\frac{\partial\vec{B}}{\partial t} \\
\vec{\nabla}\times\vec{B} &= \frac{1}{c}\frac{\partial\vec{E}}{\partial t}
\end{aligned} \tag{13.33}$$

We can define the electric and magnetic fields in terms of the potentials as follows:

$$\begin{aligned}
\vec{E} &= -\vec{\nabla}\phi - \frac{1}{c}\frac{\partial\vec{A}}{\partial t} \\
\vec{B} &= \vec{\nabla}\times\vec{A}
\end{aligned} \tag{13.34}$$

where $\vec{A}$ corresponds to the vector potential introduced in Eq. 10.78 and ϕ is known as the scalar potential. The contents of Eqs. 13.33 can be expressed equally well in terms of these gauge potentials. An interesting consequence of the definitions in Eq. 13.34 is that the electric and the magnetic fields are not sensitive to local changes or redefinitions of the gauge potentials of the following form:

$$\begin{aligned}
\delta\phi &= -\frac{1}{c}\frac{\partial\alpha(\vec{r},t)}{\partial t} \\
\delta\vec{A} &= \vec{\nabla}\alpha(\vec{r},t)
\end{aligned} \tag{13.35}$$

Namely, under the transformation in Eq. 13.35, where $\alpha(\vec{r},t)$ is an arbitrary scalar function of space-time coordinates, the $\vec{E}$ and $\vec{B}$ fields do not change:

$$\begin{aligned}
\delta\vec{E} &= -\vec{\nabla}\delta\phi - \frac{1}{c}\frac{\partial\delta\vec{A}}{\partial t} \\
&= \vec{\nabla}\frac{1}{c}\frac{\partial\alpha}{\partial t} - \frac{1}{c}\frac{\partial}{\partial t}(\vec{\nabla}\alpha) = 0 \\
\delta\vec{B} &= (\vec{\nabla}\times\delta\vec{A}) = \vec{\nabla}\times(\vec{\nabla}\alpha) = 0
\end{aligned} \tag{13.36}$$

The transformations in Eq. 13.35 are, of course, reminiscent of the gauge transformation introduced in Eq. 10.79, where $\alpha(\vec{r})$ was the position-dependent phase of the transformation. We note that, since under the gauge transformations of Eq. 13.35, the $\vec{E}$ and the $\vec{B}$ fields do not change, then Maxwell's equations must also be invariant under such redefinitions of the potentials. This invariance corresponds to the $U_Q(1)$ symmetry of the electromagnetic interactions.

An interesting consequence of this gauge invariance of Maxwell's equations is that they describe transverse electromagnetic waves propagating at the speed of light. To see this, we note from Eqs. 13.33 that

$$\vec{\nabla} \times (\vec{\nabla} \times \vec{E}) = -\frac{1}{c}\frac{\partial}{\partial t}(\vec{\nabla} \times \vec{B}) = -\frac{1}{c}\frac{\partial}{\partial t}\left(\frac{1}{c}\frac{\partial \vec{E}}{\partial t}\right)$$

or

$$\vec{\nabla}(\vec{\nabla} \cdot \vec{E}) - \vec{\nabla}^2 \vec{E} = -\frac{1}{c}\frac{\partial}{\partial t}\left(\frac{1}{c}\frac{\partial \vec{E}}{\partial t}\right)$$

or

$$\left(\vec{\nabla}^2 - \frac{1}{c^2}\frac{\partial^2}{\partial t^2}\right)\vec{E} = 0 \qquad (13.37)$$

Here we have used the property of Eq. 13.33, namely, the fact that the electric field is transverse. Equation 13.37, in fact, describes a relativistic traveling wave propagating at the speed of light. Similarly, the other pair of Maxwell's equations yield

$$\left(\vec{\nabla}^2 - \frac{1}{c^2}\frac{\partial^2}{\partial t^2}\right)\vec{B} = 0 \qquad (13.38)$$

When these waves are quantized, they correspond to massless particles (photons), which, of course, reflects the long-range nature of the Coulomb interaction.

To see that the masslessness of the gauge particle is a consequence of gauge invariance, let us note that a massive particle, of mass m, can be represented by a traveling wave of the form

$$\begin{aligned} \left(\vec{\nabla}^2 - \frac{1}{c^2}\frac{\partial^2}{\partial t^2} - \frac{m^2c^2}{\hbar^2}\right)\vec{E} &= 0 \\ \left(\vec{\nabla}^2 - \frac{1}{c^2}\frac{\partial^2}{\partial t^2} - \frac{m^2c^2}{\hbar^2}\right)\vec{B} &= 0 \end{aligned} \qquad (13.39)$$

These equations follow from a set of Maxwell-like equations of the type

$$\begin{aligned}
\vec{\nabla}\cdot\vec{E} &= -\frac{m^2c^2}{\hbar^2}\phi \\
\vec{\nabla}\cdot\vec{B} &= 0 \\
\vec{\nabla}\times\vec{E} &= -\frac{1}{c}\frac{\partial\vec{B}}{\partial t} \\
\vec{\nabla}\times\vec{B} &= \frac{1}{c}\frac{\partial\vec{E}}{\partial t} - \frac{m^2c^2}{\hbar^2}\vec{A}
\end{aligned} \tag{13.40}$$

The fact that Eq. 13.39 follows from Eq. 13.40 can be checked, as follows:

$$\vec{\nabla}\times(\vec{\nabla}\times\vec{E}) = -\frac{1}{c}\frac{\partial}{\partial t}(\vec{\nabla}\times\vec{B}) = -\frac{1}{c}\frac{\partial}{\partial t}\left(\frac{1}{c}\frac{\partial\vec{E}}{\partial t} - \frac{m^2c^2}{\hbar^2}\vec{A}\right)$$

or

$$\vec{\nabla}(\vec{\nabla}\cdot\vec{E}) - \vec{\nabla}^2\vec{E} = -\frac{1}{c}\frac{\partial}{\partial t}\left(\frac{1}{c}\frac{\partial\vec{E}}{\partial t} - \frac{m^2c^2}{\hbar^2}\vec{A}\right)$$

or

$$\vec{\nabla}\left(-\frac{m^2c^2}{\hbar^2}\phi\right) - \vec{\nabla}^2\vec{E} = -\frac{1}{c^2}\frac{\partial^2\vec{E}}{\partial t^2} + \frac{m^2c}{\hbar^2}\frac{\partial\vec{A}}{\partial t}$$

or

$$\left(\vec{\nabla}^2 - \frac{1}{c^2}\frac{\partial^2}{\partial t^2}\right)\vec{E} + \frac{m^2c^2}{\hbar^2}\left(\vec{\nabla}\phi + \frac{1}{c}\frac{\partial\vec{A}}{\partial t}\right) = 0$$

or

$$\left(\vec{\nabla}^2 - \frac{1}{c^2}\frac{\partial^2}{\partial t^2} - \frac{m^2c^2}{\hbar^2}\right)\vec{E} = 0 \tag{13.41}$$

where in the last step we have used the definition of $\vec{E}$ in Eq. 13.34. The equation for the $\vec{B}$-field can similarly be derived from the other pair of equations in 13.40. Thus, a modified set of Maxwell's equations as described in Eq. 13.40 leads to massive traveling waves, which upon quantization would yield massive particles. Unfortunately, however, unlike Maxwell's equations in 13.33, the set of equations in 13.40 depend explicitly on the gauge potentials and are therefore no longer invariant under the gauge transformation of Eq. 13.35. This brings out the intimate connection between the masslessness of gauge particles and gauge invariance, namely that gauge invariance can only hold for the case of massless gauge bosons.

The preceding analysis also points to the difficulty of generalizing the gauge principle to all forces. This is because, unlike the electromagnetic interaction, the strong and the weak forces are short ranged. If all of the forces had their origin in the existence of local symmetries, and had dynamics similar to those of Maxwell's equations, then it would follow that they should all describe long-range interactions, which is, of course, not correct. The resolution to how the weak and the strong forces can be short ranged, despite their apparent origin in a gauge principle, is quite interesting. The mechanisms responsible for the short-ranged character of the two forces are entirely different, and we discuss next how the weak forces can become short ranged.

SYMMETRY BREAKING

The meaning of symmetries can be quite subtle. As we have already seen, the invariance of the dynamical equations of a system under a set of transformations defines a symmetry of the system, whose presence can be inferred from the invariance properties of the Hamiltonian. Even if a set of dynamical equations is invariant under a given set of transformations, however, the solutions (physical states) of the system need not possess the symmetry. Thus, as a heuristic example, let us consider magnetism, which can be thought of as arising from the interaction of spins ($\vec{s}$) situated on a lattice, and which, for a ferromagnet, can be described by a Hamiltonian of the following form:

$$H = -\kappa \sum_i \vec{s}_i \cdot \vec{s}_{i+1} \tag{13.42}$$

where κ, a positive quantity, denotes the strength of coupling between spins at nearest neighbor sites. If we rotate all the spins by a constant angle, the inner product does not change, and consequently rotations correspond to a global symmetry of the Hamiltonian for a ferromagnet. On the other hand, we can recognize from the structure of the Hamiltonian that the ground state of the system—namely, the state with the lowest energy value—occurs when all the spins are parallel. Thus, a typical configuration of the spins in the ground-state can be represented as in Fig. 13.3. The ground-state configuration, therefore, picks out, at random, a preferred direction in space, and consequently breaks the rotational symmetry of the Hamiltonian.

Figure 13.3 Aligned spins in the ground state of a ferromagnet.

When the solution to a set of dynamical equations violates the symmetry that exists in the equations, we say that the symmetry of the system is "spontaneously" broken. In the case of the ferromagnet, we see that the spins in the ground state have a long-range correlation (they all point along the same direction). This is, in fact, a general feature of spontaneous breaking of symmetry, namely, when a symmetry is broken spontaneously, certain correlations become long ranged, and can be thought of as being a consequence of the presence of zero-mass particles in the quantum mechanical theory. In other words, when there is spontaneous symmetry breaking, the spectrum of states of a relativistic quantum theory develops massless particles.

A more quantitative (albeit heuristic) way to see the above result is to consider the two-dimensional classical Hamiltonian:

$$H = T + V = \frac{1}{2m}\left(p_x^2 + p_y^2\right) - \frac{1}{2}m\omega^2\left(x^2 + y^2\right) + \frac{\lambda}{4}\left(x^2 + y^2\right)^2 \qquad \text{with } \lambda > 0 \tag{13.43}$$

Except for the negative, rather than positive, sign of the the second term, this is the Hamiltonian for a two-dimensional classical anharmonic oscillator. This Hamiltonian is invariant under rotations about the z-axis, and such rotations are therefore a global symmetry of this system. We can examine the lowest energy solution, or the ground state, of this system. This solution would clearly require the kinetic energy to vanish, since that is a positive quantity. Consequently, it follows that the minima for total energy will coincide with those of the potential. The extrema of the potential can be obtained by setting the derivatives with respect to x and y to zero:

$$\begin{aligned} \frac{\partial V}{\partial x} &= x\left(-m\omega^2 + \lambda(x^2 + y^2)\right) = 0 \\ \frac{\partial V}{\partial y} &= y\left(-m\omega^2 + \lambda(x^2 + y^2)\right) = 0 \end{aligned} \tag{13.44}$$

Thus, the coordinates at the extrema of the potential satisfy the relation

$$x_{\text{min}} = y_{\text{min}} = 0 \tag{13.45}$$

or

$$x_{\text{min}}^2 + y_{\text{min}}^2 = \frac{m\omega^2}{\lambda} \tag{13.46}$$

From the form of the potential (see Fig. 13.4) it can be seen that the point $x_{\text{min}} = y_{\text{min}} = 0$ corresponds to a local maximum that is therefore unstable under perturbation. The potential has the shape of a symmetric Mexican sombrero, with a continuous set of $(x_{\min}, y_{\min})$ coordinates on the circle of Eq. 13.46 specifying the minimum (see Fig 13.4). If, for simplicity, we choose to define the minimum by

$$\begin{aligned} y_{\min} &= 0 \\ x_{\min} &= \sqrt{\frac{m\omega^2}{\lambda}} \end{aligned} \tag{13.47}$$

then this choice has the effect of picking out a preferred direction in space, and therefore of breaking the rotational invariance of the system. Small motion (oscillations) around this minimum determines its stability, which can be investigated by expanding the potential around the coordinates in Eq. 13.47:

$$V(x_{\min} + x, y) = -\frac{1}{2}m\omega^2\left((x_{\min} + x)^2 + y^2\right) + \frac{\lambda}{4}\left((x_{\min} + x)^2 + y^2\right)^2 \tag{13.48}$$

which, upon expansion up to second order in x and y, and on substitution for $x_{\min}$ using 13.47, yields

$$V\,(x_{\min} + x, y) = -\frac{m^2\omega^4}{4\lambda} + m\omega^2 x^2 + \text{higher orders} \tag{13.49}$$

Consequently, we note that the small oscillations along the x-axis are harmonic oscillations at an equivalent frequency of $\omega_x = \sqrt{2}\omega$, whereas we can think of the frequency of oscillation along the y-axis as characterized by $\omega_y = 0$.

The small oscillations around the classical ground state also bring out the essential features of correlations in the quantum mechanical ground state, as follows. From Eq. 13.49, we see that for small oscillations the

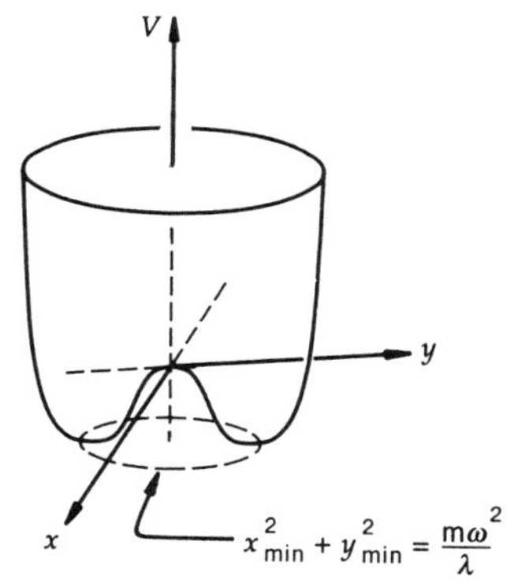

Figure 13.4 A sketch of the potential in Eq. 13.43.

Hamiltonian takes the form

$$H = \frac{p_x^2}{2m} + \frac{p_y^2}{2m} + m\omega^2 x^2 - \frac{m^2\omega^4}{4\lambda} \tag{13.50}$$

Consequently, the equation of motion along the y-axis satisfies

$$\dot{y}(t) = \text{constant} = c$$

or

$$y(t) = y(0) + ct \tag{13.51}$$

In quantum field theory, correlations are defined by the vacuum (or ground-state) expectation value of bilinear operator products involving different space-time points. Here, in analogy, we can use time correlations for $y(t)$, as follows:*

$$\begin{aligned}\langle 0|y(t)y(0)|0\rangle &= \langle 0|y(0)y(0) + cty(0)|0\rangle \\ &= \langle 0|y(0)y(0)|0\rangle\end{aligned} \tag{13.52}$$

where we have used the fact that the second term will vanish when we integrate it over all space. That is, the expectation value of $y(0)$ has to be zero for the integration in y, if the ground state has definite parity. The expectation value in Eq. 13.52 is therefore independent of time, implying a long-time correlation. This is the simple analogue to a long-distance correlation found for such systems in quantum field theories, and which is observed in the spin system of the ferromagnet.

We could have, of course, chosen any alternate solution to Eq. 13.46 such as, for example,

$$x_{\min} = y_{\min} = \sqrt{\frac{m\omega^2}{2\lambda}} \tag{13.53}$$

In fact, for any solution of Eq. 13.46, it is easy to show that we can define normal modes of oscillation such that the frequency of oscillation for one of the modes is $\sqrt{2}\omega$, while the orthogonal mode will have a vanishing frequency. Qualitatively, this can be discerned from the form of the potential: Irrespective of which point is chosen as the minimum, the motion along the valley of the potential will require no expense of energy, and will therefore correspond to the mode with zero frequency. Motion

---

*Note that $y(t)$ is in fact a coordinate operator in the quantum theory.

in the orthogonal direction will, however, require energy and therefore corresponds to a finite frequency.

Our result is a general feature of all theories that have spontaneous symmetry breaking, and, correspondingly, in such theories the quantum mechanical system will develop states of zero energy. For relativistic quantum mechanical systems, such states can be identified with massless particle states. In addition, the orthogonal mode with nonvanishing frequency corresponds to a massive particle. The massless particles, which, as we have emphasized, arise as a consequence of the spontaneous breaking of a global symmetry, are known as Nambu-Goldstone bosons, or usually termed Goldstone bosons (after Yoichiro Nambu and Jeffrey Goldstone). We should also mention that our simple example is to be regarded as purely illustrative of the basic features of spontaneous symmetry breaking, in that Goldstone bosons appear only in relativistic field theories with two or more spatial dimensions. When the spontaneous symmetry breaking is of a local rather than of a global symmetry, the Goldstone particles then become transformed into the longitudinal modes of the gauge bosons. As a consequence, the resultant gauge bosons develop mass, or equivalently, the corresponding "electric" and the "magnetic" fields lose their purely transverse character.

The preceding discussion suggests a mechanism whereby the gauge bosons of the weak interaction can develop a mass and thereby give rise to a short-ranged force. Conventionally, this is known as the Higgs mechanism (after Peter Higgs, but also discovered independently by Robert Brout and François Englert, and by Gerald Guralnik, Richard Hagen and Tom Kibble), and for the weak interactions, the massive partner of the Goldstone boson (corresponding to the mode with frequency $\sqrt{2}\omega$) is a scalar particle known as the Higgs boson. The Higgs boson has yet to be observed, and it is not clear whether it is a fundamental (structureless) particle. Thus, in this kind of scenario, we expect the local weak isospin symmetry to be spontaneously broken, and we therefore do not expect weak isospin to be a good (conserved) quantum number in the weak interactions, which agrees with observation.* In fact, even the weak hypercharge symmetry is spontaneously broken. The breaking in weak isospin and weak hypercharge, however, compensate each other such that the particular combination in Eq. 13.30, corresponding to the

---

*This statement may confuse the intrepid reader, who by applying Eq. 13.30 to the W and Z bosons will deduce that they can be regarded as $Y = 0$ objects. In fact, going further, the reader would conclude that all the transitions shown in Fig. 13.2, and the weak decays discussed in Chapter IX, conserve both weak hypercharge and weak isospin. It would, however, be incorrect to surmise that these quantum numbers are always conserved. We know that weak isospin must be a broken symmetry, because, for example, otherwise the masses of the members of the weak isospin doublets would be identical. In addition, the Higgs boson, in the context of the Standard Model, has $I = \frac{1}{2}$, but Higgs bosons interact with quarks and with W and Z bosons ($H \to W^+ + W^-$, $H \to Z + Z$, etc.). This, clearly, cannot happen if weak isospin is conserved. Consequently, the breaking of weak isospin must affect the usual fermionic transitions, and violations of weak isospin in the more common processes are expected due to higher order contributions from the Higgs sector.

electric charge symmetry, remains unbroken. Correspondingly, the photon remains massless, but the weak gauge bosons, namely the $W^{\pm}$ and the Z^0, become massive objects, with $m_{W^{\pm}} \simeq 80.6$ GeV/c^2 and $m_{Z^0} \simeq 91.2$ GeV/c^2. They are assumed to be elementary particles that can decay into lepton–antilepton or quark–antiquark pairs, as can be surmised from Fig. 13.2, and as is discussed again later in this chapter.

Returning to our example of the ferromagnet, we note that, although the ground state spontaneously breaks rotational invariance because the aligned spins pick out a preferred spatial direction, when we heat up such a system, the thermal motion randomizes the spin orientations. Above some critical high temperature or energy, the spins indeed become randomly oriented, thereby restoring rotational invariance. This feature is also found in quantum mechanical field theories with spontaneous symmetry breaking. Namely, for theories displaying spontaneous breaking of symmetry, the symmetry is indeed restored beyond a certain temperature or energy. If we extend these ideas to weak interactions, we note that above a certain energy scale the weak isospin symmetry is restored, and consequently the weak gauge bosons become massless just like the photon. As was pointed out in connection with Eq. 9.7, the strengths of the weak and the electromagnetic interactions become comparable at high momentum transfers. Both these results suggest therefore that the two forces may indeed be unified at sufficiently high energies.

QUANTUM CHROMODYNAMICS AND CONFINEMENT

As we have just seen, the short-ranged nature of the weak interactions can be argued as arising from the spontaneous breaking of the local weak isospin symmetry. However, the short-range nature of the strong nuclear force has a completely different origin. The dynamical theory of quarks and gluons that describes color interactions is known as quantum chromodynamics (QCD), and it is a gauge theory of the noncommuting color symmetry group $SU(3)$. This theory is very similar to quantum electrodynamics (QED), which describes the electromagnetic interactions of charges with photons. As we have mentioned, QED is a gauge theory of phase transformations corresponding to the commuting symmetry group $U_Q(1)$. Being a gauge theory of color symmetry, QCD also contains massless gauge bosons (namely gluons) that have properties similar to photons.

There are, however, essential differences between the two theories, which arise because of the different nature of the two symmetry groups. As we noted earlier, the photon, which is the carrier of the force between charged particles, is itself charge–neutral. As a result, the photon does not interact with itself. In contrast, the gluon, which is the mediator of color interactions, also carries color charge, and consequently has self-interactions. Another important consequence of the non-Abelian nature of the color symmetry is related to how color–neutral states can be formed. Consider, for example, a red-colored quark. We can obtain a color–neutral system by combining the red quark with an anti-red ($\overline{\text{red}}$) antiquark. This is very much how electric charges would add, namely,

$$\text{red} + \overline{\text{red}} = \text{color neutral} \tag{13.54}$$

Since three quarks with distinct colors can also yield a color neutral baryon, however, there must therefore be an alternative way of obtaining a color–neutral combination for three colored quarks, and this must simply be

$$\text{red} + \text{blue} + \text{green} = \text{color–neutral} \tag{13.55}$$

which is clearly different from the way electric charges add together.

This difference between color charge and electric charge has great physical consequences. For example, a classical test particle carrying positive electric charge polarizes a dielectric medium by creating pairs of oppositely charged particles (dipoles). Due to the nature of Coulomb interactions, the negatively charged particles so produced will be attracted toward the test particle, while the positively charged particles will be repelled (see Fig. 13.5). As a consequence, the charge of the test particle is shielded and the effective charge seen far away is smaller than the actual charge carried by the test particle. (Recall that the electric field in a dielectric medium is reduced relative to that in vacuum by the value of the dielectric constant of the medium.) In fact, the effective charge depends on the distance scale at which we probe the test particle. The magnitude of the charge increases as we probe it at ever smaller distances, and only asymptotically (at largest momentum transfers) do we obtain the true charge of the test particle. Since the distance probed is inversely proportional to the momentum transfer, one conventionally says that the effective electric charge, or the strength of the electromagnetic interaction, increases with the momentum transfer, and, as we have just argued, this phenomenon is a consequence of the screening of the electric charge in a dielectric medium. Because of the presence of quantum fluctuations, a similar effect arises for charged particles in vacuum, the impact of which is that the fine structure constant

Figure 13.5 Polarization of a dielectric medium around a positive electric charge, and the effective value of the charge as a function of probing distance and of momentum transfer.

$\alpha = e^2/\hbar c$ increases, albeit only slightly, with momentum transfer. (This has been confirmed in high-energy e^+e^- scattering.)

In contrast, a test particle carrying color charge polarizes the medium in two different ways. First, just as in the case of QED, it can create pairs of particles with opposite color charge. But it can also create three particles of distinct color, while still maintaining overall color neutrality. Consequently, for the color force, the effect of color charge on a polarized medium is quite different and more complicated. A detailed analysis of QCD reveals that the color charge of the test particle is, in fact, anti-screened. In other words, far away from the test particle, the magnitude of the measured color charge is larger than actually carried by the test particle. In fact, as we probe deeper, the magnitude of this charge decreases. Thus, the qualitative dependence of the color charge on probing distance, or on the probing momentum transfer, is exactly opposite of that found for electromagnetic interactions (see Fig. 13.6). This implies that the strength of the strong interactions decreases with increasing momentum transfer, and vanishes asymptotically. Conventionally, this is referred to as asymptotic freedom, and simply refers to the fact that at infinite energies quarks will behave essentially as free particles, since the effective strength of the coupling for interactions will vanish in this limit. (The asymptotic freedom of QCD was discovered independently by David Politzer, by David Gross and Frank Wilczek, and by Gerard t'Hooft.) This principle has the additional implication that, in very high-energy collisions, hadrons consist of quarks that move inside them as if they were free and essentially independent particles. This simple model of high-energy hadrons is known as the *parton model*. It has been tested extensively and found to agree with a variety of high-energy scattering experiments.

The very fact that the strength of coupling of QCD decreases at high energies is quite important. In fact, it means that the effect of color interactions can be calculated perturbatively at small distances or at large momentum transfers. Consequently, the predictions of QCD can be tested in high-energy experiments. At present, all known predictions of QCD are in excellent agreement with the data (see later in this chapter).

At low energies, color interactions become stronger, thereby making perturbative calculations unreliable. This, however, also points to the fact

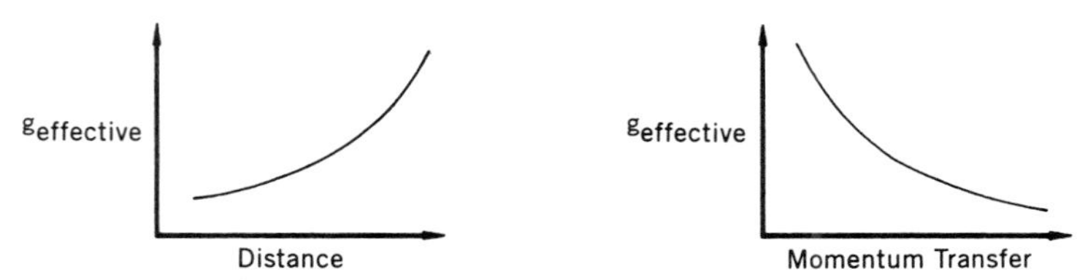

Figure 13.6 The effective value of color charge as a function of probing distance and of momentum transfer.

that as color couplings increase, the quarks can form bound states, which we know as the hadrons. In fact, quarks alone cannot account for the properties of hadrons. As inferred from high-energy collisions, quarks carry only about one-half of the momentum of the hadrons, the rest has to be attributed to the presence of other pointlike constituents that appear to be electrically neutral, and have spin $J = 1$. These constituents can be identified with the gluons. For consistency, we must therefore modify the interpretation of the quark model of hadrons from one where hadrons are composed only of quarks, to one in which quarks provide just the effective "valence" properties of hadrons.

There have been many attempts to understand the low-energy, nonperturbative behavior of QCD. The present qualitative picture can be summarized best by a phenomenological linear potential between quarks and antiquarks of the form

$$V(r) \propto kr \tag{13.56}$$

This kind of picture works particularly well for the heavier quarks. Intuitively, we can think of the $q\overline{q}$ system as being connected by a string. As the $q\overline{q}$ pair is forced to separate, the potential between the two constituents increases. At some separation length, it becomes energetically more favorable for a $q\overline{q}$ pair to split into two $q\overline{q}$ pairs. Pictorially, we can describe the process as shown in Fig. 13.7. In other words, the strong color attraction increases with separation distance between the quarks, and therefore precludes the possibility of observing an isolated quark.* This effect, known as *confinement*, is, of course, consistent with observation. That is, all observed particles appear to be color neutral, and there has never been any evidence for the production of an isolated quark or gluon with color charge. When additional quarks are produced in high-energy collisions, they are always found in states whose total color adds up to zero (i.e., color neutral). As these quarks leave the region of their production, they "dress" themselves (become converted) into hadrons, and their presence can be inferred from a "jet" of particles that is formed from their initial energy. Similarly, gluons emitted in hadronic interactions

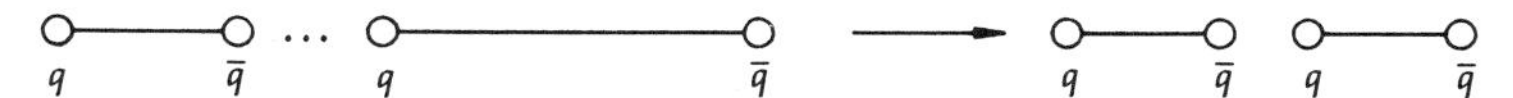

Figure 13.7 Creation of a new $q\overline{q}$ pair from the vacuum when the separation between the original $q\overline{q}$ pair is increased.

*The splitting of a $q\overline{q}$ pair is similar to what happens when a bar magnet is cut into two. The net result is two separate bar magnets rather than a north and a south pole.

also become dressed into hadrons and produce jets of particles as they leave the point of collision. While we currently believe in the confinement of quarks and gluons, a detailed proof of this requirement on the basis of QCD is still lacking.

In the context of the Standard Model, the strong nuclear force between hadrons can be thought of as a residual color Van der Waals force, analogous to the Van der Waals force that describes the residual electromagnetic interactions of charge–neutral molecules. Namely, just as the Van der Waals force reflects the presence of charged atomic constituents that can interact through the Coulomb force, the strong nuclear force reflects the presence of far more strongly interacting color objects that are present within hadrons. The Van der Waals force, as we know, falls off far more rapidly with distance than the Coulomb force, which suggests that a similar effect could be expected for the case of color, which would explain the origin of the short-range nature of the strong force between hadrons, both within as well as outside of nuclei.

QUARK–GLUON PLASMA

As we have already argued, quarks are confined within hadrons. If we increase the temperature of our hadronic system, however, the increased random thermal motion may eventually lead to a complete disintegration of the hadron. If this happens, then quarks and gluons can appear as free particles in a new transformed kind of matter known as the quark–gluon plasma phase. This phase is quite similar to the plasma state of charged particles that exists inside the sun and the stars, where electrons and protons from ionized hydrogen atoms move about freely. The best theoretical evidence that, as the temperature increases, such a transition between the confined and the deconfined phases of quarks takes place comes from extensive computer simulations based on QCD. Such a quark–gluon plasma phase of matter could have been formed right after the big bang, when the temperature in the universe was quite high. In this type of a phase, we would expect to have a large number of rapidly moving charged quarks, which would scatter and therefore radiate photons, leading to enhanced direct single-photon production. Also, because of the high temperatures (or high energies), the production of quarks would not be limited to the low-mass flavors, and therefore production of a quark–gluon phase would also lead to an increase in the production of quarks with more exotic flavors, such as strangeness and charm. Experimental verification of such signals in high-energy interactions is an interesting area of research, and has led to the proposed study of heavy-ion collisions at the Relativistic Heavy Ion Collider (RHIC), under construction at Brookhaven National Laboratory. These collider experiments will study interactions of large-A nuclei, each with energies of several hundred GeV per nucleon. The energy and matter densities in these experiments are expected to be large enough to observe the transformation of normal nuclei into free quark–gluon systems. (RHIC is scheduled for completion in the late 1990s.) The attempts to test such ideas are quite challenging. The theoretical properties of the quark–gluon plasma phase are not as yet completely understood, and furthermore, the experimental signatures for the presence of the quark–gluon plasma are also not entirely clear. In spite of this uncertainty, or perhaps because of it, it is exceedingly interesting to learn whether such states of matter can be produced.

COMMENTS ON PHENOMENOLOGY AND COMPARISON WITH DATA

As example of the kind of agreement that has been observed between expectations from QCD and collisions studied at high energies, we show in Figs. 13.8 and 13.9, respectively, the data and theoretical predictions for production of W and Z bosons and for production of particle jets (quarks and gluons that form, or evolve into, color–neutral particles) in antiproton–proton collisions. The primary uncertainty in the theory (displayed as the allowed regions between the two sets of smooth curves in Fig. 13.8) stems from the inability to predict the content and the momentum distributions of constituents that are bound within hadrons. This is an issue related to confinement and interactions of quarks and gluons at low momentum transfer, which cannot be calculated reliably in perturbation theory. Such

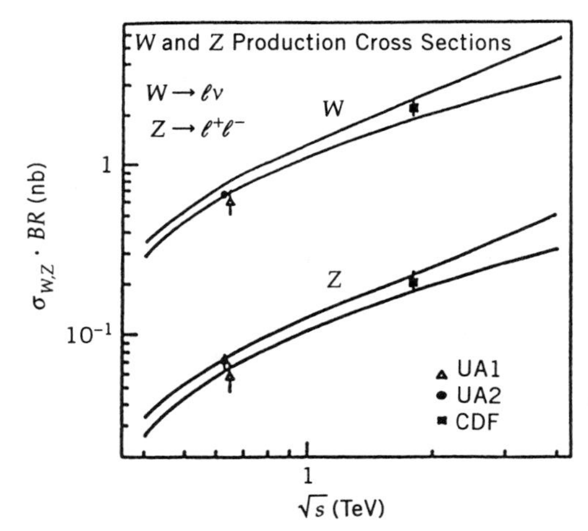

Figure 13.8 The cross section for W and Z production in $\bar{p}p$ collisions compared to theoretical predictions based on the Standard Model.

(From A. G. Clark. 1991. *Techniques and Concepts of High Energy Physics VI*, T. Ferbel, Ed. Plenum Press, New York.)

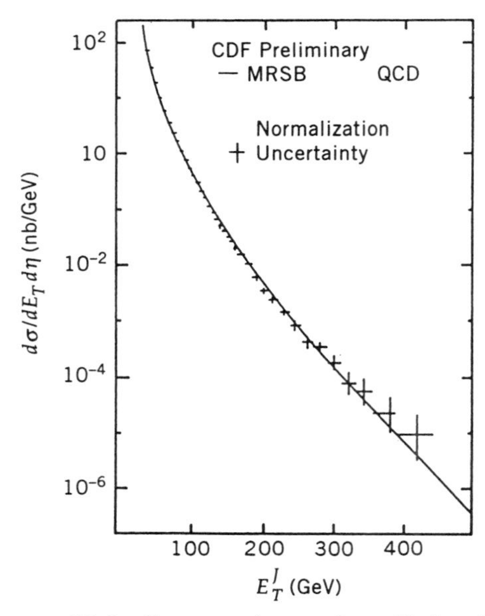

Figure 13.9 Cross section and prediction from QCD for production of particle jets at large momentum transfers in $\overline{p}p$ collisions at $\sqrt{s} \sim 1.8$ TeV.

(From A. G. Clark. 1991. *Techniques and Concepts of High Energy Physics VI*, T. Ferbel, Ed. Plenum Press, New York.)

information has to be extracted from other reactions (e.g., from electron scattering off protons) and then applied to predict results for collisions of partons located within separate hadrons. Thus, for the case of W production, the main contribution to the yield arises from the interaction of $\overline{u}$ (or $\overline{d}$) quarks in the antiproton that fuse with d (or u) quarks within the proton to produce a W^- (or W^+) and remnant jets of particles. For the case of jet production, any parton in one of the interacting hadrons can scatter elastically off any parton in the other hadron, and then both partons can evolve into jets. Clearly, the scattered partons can appear at large angles relative to the collision axis, while the other (unscattered) constituents will evolve into color–neutral states at small angles along the collision axis. In fact, since momentum must be conserved in the direction transverse to the collision axis, we would expect the scattered-parton jets to be emitted back to back. Two typical events of this kind are shown in Fig. 13.10. This kind of display is referred to as a *lego plot*. The height of any entry is proportional to the energy observed in that region of coordinates. The axes correspond to the azimuth (ϕ) around the collision axis, and the polar angle θ relative to the collision's axis. The jets are observed at 180° relative to each other in azimuth, reflecting transverse momentum conservation in collisions between constituents inside the hadrons.

We showed in Fig. 13.2 how the W and Z bosons can produce transitions between members of the same weak isospin doublet. However,

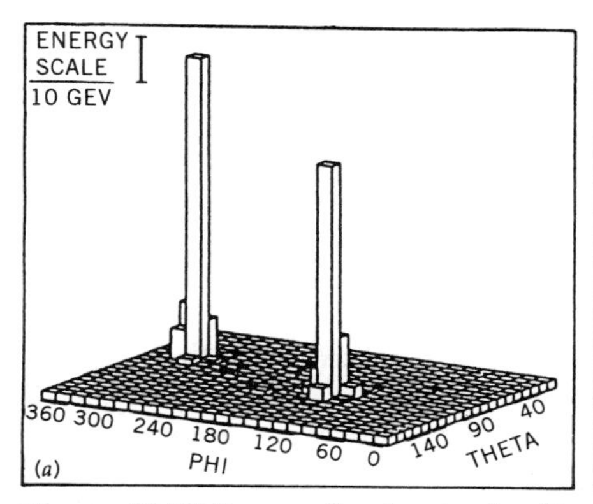

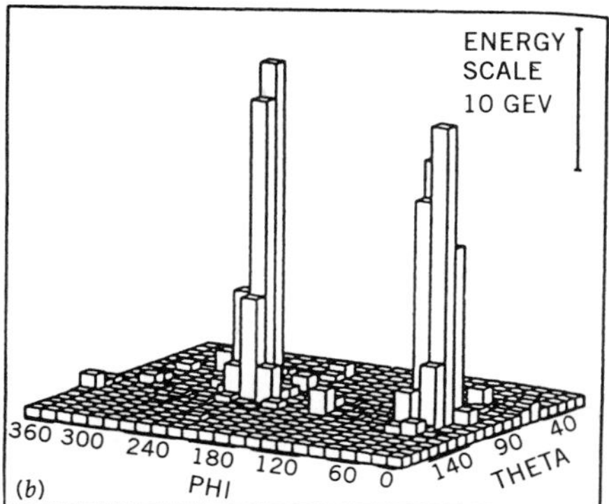

Figure 13.10 Energy flowing in the direction transverse to the collision axis for production of particle jets in $\bar{p}p$ collisions at $\sqrt{s} \sim 600$ GeV.

(From L. DiLella. 1987. *Techniques and Concepts of High Energy Physics IV*, T. Ferbel, Ed. Plenum Press, New York.)

if, in fact, W and Z bosons could not provide transitions among particles belonging to different multiplets, it would clearly present a great puzzle concerning the origin of $|\Delta S| = 1$ strangeness changing weak decays. The solution to this issue comes from our earlier observation that strangeness is a quantum number that is not conserved in weak interactions. Consequently, the eigenstates of the weak Hamiltonian are different from those of the strong Hamiltonian, and, in particular, do not have unique strangeness. In analogy with our analysis of the K^0–$\overline{K^0}$ system, we can try to redefine the quark doublet eigenstates of the weak Hamiltonian as mixed states of the doublets of Eq. 13.2. Before the discovery of the charm quark, and based on the experimental results available at that time, Nicola Cabibbo showed that all data was consistent with altering the doublet corresponding to the first family of quarks as follows:

$$\begin{pmatrix} u \\ d \end{pmatrix} \longrightarrow \begin{pmatrix} u \\ d' \end{pmatrix} \tag{13.57}$$

where the newly defined state d' is a mixture of d and s quarks,

$$d' = \cos\theta_c \; d + \sin\theta_c \; s \tag{13.58}$$

This kind of state clearly does not have a unique strangeness quantum number, and if the weak gauge bosons can give rise to transitions within the u, d' multiplet, then they can, in fact, induce strangeness-changing processes. The angle θ_c parameterizing the mixing between the d and s quarks in Eq. 13.58 is commonly called the Cabibbo angle, and its value determines the relative rates for processes such as

$$
\begin{aligned}
W^+ &\longrightarrow u\bar{s} \\
W^+ &\longrightarrow u\bar{d} \\
Z^0 &\longrightarrow u\bar{u} \\
Z^0 &\longrightarrow d\bar{s}
\end{aligned}
\tag{13.59}
$$

The Cabibbo angle can be determined experimentally from the preceding transitions, and has the value $\sin\theta_c = 0.23$. Figure 13.11 shows how the decay of a K^0 into a π^+ and a π^- can now be described in the Standard Model.

While the Cabibbo hypothesis accommodated most of the $W^\pm$-induced decays, certain strangeness-changing processes, particularly involving leptonic decay modes of the K^0, remained puzzling. For example, it was known that

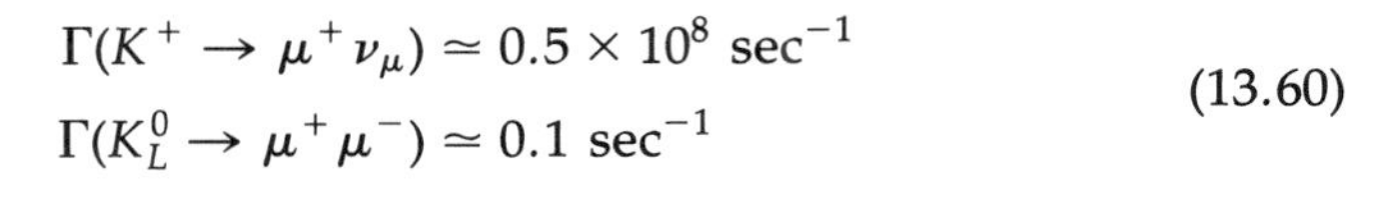

$$
\begin{aligned}
\Gamma(K^+ \to \mu^+\nu_\mu) &\simeq 0.5 \times 10^8 \text{ sec}^{-1} \\
\Gamma(K_L^0 \to \mu^+\mu^-) &\simeq 0.1 \text{ sec}^{-1}
\end{aligned}
\tag{13.60}
$$

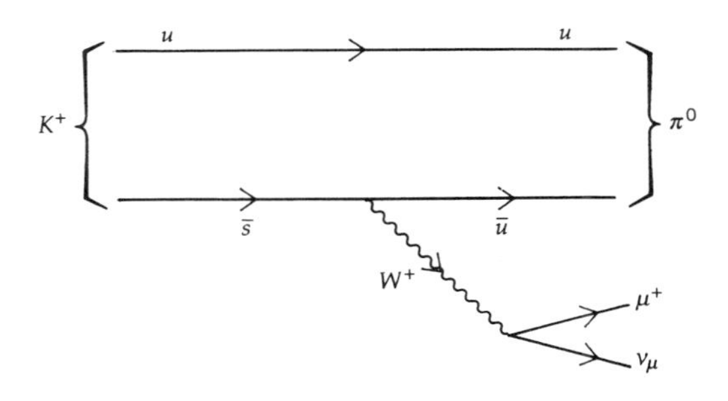

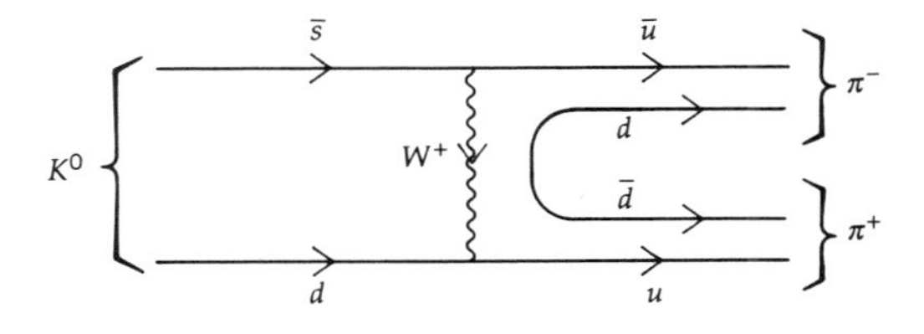

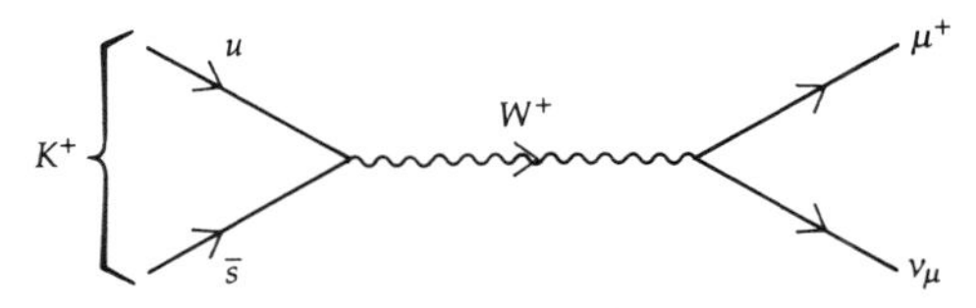

Figure 13.11 Strangeness-changing transitions in the Standard Model. A $d\bar{d}$ pair produced from the vacuum (in the middle graph) combines with the other quarks to produce a π^+ and a π^- for the final state of the K^0 decay.

leading to

$$\frac{\Gamma(K_L^0 \rightarrow \mu^+\mu^-)}{\Gamma(K^+ \rightarrow \mu^+\nu_\mu)} \simeq 10^{-9} \tag{13.61}$$

The smallness of this ratio could not be explained within the framework of the Cabibbo analysis. In fact, further investigation of this problem led Sheldon Glashow, John Illiopoulos, and Luciano Maiani to propose the existence of a fourth quark (charm quark) in a doublet structure of the type given in Eq. 13.57:

$$\begin{pmatrix} c \\ s' \end{pmatrix} \tag{13.62}$$

with

$$s' = -\sin\theta_c\, d + \cos\theta_c\, s$$

This, indeed, leads to a resolution of all the leptonic decay modes of the strange mesons, and this proposal is commonly referred to as the "GIM mechanism."

The ideas of Cabibbo and GIM can be summarized by saying that for the two doublets

$$\begin{pmatrix} u \\ d' \end{pmatrix} \quad \text{and} \quad \begin{pmatrix} c \\ s' \end{pmatrix}$$

the weak eigenstates are related to the eigenstates of the strong Hamiltonian by an orthogonal matrix:

$$\begin{pmatrix} d' \\ s' \end{pmatrix} = \begin{pmatrix} \cos\theta_c & \sin\theta_c \\ -\sin\theta_c & \cos\theta_c \end{pmatrix} \begin{pmatrix} d \\ s \end{pmatrix} \tag{13.63}$$

Today, we believe that there are, in fact, three doublets of quarks of the form

$$\begin{pmatrix} u \\ d' \end{pmatrix} \quad \begin{pmatrix} c \\ s' \end{pmatrix} \quad \text{and} \quad \begin{pmatrix} t \\ b' \end{pmatrix} \tag{13.64}$$

In this case, the relation between the three states d', s', and b' and the eigenstates d, s, b is more complicated and involves a 3×3 matrix that is known as the Kobayashi–Maskawa matrix. With this generalization, the Standard Model is now in good agreement with all the known decays.

(Note that only quarks with same electric charge can mix, and, consequently, b' would be a superposition of only the d, s, and b quarks.)

Finally, we wish to comment on the connection between electrodynamics and the weak interactions. Sheldon Glashow and Abdus Salam and Steven Weinberg, independently formulated the Electroweak Model, which is one of the cornerstones of the Standard Model. The Electroweak Theory relates the strengths of the electromagnetic and weak interactions of the fundamental particles through the weak mixing angle, θ_W, and through the masses of the gauge bosons. In particular, we can write the following relations for the parameters:

$$\sin^2\theta_W = \frac{\pi\alpha}{\sqrt{2}G_F}\frac{1}{m_W^2} = 1 - \frac{m_W^2}{m_Z^2} = 0.23 \qquad (13.65)$$

where α is the fine structure "constant" and G_F is the weak coupling constant. The value of θ_W has been measured in different scattering experiments, and the masses of the W and Z gauge bosons have been measured in $\overline{p}p$ and in e^+e^- collisions. All current experimental data are consistent with Relation (13.65) and with all the other predictions of the electroweak theory of the Standard Model.

Problems

XIII.1 In discussing weak decays proceeding through W or Z bosons, we have focussed primarily on the fundamental transitions among quarks and leptons. However, such decays often involve hadrons that contain "spectator" quarks, in addition to those quarks that are involved in the weak interaction per se. For example, Fig. 13.11 shows a diagram for the decay of a K^0 into a $\pi^+\pi^-$ pair. Using similar "quark-line" diagrams, draw processes for the following decays: (a) $K^+ \to \pi^+ + \pi^0$, (b) $n \to p + e^- + \overline{\nu}_e$, (c) $\pi^+ \to \mu^+ + \nu_\mu$, (d) $K^0 \to \pi^- + e^+ + \nu_e$.

XIII.2 Draw quark-line diagrams for the following reactions: (a) $\pi^- + p \to \Lambda^0 + K^0$, (b) $\pi^+ + p \to \Sigma^+ + K^+$, (c) $\pi^+ + n \to \pi^0 + p$, (d) $p + p \to \Lambda^0 + K^+ + p$, (e) $\overline{p} + p \to K^+ + K^-$.

XIII.3 Draw quark-line diagrams for the following weak interactions, and include any required intermediate W or Z bosons: (a) $\nu_e + n \to \nu_e + n$, (b) $\overline{\nu}_\mu + p \to \mu^+ + n$, (c) $\pi^- + p \to \Lambda^0 + \pi^0$.

XIII.4 According to the quark model, wave functions of baryons are antisymmetric in color. Construct a wave function for the Δ^{++} that is explicitly antisymmetric under the exchange of any two of its quark constituents in color space.

Suggested Readings

Frauenfelder, H., and E. M. Henley. 1991. *Subatomic Physics*. Englewood Cliffs, N.J.: Prentice-Hall.

Goldstein, H. 1980. *Classical Mechanics*. Reading, Mass.: Addison-Wesley.

Griffiths, D. 1987. *Introduction to Elementary Particles*. New York: Wiley.

Sakurai, J. J. 1964. *Invariance Principles and Elementary Particles*. Princeton, N.J.: Princeton Univ. Press.

Williams, W. S. C. 1991. *Nuclear and Particle Physics*. London/New York: Oxford Univ. Press.

Also, see standard texts on quantum mechanics, e.g., Das, A., and A. C. Melissinos. 1986. *Quantum Mechanics*. New York: Gordon & Breach.

Chapter XIV

BEYOND THE STANDARD MODEL

INTRODUCTORY REMARKS

The Standard Model of fundamental interactions—namely of the strong, weak, and electromagnetic interactions—is a gauge theory involving quarks and leptons based on the symmetry group $SU_{\text{color}}(3) \times SU_L(2) \times U_Y(1)$. As we have argued, the weak isospin and hypercharge symmetries, or the symmetry groups $SU_L(2)$ and $U_Y(1)$, are spontaneously broken. As a result, the weak gauge bosons become massive and the symmetry at low energy reduces to the gauge symmetry of electromagnetism and of the color symmetry, namely $SU_{\text{color}}(3) \times U_Q(1)$. Ignoring some of the finer technical points, this is essentially the spirit of the Standard Model. Theoretically, the Standard Model leads to many interesting perturbative predictions, and all of them appear to hold true experimentally. In fact, the agreement between experiment and theory appears to be remarkable (see previous chapter). Thus, it seems reasonable to conclude that the Standard Model leads to the correct description of the fundamental interactions at low energies. The standard model, however, has many parameters (e.g., masses of the leptons, quarks, gauge bosons, Higgs, various couplings). Also, it does not incorporate gravity, which is another fundamental force, and it does not address various questions of aesthetics. Therefore, although from the experimental point of view there does not seem to be a reason to look beyond the Standard Model, there are many theoretical reasons to do just that. In this chapter, we describe various attempts to look beyond the Standard Model.

GRAND UNIFICATION

When we examine the properties of the quark and the lepton multiplets (families), it appears that the electric charges of all fundamental particles can be regarded as quantized in units of $\frac{1}{3}e$. We already know from the properties of angular momentum that the angular momentum quantum number is also quantized—in units of $\hbar/2$. However, the quantization of this quantum number arises because the angular momentum algebra is noncommutative. In other words, it is a general feature of noncommutative (non-Abelian) symmetry groups that they give rise to conserved charges that have discrete, quantized values. In contrast, the symmetry that gives rise to a conserved electric charge corresponds to a simple phase transformation that is described by the commuting $U_Q(1)$ symmetry group. This symmetry group does not require the corresponding conserved charge to take on quantized values, and consequently, within the framework of the Standard Model, the quantization of electric charge must be regarded as a great mystery. If, however, for some reason all these symmetries—namely, the $U_Y(1)$, $SU_L(2)$, and $SU_{\text{color}}(3)$ groups—were part of a larger noncommutative symmetry group, then that could explain the origin of quantization of electric charge.

In addition, there appears to be a certain phenomenological symmetry between the quarks and the leptons. Namely, for every lepton family, there exists a family of quarks with three different colors.

$$\begin{gathered}
\begin{pmatrix} \nu_e \\ e^- \end{pmatrix} \longleftrightarrow \begin{pmatrix} u^a \\ d^a \end{pmatrix} \\
\begin{pmatrix} \nu_\mu \\ \mu^- \end{pmatrix} \longleftrightarrow \begin{pmatrix} c^a \\ s^a \end{pmatrix} \\
\begin{pmatrix} \nu_\tau \\ \tau^- \end{pmatrix} \longleftrightarrow \begin{pmatrix} t^a \\ b^a \end{pmatrix}
\end{gathered} \tag{14.1}$$

Although the top quark has yet to be detected, the internal consistency of the Standard Model suggests that it must exist. In fact, the current prediction is that the mass of the top quark is about 130 ± 30 GeV/c^2, and it is expected to be observed in the near future. The origin of the quark–lepton symmetry can be understood if we assume that quarks and leptons simply correspond to different states of the same particle. For example, we can think of each quark as having four colors, where the fourth color corresponds to the lepton quantum number. Such an assumption would then lead naturally to having the number of lepton families equal the number of quark families.

The idea that quarks and leptons are different manifestations of the same particle leads to a very interesting parallel suggestion. Namely, that the interactions of leptons and quarks, which appear to be rather different—leptons interacting weakly through the Z^0 and $W^{\pm}$ bosons, and quarks interacting strongly through the color gluons—might be related. That is, it would only make sense for the particles to be grouped together if the strong and the weak forces also correspond to different manifestations of a single fundamental force. This is a simplification that most physicists would find very appealing. That is, if the three fundamental forces were, in fact, different manifestations of one truly basic force, it would lead to a simplicity and beauty of the fundamental laws of nature that would be esthetically pleasing. This is the concept known as grand unification, and it can be examined in theories reaching beyond the Standard Model.

As we have discussed earlier, the strengths of coupling for the three forces are rather different. It is therefore not a priori clear in what way the different forces can be regarded as separate manifestations of a single force. Here, our observations in Chapters IX and XIII concerning the dependence of coupling constants on the momentum scale (or distance) become crucial. We know that electric charge grows with momentum transfer, whereas the charges associated with non-Abelian symmetries, such as color, decrease with momentum. It is therefore conceivable that, at some large energy scale, the three coupling strengths can become equal. In such a situation, the three kinds of interactions could not be distinguished, and could therefore be described by a single force that would operate beyond the unifying energy scale. At low energies, the single force would simply separate into the three of the four known fundamental forces of nature.

To understand how such a separation of forces can be achieved, we must recognize that, in order to incorporate quarks and leptons into a single family, we must also enlarge the overall symmetry group. (This also leads more naturally to the quantization of electric charge.) There are several symmetry groups, with different degrees of complexity, that can be used to implement the ideas of grand unification. The simplest, however, is a symmetry group known as $SU(5)$, which is analogous to isospin and corresponds to a rotation in an internal space of five dimensions. In this particular model (suggested by Howard Georgi and Sheldon Glashow), it is assumed that the symmetry of the fundamental interactions beyond the unifying energy scale corresponds to a local symmetry based on this larger $SU(5)$ group. At the unification scale, however, the local symmetry breaks down spontaneously to the low-energy symmetry group of the standard model—namely, $SU_{\text{color}}(3) \times SU_L(2) \times U_Y(1)$—which subsequently breaks down spontaneously to an even lower symmetry group of $SU_{\text{color}}(3) \times U_Q(1)$ at the weak energy scale. This can explain how a single force at very high energies can manifest itself as three separate forces at low energies.

Detailed analysis suggests that whereas the weak scale corresponds to about 10^2 GeV, the unification scale is close to 10^{15} GeV.

For the case of $SU(5)$, a family of quarks and leptons can be incorporated consistently into a five-dimensional multiplet and a ten-dimensional antisymmetric matrix multiplet. The five-dimensional multiplet consists of right-handed particles, whereas the ten-dimensional multiplet contains only left-handed particles. Explicitly, the multiplets take the form:

$$\begin{pmatrix} d^{\text{red}} \\ d^{\text{blue}} \\ d^{\text{green}} \\ e^+ \\ \overline{\nu}_e \end{pmatrix}_R \quad \begin{pmatrix} 0 & \overline{u}^{\text{ green,}} & \overline{u}^{\text{ blue,}} & u^{\text{ red}} & d^{\text{ red}} \\ & 0 & \overline{u}^{\text{ red,}} & u^{\text{ blue}} & d^{\text{ blue}} \\ & & 0 & u^{\text{ green}} & d^{\text{ green}} \\ & & & 0 & e^+ \\ & & & & 0 \end{pmatrix}_L \tag{14.2}$$

The particle representations in other unifying groups are more complex. Generically, however, we can represent a multiplet of a unifying group as

$$\begin{pmatrix} q \\ \ell \end{pmatrix} \tag{14.3}$$

Namely, each multiplet will necessarily contain both quarks and leptons.

As we discussed previously, the gauge bosons provide transitions among the members of a given multiplet. Thus, from the structure of Eq. 14.3, we would conclude that, in theories with grand unification, there will necessarily be transitions between quarks and leptons caused by the new massive gauge bosons of $SU(5)$. This, of course, implies that in such theories baryon and lepton numbers need not be conserved, and, as a consequence, the proton will decay. For the case of $SU(5)$, for example, the process shown in Fig. 14.1 corresponds to a proton decaying into a π^0 and e^+:

$$p \longrightarrow \pi^0 + e^+ \tag{14.4}$$

We considered this decay in Chapter IX in the context of baryon number conservation. The lifetime of the proton can, in fact, be calculated in grand

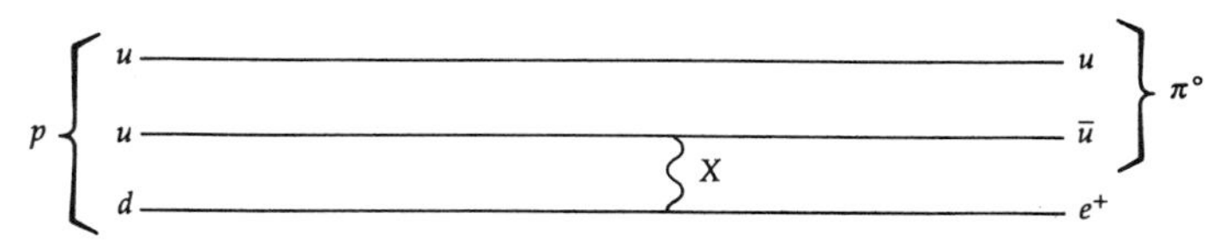

Figure 14.1 Mechanism for proton decay involving the X gauge boson of $SU(5)$.

unified models. From the observed stability of the universe, we would, of course, expect the proton to be long-lived. In fact, an impressive set of experiments has determined the lifetime of the proton to be

$$\tau_p > 10^{31} \text{ years} \tag{14.5}$$

This disagrees with the shorter lifetime expected from calculations using the simplest grand unified theory (GUT) based on the symmetry group $SU(5)$, and therefore rules out such a theory. There are, however, other models of grand unification (with more parameters and greater flexibility) that are consistent with the limit of Eq. 14.5.

The ideas of grand unification have also had considerable influence on the study of cosmology. This application rests on the fact that when a symmetry is spontaneously broken, the system undergoes a phase transition. The simplest way to see this is to consider, again, the example of the ferromagnet. At very high temperatures, the thermal motion orients the spins in a random fashion and therefore prevents any order in the system. Rotations are consequently a symmetry of the system. As the temperature drops, the thermal motion diminishes, leading to a ground state where the spins are aligned, and rotational symmetry is spontaneously broken.

Consequently, we see that, at a lower temperature, the system makes a transition to an ordered phase. Such concepts can also be applied to the evolution of our physical universe. In particular, immediately after the big bang, the temperature in the universe was exceedingly high, and assuming the validity of grand unification, we expect that the larger symmetry of the grand unified theories was a true symmetry of the universe. (Higher temperatures, of course, also correspond to higher energies.) As the universe expanded and cooled below the value corresponding to the unification scale, the symmetry of the system reduced to that of the Standard Model. In other words, there was a phase transition. Such transitions are normally exothermic, and consequently, energy was released in the process. Recalling that energy is the source of the gravitational force, we would infer that such a phase transition would have influenced the dynamical evolution of our universe. In fact, using concepts of particle physics, it can be shown explicitly that such a phase transition would have led to an epoch where the universe expanded exponentially, and much faster than predicted by previous cosmological models. This conjecture (made by Alan Guth) has the added benefit that it solves several other important problems in standard cosmology.

Grand unified theories can also provide a natural explanation for the baryon asymmetry in the universe. The argument is based on the ratio of the observed number of baryons to the number of photons in our universe, which has the value:

$$\frac{n_B}{n_\gamma} \simeq 10^{-9}$$

The photons (mostly from the 3 K "background radiation") have typical energies of about 10^{-4} eV. This means that the energy in our universe is mainly in the form of matter (or "matter dominated"). As we have seen, grand unified theories can lead to baryon nonconservation through processes such as proton decay. Furthermore, if CP violation is built into such theories, they can then generate a baryon asymmetry and can be used to calculate the n_B/n_γ ratio, which comes out quite close to the observed value. It is worth emphasizing that models based on grand unified theories are the only ones where the ratio n_B/n_γ is calculable, and can be made consistent with the observed value, thereby providing a possible understanding of the origin of the baryon asymmetry in the universe.

SUPERSYMMETRY

So far our discussions of symmetry have been restricted to transformations that relate similar kinds of particles. For example, a rotation can take a spin "up" electron to a spin "down" electron state. An isospin rotation can take a proton state to a neutron state, or a π^+ meson state to a π^0 meson state, and so on. Namely, the conventional symmetry transformations rotate bosonic states to bosonic ones and fermionic states to fermionic ones. A novel form of symmetry transformation would be one that can rotate a bosonic state to a fermionic one, and vice versa. If this were possible, it would imply that bosons and fermions are merely different manifestations of the same state, and in some sense would correspond to an ultimate form of unification. For a long time, it was believed that such a symmetry transformation was not possible to implement in physical theories. At present, however, we know that such transformations can be defined, and, in fact, there exist theories that are invariant under such transformations. These transformations are known as *supersymmetry* (SUSY) *transformations*, and the corresponding theories invariant under such transformations are called *supersymmetric theories*.

To get a qualitative understanding of supersymmetry, let us consider a simple quantum mechanical example. For a bosonic harmonic oscillator in one dimension, we can write the Hamiltonian in terms of creation and annihilation operators as follows:

$$H_B = \frac{\hbar\omega}{2}\left(a_B a_B^\dagger + a_B^\dagger a_B\right) \tag{14.6}$$

Here a_B and $a_B^\dagger$ lower and raise, respectively, the number of quanta in a state, and they satisfy the commutation relations

$$\begin{aligned} [a_B, a_B] &= 0 = \left[a_B^\dagger, a_B^\dagger\right] \\ \left[a_B, a_B^\dagger\right] &= 1 \end{aligned} \tag{14.7}$$

We can also rewrite the Hamiltonian of Eq. 14.6 in the more familiar form

$$H_B = \hbar\omega\left(a_B^\dagger a_B + \tfrac{1}{2}\right) \tag{14.8}$$

The energy spectrum of this Hamiltonian is easy to obtain, and the quantum states, with their corresponding energy values, take the form

$$|n_B\rangle \longrightarrow E_{n_B} = \hbar\omega\left(n_B + \tfrac{1}{2}\right), \qquad n_B = 0, 1, 2, \ldots \tag{14.9}$$

We note, in particular, that the ground state energy of the system has the value

$$E_0 = \frac{\hbar\omega}{2} \tag{14.10}$$

Quantum mechanically, we can also have an oscillator satisfying Fermi-Dirac statistics. In this case, the Hamiltonian has the form

$$H_F = \frac{\hbar\omega}{2}\left(a_F^\dagger a_F - a_F a_F^\dagger\right) \tag{14.11}$$

Being fermionic operators, a_F and $a_F^\dagger$ satisfy the anticommutation relations

$$\begin{aligned} a_F^2 &= 0 = \left(a_F^\dagger\right)^2 \\ a_F a_F^\dagger + a_F^\dagger a_F &= 1 \end{aligned} \tag{14.12}$$

Using Eq. 14.12, we can also rewrite the Hamiltonian for the fermionic oscillator as follows:

$$H_F = \hbar\omega\left(a_F^\dagger a_F - \tfrac{1}{2}\right) \tag{14.13}$$

Such a system has only two energy eigenstates and the corresponding energy eigenvalues are given by

$$|n_F\rangle \longrightarrow E_{n_F} = \hbar\omega\left(n_F - \tfrac{1}{2}\right), \qquad n_F = 0, 1 \tag{14.14}$$

The simplicity of this spectrum is purely a consequence of Fermi-Dirac statistics, according to which any physical state can either have one fermionic quantum ($n_F = 1$) or be the empty (bosonic) ground state containing no fermions ($n_F = 0$).

If we now consider a mixed bosonic and fermionic oscillator of the same frequency, then we can write the Hamiltonian as

$$H = H_B + H_F = \hbar\omega\left(a_B^\dagger a_B + a_F^\dagger a_F\right) \tag{14.15}$$

Very roughly speaking, this Hamiltonian is invariant under an interchange of bosons and fermions:

$$\text{``}a_B\text{''} \longleftrightarrow \text{``}a_F\text{''} \tag{14.16}$$

The simplest way to see this (see Eqs. 14.9 and 14.14) is to note that the energy spectrum of this system can be represented as

$$|n_B, n_F\rangle \longrightarrow E_{n_B,n_F} = \hbar\omega(n_B + n_F) \qquad \begin{aligned} n_F &= 0, 1 \\ n_B &= 0, 1, 2, \ldots \end{aligned} \tag{14.17}$$

Thus a bosonic state $|n_B, n_F = 0\rangle$ and a fermionic state $|n_B - 1, n_F = 1\rangle$ (with $n_B \geq 1$) are degenerate in energy, with eigenvalue

$$E = \hbar\omega n_B \tag{14.18}$$

This degeneracy is a consequence of the invariance of the Hamiltonian in Eq. 14.15 under supersymmetry. Without going into details, let us simply note that generators of supersymmetry transformations do, in fact, exist in this case, and can be identified with

$$\begin{aligned} Q_F &= a_B^\dagger a_F \\ Q_F^\dagger &= a_F^\dagger a_B \end{aligned} \tag{14.19}$$

which can be shown to satisfy the anticommutation relations

$$[Q_F, Q_F^\dagger]_+ = Q_F Q_F^\dagger + Q_F^\dagger Q_F = \frac{H}{\hbar\omega} \tag{14.20}$$

The operators Q_F and $Q_F^\dagger$, in analogy with raising and lowering operators of angular momentum, transform a bosonic state to a fermionic one of same energy, and vice versa. (It is worth pointing out that the ground state for the supersymmetric oscillator, as can be seen from Eq. 14.17, has zero energy. This is, in fact, a general property of all supersymmetric theories, and has bearing on the nature of spontaneous symmetry breaking in such theories.)

Supersymmetry is not only a beautiful concept, but it also solves many technical difficulties of grand unified theories. For example, we have noted that the unification scale is about 10^{15} GeV, whereas the observed gauge bosons, leptons, and hadrons have much smaller masses. Without supersymmetry, it is exceedingly difficult to understand why the observed particles are so light, when the scale of their unifying interaction is 10^{15} GeV. The presence of supersymmetry can rather naturally prevent our fundamental particles from becoming far more massive. There are many other reasons that have prompted the examination of supersymmetric grand unified theories. For simplest supersymmetric GUTs, the calculated proton lifetime again turns out to be inconsistent with the experimental limit in Eq. 14.5, but there are other supersymmetric models where this is not the case, and such models are therefore viable. The main current difficulty with accepting supersymmetric theories lies in the fact that they suggest the doubling of the spectrum of the fundamental particles. Namely, these theories require bosonic partners for all fermions, and vice versa (in analogy with the two degeneracies of Eq. 14.18). Unfortunately, such particles (or supersymmetric partners) have not as yet been observed. It is expected, however, that if supersymmetry is a true symmetry of nature, these new SUSY particles will be detected in experiments at the next-generation colliders such as the LHC.

Finally, a brief comment on the nature of spontaneous symmetry breaking at the weak scale, which remains an interesting and open question. We have noted that when a symmetry is spontaneously broken, there arise fundamental massive particles known as Higgs bosons. The Higgs boson for the case of electroweak symmetry breaking has yet to be observed. There is an alternative scenario for symmetry breaking where the breaking is induced by composite rather than fundamental bosons. "Technicolor" theories are those in which the symmetry is spontaneously broken by a composite state consisting of a fermion–antifermion pair. These theories have an additional symmetry group known as the technicolor group, where new quarks, carrying new technicolor charges, form bound states that spontaneously break the low-energy symmetry of the standard model. At present, however, there is no evident technicolor theory that reduces in a natural way to the $SU_{\text{color}}(3) \times SU_L(2) \times U_Y(1)$ structure of the standard model.

SUPERGRAVITY, SUPERSTRINGS

Grand unified theories—the standard or the supersymmetric kind—are not complete because they leave out one of the four fundamental forces, namely gravitation. As we have seen, this force is very weak, and can be ignored for interactions involving the current sub-TeV energy scales. However, from the form of the gravitational potential energy, namely,

$$V_{\text{grav}}(r) = G_N \frac{m^2}{r} \tag{14.21}$$

we know that, at very small distances, this force can be expected to become appreciable. In fact, for distances of the order of the Planck length of $\sim 10^{-33}$ cm, or for equivalent energy scales of the order of 10^{19} GeV, the effects of the gravitational interaction cannot be neglected. This can be seen heuristically in the following way. Considering two relativistic particles with energy $E = pc$, we can generalize the relation of Eq. 14.21 to

$$V_{\text{grav}} = \frac{G_N(E/c^2)^2}{r} \tag{14.22}$$

Using the uncertainty principle, we can write

$$r \simeq \frac{\hbar}{p} = \frac{\hbar c}{pc} = \frac{\hbar c}{E} \tag{14.23}$$

and we can therefore express the potential energy as

$$V_{\text{grav}} \simeq \frac{G_N}{\hbar c} \cdot E \cdot \left(\frac{E}{c^2}\right)^2 \tag{14.24}$$

From Eq. 14.24, we can deduce the energy scale at which the gravitational potential energy can no longer be neglected. This should correspond to the range when $V \sim E$:

$$V_{\text{grav}} \simeq \frac{G_N}{\hbar c} \cdot E \cdot \left(\frac{E}{c^2}\right)^2 \simeq E$$

or

$$\left(\frac{E}{c^2}\right)^2 \simeq \frac{\hbar c}{G_N} \simeq \frac{6}{6.7} \times 10^{39}(\text{GeV}/c^2)^2$$

or

$$E \simeq 10^{19} \text{ GeV} \tag{14.25}$$

At such energies, the effects of gravitation become appreciable. As already mentioned, this energy scale corresponds to a length scale of $\sim 10^{-33}$ cm. Since the unification scale is about 10^{15} GeV, which is relatively close to the Planck scale, a consistent description of the fundamental interactions at such energies should therefore include gravity.

The primary reason why gravity is avoided in all such considerations is because Einstein's theory of gravity does not readily lend itself to quantization. In fact, if we naively quantize Einstein's theory, it yields divergent results for the calculation of any cross section. Divergences are not unheard-of features of quantum field theories. In fact, any relativistic quantum mechanical field theory leads to divergences. In all these theories, however, there is a systematic procedure for extracting meaningful physical quantities from these seemingly infinite results. This procedure, referred to as renormalization, unfortunately fails for Einstein's gravity.

Supersymmetric theories, on the other hand, are known to have much better divergence behavior. Consequently, it is natural to consider supersymmetrizing Einstein's theory of gravity to see if this improves the divergence properties. (Supersymmetrizing Einstein's theory does not affect the desirable classical predictions because any such modification to the theory proceeds through the addition of supersymmetric fermionic partners of gravity, which have no classical analogue, and consequently do not contribute in the classical limit.) Supersymmetrized quantum gravity is found to have a local supersymmetry invariance. (This is related to the fact that Einstein's gravity has no preferred reference frame.) This gauge theory of supersymmetry is known as *supergravity* and it has, indeed, better divergence structure than ordinary quantum gravity theories. Even the most sophisticated form of supergravity theory does not, however, appear to be free of divergences. In addition, these theories do not appear to reduce naturally to the Standard Model.

The origin of divergences in relativistic quantum theories can be traced primarily to the presumed local nature of the interactions. That is, when a charged particle emits a photon, or when any other interaction takes place, it is always assumed to occur at a specific space-time point. In such theories, there are therefore no uncertainties in the positions of interaction points. This is essentially because we are dealing with point particles. As a consequence, such interactions have $\Delta x = 0$, and the uncertainty principle therefore requires infinite uncertainties in the transferred momentum. This is the origin of divergences in ordinary quantum theories. The simplest way to remedy this problem is to regard the fundamental constituents not as point particles, but rather as objects of infinitesimal size (of the order of 10^{-33} cm). In such a case, the interaction vertices are no longer completely localized (see Fig. 14.2), the momentum transfers consequently have a finite uncertainty, and divergences disappear. Theories that describe particles as objects of infinitesimal linear extension are known as *string theories*, and they appear to incorporate gravity in a natural manner. This possibility therefore offers the only currently known prescription for a fully quantized theory of gravity that does not suffer from problems of divergence. There are two kinds of string theories: bosonic string theories and superstring (supersymmetric string) theories. They are quite elegant and incorporate many interesting symmetries. However, they can be formulated in a consistent manner only in 10 (for superstrings) or 26 (for

Figure 14.2 Difference between a pointlike (left) and a stringlike (right) interaction.

bosonic strings) space-time dimensions. Of the two possibilities, superstrings appear to be more interesting. The problem with string theories, however, is that it is not clear at the moment how one can reduce them down to four space-time dimensions in a natural manner. Much more work remains to be done in this field before clean experimental tests can be formulated, and this continues to be an active area of research.

Suggested Readings

Georgi, H. 1981. A unified theory of elementary particles and forces. Sci. Am. **244**(4):48.

Green, M. B. 1986. Superstrings. Sci. Am. **255**(3):48.

Weinberg, S. 1981. The decay of the proton. Sci. Am. **244**(6):64.

Appendix A

SPECIAL RELATIVITY

Essentially all of particle physics and many areas of nuclear physics deal with particles that travel at relativistic velocities, namely velocities that are close to the speed of light (c). In this appendix we therefore summarize some of the basic concepts and results of special relativity that are needed for interpreting relativistic processes.

Starting off with the assumption that the laws of physics do not depend on the relative motion of observers at rest in different inertial frames, and that the speed of light (in vacuum) is a constant of nature that is independent of the inertial frame, Albert Einstein showed that the space-time coordinates of an event observed in two such frames can be related through the Lorentz transformation. That is, for two inertial frames that move with a relative velocity $v = v_z = \beta c$ with respect to each other, the relationship between the coordinates of any event in the two frames can be expressed as follows:

$$\begin{aligned} ct' &= \gamma(ct - \beta z) \\ x' &= x \\ y' &= y \\ z' &= \gamma(z - \beta ct) \end{aligned} \tag{A.1}$$

where we have chosen to define the z-axis as the direction of relative motion of our two coordinate frames (with primed and unprimed coordinates), and $\gamma = (1 - \beta^2)^{-1/2}$. The relations given in Eq. A.1 can be written in the form of a matrix, as follows:

$$\begin{pmatrix} ct' \\ x' \\ y' \\ z' \end{pmatrix} = \begin{pmatrix} \gamma & 0 & 0 & -\beta\gamma \\ 0 & 1 & 0 & 0 \\ 0 & 0 & 1 & 0 \\ -\beta\gamma & 0 & 0 & \gamma \end{pmatrix} \begin{pmatrix} ct \\ x \\ y \\ z \end{pmatrix} \tag{A.2}$$

The inverse transformation involves just a change in the sign of v (and, therefore, β)

$$\begin{pmatrix} ct \\ x \\ y \\ z \end{pmatrix} = \begin{pmatrix} \gamma & 0 & 0 & \beta\gamma \\ 0 & 1 & 0 & 0 \\ 0 & 0 & 1 & 0 \\ \beta\gamma & 0 & 0 & \gamma \end{pmatrix} \begin{pmatrix} ct' \\ x' \\ y' \\ z' \end{pmatrix} \tag{A.2'}$$

For a general Lorentz transformation, the matrix connecting the coordinates of the two reference frames will be more complicated. But since the basic properties of the transformations do not change, we continue to use this simple matrix for our discussion.

The four coordinates ($x^0 = ct, x^1 = x, x^2 = y, x^3 = z$), or $(x^0, \vec{x})$, are referred to as the components of a space-time "four-vector" x. Now, just as the scalar or dot product of any two common "three-vectors" $\vec{S}$ and $\vec{R}$, namely $\vec{S} \cdot \vec{R}$, is invariant (remains the same) under a rotation of coordinates, so is the following "contraction" of any two four-vectors x and y invariant under a Lorentz transformation (which consists of rotations and "boosts"):

$$x \cdot y = x^0 y^0 - x^1 y^1 - x^2 y^2 - x^3 y^3 = x^0 y^0 - \vec{x} \cdot \vec{y} \tag{A.3}$$

Similarly, the momentum vector $\vec{P}$ and the energy E of any particle, also define a four-vector p, commonly referred to as the energy-momentum four-vector:

$$p = (E, c\vec{P}) = (E, cP_x, cP_y, cP_z) = (p^0, p^1, p^2, p^3) \tag{A.4}$$

Although the individual components of such energy-momentum four-vectors will be different in different inertial frames, they can be related to one another through the same Lorentz transformation that relates the coordinates, namely,

$$\begin{pmatrix} E' \\ cP'_x \\ cP'_y \\ cP'_z \end{pmatrix} = \begin{pmatrix} \gamma & 0 & 0 & -\beta\gamma \\ 0 & 1 & 0 & 0 \\ 0 & 0 & 1 & 0 \\ -\beta\gamma & 0 & 0 & \gamma \end{pmatrix} \begin{pmatrix} E \\ cP_x \\ cP_y \\ cP_z \end{pmatrix} \tag{A.5}$$

Again, given any two energy-momentum four-vectors $p = (p^0, c\vec{P})$ and $q = (q^0, c\vec{Q})$, the quantity $p \cdot q = (p^0 q^0 - c^2 \vec{P} \cdot \vec{Q})$ will be independent of any Lorentz frame, namely a constant, or a relativistic invariant. In particular, $p \cdot p$ will also be an invariant, and for a particle with energy E and momentum $\vec{P}$, we note that

$$p \cdot p = E^2 - c^2 |\vec{P}|^2 = \text{constant} \tag{A.6}$$

Since this quantity is independent of reference frame, we can, in particular, consider its value in the rest frame of the particle ($\vec{P} = 0$), where we can identify it with the square of the rest energy, namely,

$$p \cdot p = E_{\text{rest}}^2 = (Mc^2)^2 \tag{A.6$'$}$$

where M is the rest mass of the particle.

For any particle moving with a velocity $\vec{v} = \vec{\beta} c$ relative to a stationary observer, the relativistic momentum and energy can be written as follows:

$$\begin{aligned} \vec{P} &= M\gamma\vec{v} = M\gamma\vec{\beta}c \\ E &= M\gamma c^2 \end{aligned} \tag{A.7}$$

From which it follows that

$$\vec{\beta} = \frac{c\vec{P}}{E} \quad \text{and} \quad \gamma = \frac{E}{Mc^2} \tag{A.8}$$

The total energy E can also be decomposed into the frame-independent rest energy (Mc^2) and the relativistic kinetic energy T as

$$E = T + Mc^2 \tag{A.9}$$

Consequently, the kinetic energy of any particle can be expressed in terms of the momentum observed in any rest frame as follows:

$$T = \sqrt{(Mc^2)^2 + c^2|\vec{P}|^2} - Mc^2 \tag{A.10}$$

or, equivalently,

$$c|\vec{P}| = \sqrt{T^2 + 2Mc^2 T} \tag{A.10$'$}$$

Because the sum or difference of any two four-vectors is also a four-vector, it follows that the "square" of the sum or difference of any number of four-vectors is also a Lorentz invariant quantity. In particular, for any set of four-vectors $q_i = (q_i^0, q_i^1, q_i^2, q_i^3)$, if we define the sum of the q_i as a four-vector $q = \sum_i q_i$, then the square of the four-vector q^2 is an invariant quantity

$$q^2 = \left(\sum_i q_i^0\right)^2 - \left(\sum_i q_i^1\right)^2 - \left(\sum_i q_i^2\right)^2 - \left(\sum_i q_i^3\right)^2 \quad \text{(A.11)}$$

For example, when the q_i represent the energy-momentum four-vectors of some group of particles, the quantity q^2 is the square of the rest energy of the entire system. This is the quantity we call s in Chapter I.

An unstable relativistic particle of energy E and momentum $\vec{P}$, that has a mean life τ in its own rest frame, will have an observed mean life given by the appropriate Lorentz transformation for a time interval; namely, if a particle is produced at rest in the laboratory, its point of production will coincide with its point of decay ($x_2 = x_1, y_2 = y_1, z_2 = z_1$) and its decay time ($t_2 - t_1$) will be characterized by its mean life τ. If the particle has velocity $v = \beta c = |\vec{P}|/E$ in the laboratory, however, its mean life will be dilated. The time interval in the laboratory is related to the one in the particle's rest frame through the first of the relations in Eq. A.1:

$$t_2' - t_1' = \gamma(t_2 - t_1) - \frac{\beta}{c}(z_2 - z_1) \quad \text{(A.12)}$$

In the rest frame of the particle $z_2 = z_1$, and consequently the mean life observed in the laboratory ($\tau' = t_2' - t_1'$), will be given by

$$\tau' = \gamma\tau \quad \text{(A.13)}$$

Appendix B

SPHERICAL HARMONICS

The spherical harmonic functions $Y_{\ell,m}(\theta,\phi)$ are eigenstates of both the square of the angular momentum operator L^2, as well as of L_z, the projection of $\vec{L}$ on some specific axis z (see Eq. 3.26):

$$\begin{aligned} L^2 Y_{\ell,m}(\theta,\phi) &= \hbar^2 \ell(\ell+1) Y_{\ell,m}(\theta,\phi) \\ L_z Y_{\ell,m}(\theta,\phi) &= \hbar m Y_{\ell,m}(\theta,\phi) \end{aligned} \tag{B.1}$$

The $Y_{\ell,m}(\theta,\phi)$ are products of periodic functions of θ and of ϕ that are often encountered in quantum mechanics and in other areas where we seek solutions to problems with spherical symmetry. The $Y_{\ell,m}(\theta,\phi)$ can be written in terms of associated Legendre polynomials $P_{\ell,m}(\cos\theta)$ and exponentials of ϕ, as follows:

$$Y_{\ell,m}(\theta,\phi) = \sqrt{\frac{2\ell+1}{4\pi}\frac{(\ell-m)!}{(\ell+m)!}} P_{\ell,m}(\cos\theta) e^{im\phi} \tag{B.2}$$

where the associated Legendre functions are given by

$$P_{\ell,m}(x) = \frac{(-1)^m}{2^\ell \ell!}(1-x^2)^{m/2}\frac{d^{\ell+m}}{dx^{\ell+m}}(x^2-1)^\ell \tag{B.3}$$

where $x = \cos\theta$. The $P_{\ell,m}(x)$ are defined such that the spherical harmonics obey the following normalization relation over the full solid angle:

$$\int_{\phi=0}^{2\pi}\int_{\theta=0}^{\pi} Y^*_{\ell' m'}(\theta,\phi) Y_{\ell,m}(\theta,\phi)\sin\theta\, d\theta\, d\phi = \delta_{\ell'\ell}\delta_{m'm} \tag{B.4}$$

where the δ_{nm} are the Kronecker symbols (see Eq. 10.21). Also, the following general relation holds for the $Y_{\ell,m}(\theta,\phi)$:

$$Y^*_{\ell,m}(\theta,\phi) = (-1)^m Y_{\ell,-m} \tag{B.5}$$

Some of the low-order spherical harmonics are

$$\begin{aligned}
Y_{0,0}(\theta,\phi) &= \frac{1}{\sqrt{4\pi}}, & Y_{1,1}(\theta,\phi) &= -\sqrt{\frac{3}{8\pi}}\sin\theta e^{i\phi} \\
Y_{1,0}(\theta,\phi) &= \sqrt{\frac{3}{4\pi}}\cos\theta, & Y_{1,-1}(\theta,\phi) &= \sqrt{\frac{3}{8\pi}}\sin\theta e^{-i\phi} \\
Y_{2,2}(\theta,\phi) &= \sqrt{\frac{15}{32\pi}}\sin^2\theta' e^{2i\phi}, & Y_{2,1}(\theta,\phi) &= -\sqrt{\frac{15}{8\pi}}\sin\theta\cos\theta e^{i\phi} \\
Y_{2,0}(\theta,\phi) &= \sqrt{\frac{5}{4\pi}}\left(\frac{3}{2}\cos^2\theta - \frac{1}{2}\right)
\end{aligned} \tag{B.6}$$

SPHERICAL BESSEL FUNCTIONS

The spherical Bessel functions $j_\ell(x)$ arise in solving the radial Schrödinger equation in spherical coordinates. These functions are related to the ordinary Bessel functions $J_n(x)$ that are usually encountered in systems that possess cylindrical symmetry. The relation between the two type of functions are

$$j_\ell(x) = \sqrt{\frac{\pi}{2x}}\, J_{\ell+1/2}(x) \tag{C.1}$$

The more standard Bessel functions are given by the expansion

$$J_n(x) = \sum_{\lambda=0}^{\infty} \frac{(-1)^\lambda (x/2)^{n+2\lambda}}{\Gamma(\lambda+1)\Gamma(\lambda+n+1)} \tag{C.2}$$

where Γ refers to the factorial function (Gamma function).

Using identities to relate Γ functions of different argument, it can be shown that the series obtained by substituting Eq. C.2 into Eq. C.1 can be identified with expansions of simple periodic functions. In particular, it follows that some of the lowest order spherical Bessel functions can be written as

$$j_0(x) = \frac{\sin x}{x}, \qquad j_1(x) = \frac{\sin x}{x^2} - \frac{\cos x}{x}$$
$$j_2(x) = \left(\frac{3}{x^3} - \frac{1}{x}\right)\sin x - \frac{3\cos x}{x^2}, \qquad \text{etc.} \tag{C.3}$$

All the $j_\ell(x)$ are well-behaved near $x = 0$. In fact, all but the $\ell = 0$ function vanish at the origin, and $j_0(0) = 1$. The solutions of the radial Schrödinger equation that are singular at the origin are given by the Neumann functions, but such functions are not normalizable and therefore do not correspond to physical solutions for bound quantum mechanical systems.

Appendix D

BASICS OF GROUP THEORY

A group consists of a set of elements (objects, quantities)—finite or infinite in number—with a rule for combining the elements ("multiplication" rule) such that the set is "closed" under multiplication. Thus, if G represents a group, with $(g_1, g_2, \ldots, g_n)$ as its elements, then the combination of any two of its elements g_i and g_j, denoted by $g_i \bullet g_j$, also belongs to the group. (Mathematically, $g \in G$ stands for the statement g belongs to G.) It should be understood that the combination rule for the elements (namely, the multiplication rule) does not necessarily have to be an ordinary product of the elements. It can also be any other operation such as addition.

The set of elements have to satisfy several other properties in order to define a group. These are:

1. The combination of the elements must be associative, namely,

$$g_1 \bullet (g_2 \bullet g_3) = (g_1 \bullet g_2) \bullet g_3 \in G \tag{D.1}$$

2. There must be an identity element of the group, denoted as I, such that combining any element with the identity gives back the same element.

$$g \bullet I = g = I \bullet g \tag{D.2}$$

3. For every element $g \in G$, there must exist a unique inverse element $g^{-1} \in G$, such that

$$g \bullet g^{-1} = I = g \bullet g^{-1} \tag{D.3}$$

As a simple example of a group, let us assume that G consists of all the real numbers, both positive and negative. In this case, we can define

$$g_1 \bullet g_2 = g_1 + g_2 \tag{D.4}$$

With this combination formula, it is clear that the sum of any two real numbers is again a real number, and G is therefore closed under multiplication. We also recognize that ordinary addition is associative. Therefore, for this case,

$$g_1 \bullet (g_2 \bullet g_3) = g_1 \bullet (g_2 + g_3) = g_1 + g_2 + g_3 = (g_1 \bullet g_2) \bullet g_3 \tag{D.5}$$

Furthermore, we can identify the identity element with zero, so that

$$g \bullet I = g + 0 = g = I \bullet g \tag{D.6}$$

Finally, for any real number g, we can identify its inverse as $g^{-1} = -g$ such that

$$g \bullet g^{-1} = (g + g^{-1}) = (g - g) = 0 = I = g^{-1} \bullet g \tag{D.7}$$

This shows that the set of all real numbers defines a group, with ordinary addition representing the multiplication rule.

Let us next consider the set of all real phases, and denote them as

$$G = \{U(\alpha) = e^{i\alpha}, \quad \alpha \text{ real and in the range } -\infty \leq \alpha \leq \infty\} \tag{D.8}$$

The elements of this set are labeled by a continuous parameter, α, and consequently this defines a continuous set. Note that if we choose ordinary multiplication to be the combination formula, then

$$U(\alpha) \bullet U(\beta) = U(\alpha)U(\beta) = e^{i\alpha}e^{i\beta} = e^{i(\alpha+\beta)} = U(\alpha + \beta) \tag{D.9}$$

That is, we see that the combination is a phase, and therefore belongs to the set. The set G is consequently closed under multiplication. The ordinary products are, of course, associative, and we have

$$\begin{aligned} U(\alpha) \bullet (U(\beta) \bullet U(\gamma)) &= U(\alpha)(U(\beta)U(\gamma)) = (U(\alpha)U(\beta))U(\gamma) \\ &= e^{i(\alpha+\beta+\gamma)} = U(\alpha + \beta + \gamma) \in G \end{aligned} \tag{D.10}$$

For the identity element, we can choose the element with zero phase to correspond to $I = 1$, so that

$$U(\alpha) \bullet I = e^{i\alpha} \times 1 = e^{i\alpha} = U(\alpha) = I \bullet U(\alpha) \qquad \text{(D.11)}$$

Furthermore, given a phase $U(\alpha)$, we can identify its inverse with $U^{-1}(\alpha) = U(-\alpha)$, such that

$$\begin{aligned} U(\alpha) \bullet U^{-1}(\alpha) &= U(\alpha)U^{-1}(\alpha) = U(\alpha)U(-\alpha) \\ &= e^{i\alpha}e^{-i\alpha} = 1 = I = U^{-1}(\alpha) \bullet U(\alpha) \end{aligned} \qquad \text{(D.12)}$$

Thus, the set of all real and continuous phases defines a group. For this case, the adjoint (complex conjugate) element is also the inverse element,

$$U^{\dagger}(\alpha) = e^{-i\alpha} = U(-\alpha) = U^{-1}(\alpha) \qquad (\alpha \text{ is real}) \qquad \text{(D.13)}$$

and such groups are referred to as unitary groups. Furthermore, because the group elements in this case are completely defined by a single parameter, the group is denoted as the group $U(1)$, or the unitary group in one dimension.

It is important to recognize that, in general, the combination rule for the elements of any group need not be commutative. That is,

$$g_1 \bullet g_2 \neq g_2 \bullet g_1 \qquad \text{(D.14)}$$

In our simple example, however, we see that

$$U(\alpha) \bullet U(\beta) = U(\alpha)U(\beta) = U(\alpha + \beta) = U(\beta) \bullet U(\alpha) \qquad \text{(D.15)}$$

Thus, we say that the group $U(1)$ is commutative or Abelian.

In a similar manner, it can be shown that the set of all (2×2) unitary matrices, with determinant (det) equaling unity, define a group that has ordinary matrix multiplication representing the combination rule for the elements. This kind of group is known as the special (det = 1) unitary group in two dimensions, and is denoted by $SU(2)$. An element of such a group can also be represented as a phase of the form

$$U(\vec{\alpha}) = e^{iT(\vec{\alpha})} \qquad \text{(D.16)}$$

where $\vec{\alpha}$ denotes a vector parameter that labels the phase, and $T(\vec{\alpha})$ corresponds to a 2×2 matrix.* Note that for $U(\vec{\alpha})$ to be unitary, we must

*The effect of the exponentiated matrix is equivalent to the normal series expansion of the operator: $e^{iT(\vec{\alpha})} = I + iT(\vec{\alpha}) - (1/2!)T^2(\vec{\alpha}) - (i/3!)T^3(\vec{\alpha}) + \cdots$

have

$$U^{\dagger}(\vec{\alpha}) = U^{-1}(\vec{\alpha})$$

or

$$e^{-iT^{\dagger}(\vec{\alpha})} = e^{-iT(\vec{\alpha}))} \tag{D.17}$$

We conclude, therefore, that the matrices $T(\vec{\alpha})$ must be Hermitian. Furthermore, $\det U(\vec{\alpha})$ can equal unity only if the matrices $T(\vec{\alpha})$ are traceless. This can be seen by noting that, for any matrix A, we can write, in general,

$$\det A = e^{\mathrm{Tr}\ \ln A} \tag{D.18}$$

where Tr refers to the trace of a matrix. Thus, requiring

$$\det U(\vec{\alpha}) = 1$$

or

$$e^{\mathrm{Tr}\ \ln U(\vec{\alpha})} = 1$$

or

$$e^{i\ \mathrm{Tr}\ T(\vec{\alpha})} = 1 \tag{D.19}$$

Since this has to be true for any arbitrary vector $\vec{\alpha}$, we conclude that

$$\mathrm{Tr}\ T(\vec{\alpha}) = 0 \tag{D.20}$$

We know that there are only three linearly independent, Hermitian, and traceless 2×2 matrices, namely the Pauli matrices

$$\sigma_1 = \begin{pmatrix} 0 & 1 \\ 1 & 0 \end{pmatrix} \quad \sigma_2 \begin{pmatrix} 0 & -i \\ i & 0 \end{pmatrix} \quad \sigma_3 = \begin{pmatrix} 1 & 0 \\ 0 & -1 \end{pmatrix} \tag{D.21}$$

Therefore, for a general element of $SU(2)$, we can write

$$U(\vec{\alpha}) = e^{iT(\vec{\alpha})} = e^{i\sum_{j=1}^{3} \alpha_j T_j} \tag{D.22}$$

where, conventionally, we identify

$$T_j = \tfrac{1}{2}\sigma_j \tag{D.23}$$

The phases or elements of $SU(2)$, therefore, are labeled by three continuous parameters $\alpha_1, \alpha_2, \alpha_3$. Furthermore, we note that, because the matrix product is not commutative, for this case we have

$$U(\vec{\alpha})U(\vec{\beta}) = e^{i\sum_{j=1}^{3}\alpha_j T_j} e^{i\sum_{k=1}^{3}\beta_k T_k} \neq U(\vec{\beta})U(\vec{\alpha}) \qquad \text{(D.24)}$$

The group $SU(2)$ is therefore noncommutative or non-Abelian. The properties of the group can be completely determined, however, once we know the properties of the matrices T_j. These matrices, as we know, satisfy the commutation relations

$$[T_j, T_k] = \left[\tfrac{1}{2}\sigma_j, \tfrac{1}{2}\sigma_k\right] = i\epsilon_{jk\ell}\tfrac{1}{2}\sigma_\ell = i\epsilon_{jk\ell}T_\ell \qquad \text{(D.25)}$$

(Where $\epsilon_{jk\ell}$ is the antisymmetric Levi-Civita symbol introduced in our discussion of continuous symmetries in Chapter X.) This is known as the Lie algebra for the group $SU(2)$.

Similarly, one can show that the set of all 3×3 unitary matrices with determinant 1 constitute a group known as $SU(3)$. The set of all 3×3 real orthogonal matrices with determinant 1 define a group known as $SO(3)$, and so on. The properties of these groups are fully determined once their Lie algebras are specified.

Appendix **E**

TABLE OF PHYSICAL CONSTANTS*

Constant	*Symbol*	*Value*
Avogadro's number	A_0	6.022137×10^{23} mole^{-1}
Boltzmann's constant	k	8.61739×10^{-5} eV/K
Electron charge	e	4.803207×10^{-10} esu
Mass of electron	m_e	0.5109991 MeV/c^2
		9.109390×10^{-28} gm
Fermi's constant	$G_F/(\hbar c)^3$	1.16639×10^{-5} GeV^{-2}
Fine structure constant	$\alpha = \dfrac{e^2}{\hbar c}$	1/137.035990
Speed of light	c	$2.99792458 \times 10^{10}$ cm/sec
Newton's gravitational constant	G_N	6.6726×10^{-8} cm^3/gm-sec^2
		$6.7071 \times 10^{-39}\ \hbar c(\text{GeV}/c^2)^{-2}$
("Reduced" Planck's constant) · c	$\hbar c$	197.32705 MeV fm

*Review of particle properties, Phys. Rev., Part 2, Vol. D45, June 1992. See also the latest *CRC Handbook*. The physical constants are uncertain only in the last significant figure given in the table.

INDEX*

*p refers to problem and f to figure